Student Solutions Manual

SEARS AND ZEMANSKY'S

UNIVERSITY PHYSICS

TENTH EDITION

VOLUME 1

YOUNG & FREEDMAN

A. LEWIS FORD
TEXAS A&M UNIVERSITY

▲ ADDISON-WESLEY

An imprint of Addison Wesley Longman, Inc.

San Francisco • Reading, Massachusetts
New York • Harlow, England • Don Mills, Ontario
Sydney • Mexico City • Madrid • Amsterdam

0-201-64394-4
5 6 7 8 9 10—CRS—03 02 01

Addison Wesley Longman, Inc.
1301 Sansome Street
San Francisco, California 94111

PREFACE

This *Student Solutions Manual*, Volume 1, contains detailed solutions for approximately one third of the Exercises and Problems in Chapters 1 through 21 of the Tenth Edition of *University Physics*. The Exercises and Problems included in this manual are selected solely from the odd-numbered Exercises and Problems in the text (for which the answers are tabulated in the back of the textbook). The Exercises and Problems included were not selected at random but rather were carefully chosen to include at least one representative example of each problem type. The remaining Exercises and Problems, for which solutions are not given here, constitute an ample set of problems for you to tackle on your own. In addition, there are the Challenge Problems in the text for which no solutions are given here.

This manual greatly expands the set of worked-out examples that accompanies the presentation of physics laws and concepts in the text. This manual was written to provide you with models to follow in working physics problems. The problems are worked out in the manner and style in which you should carry out your own problem solutions.

The author will gratefully receive comments as to style, points of physics, errors, or anything else relating to this manual. The *Student Solutions Manual* Volumes 2 and 3 companion volume is also available from your college bookstore.

A. Lewis Ford
Physics Department
Texas A&M University
College Station TX 77843
Email: ford@physics.tamu.edu

CONTENTS

CHAPTER 1
UNITS, PHYSICAL QUANTITIES, AND VECTORS

Exercises 1, 7, 9, 15, 21, 23, 33, 35, 37, 39, 41, 43, 45, 49
Problems 53, 57, 59, 61, 69, 73, 75

Exercises

1-1   1.00 in. = 2.54 cm

1.00 mi = 1.00 mi(5280 ft/1 mi)(12 in./1 ft)(2.54 cm/1 in.)(1 m/10^2 cm)
(1 km/10^3 m) = 1.61 km

Note that the units all cancel to give the desired units of the answer.

1-7   **a)** 1450 mi/h = (1450 mi/h)(1.609 km/1 mi) = 2.33×10^3 km/h

b) 2.33×10^3 km/h = $(2.33 \times 10^3$ km/h)(10^3 m/1 km)(1 h/3600 s) = 647 m/s

1-9   15.0 km/L = (15.0 km/L)(0.6214 mi/1 km)(3.788 L/1 gal) = 35.3 mi/gal

1-15   **a)** The volume of a disk of diameter d and thickness t is $V = \pi(d/2)^2 t$.
The average volume is $V = \pi(8.50 \text{ cm}/2)^2(0.050 \text{ cm}) = 2.837$ cm^3. But t is given
to only two significant figures so the answer should be expressed to two significant
figures: $V = 2.8$ cm^3.

We can find the uncertainty in the volume as follows. The volume could be as large
as $V = \pi(8.52 \text{ cm}/2)^2(0.055 \text{ cm}) = 3.1$ cm^3, which is 0.3 cm^3 larger than the average
value. The volume could be as small as $V = \pi(8.52 \text{ cm}/2)^2(0.045 \text{ cm}) = 2.5$ cm^3,
which is 0.3 cm^3 smaller than the average value. The uncertainty is ± 0.3 cm^3, and
we express the volume as $V = 2.8 \pm 0.3$ cm^3.

b) The ratio of the average diameter to the average thickness is 8.50 cm/0.050 cm =
170. By taking the largest possible value of the diameter and the smallest possible
thickness we get the largest possible value for this ratio: 8.52 cm/0.045 cm = 190.
The smallest possible value of the ratio is 8.48/0.055 = 150. Thus the uncertainty
is ± 20 and we write the ratio as 170 ± 20.

1-21   I **estimate** that my scalp's area is about that of a 10 in. diameter circle:

$d = 10$ in.(2.54 cm/1 in.)(10 mm/1 cm) = 250 mm. The estimated area is thus
$A = \pi r^2$ with $r = d/2 = 125$ mm: $A = \pi(125 \text{ mm})^2 = 5 \times 10^4$ mm^2.

I further **estimate** that on my scalp there are 5 hairs per mm^2. The number of hairs on my head is thus estimated to be about $(5 \text{ hairs/mm}^2)(5 \times 10^4 \text{ mm}^2) =$ 2×10^5 hairs.

1-23   Estimate that the pile is 18 in. × 18 in. × 5 ft 8 in.,

so the volume of gold in the pile is $V =$18 in. × 18 in. × 68 in. = 22,000 in.^{3}.

Convert to cm^3: $V = 22{,}000$ in.^{3} $(1000 \text{ cm}^3/61.02 \text{ in.}^3) = 3.6 \times 10^5 \text{ cm}^3$.

From Example 1-4 the density of gold is 19.3 g/cm^3, so the mass of this volume of gold is $m = (19.3 \text{ g/cm}^3)(3.6 \times 10^5 \text{ cm}^3) = 7 \times 10^6$ g.

Convert to ounces: $m = 7 \times 10^6$ g (1 ounce/30 g) $= 2 \times 10^5$ ounces.

Also from Example 1-4, the monetary value of one ounce is $400, so the gold has a value of ($400/ounce)$(2 \times 10^5$ ounces) $= \$8 \times 10^7$, or about 100×10^6 (one hundred million dollars).

1-33

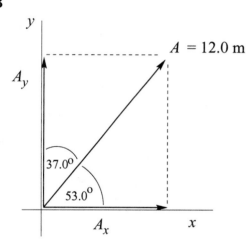

$A_x = A \cos 53.0°$
$A_x = (12.0 \text{ m}) \cos 53.0° = +7.22 \text{ m}$

$A_y = A \sin 53.0°$
$A_y = (12.0 \text{ m}) \sin 53.0° = +9.58 \text{ m}$

(The sketch shows that A_x and A_y should both be positive, in agreement with our calculation.)

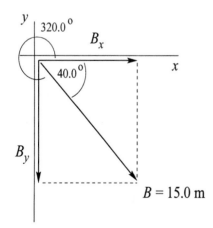

$B_x = B \cos 320.0°$
$B_x = (15.0 \text{ m}) \cos 320.0° = +11.5 \text{ m}$

$B_y = B \sin 320.0°$
$B_y = (15.0 \text{ m}) \sin 320.0° = -9.64 \text{ m}$

(The sketch shows that B_x is positive and B_y is negative, in agreement with our calculation.)

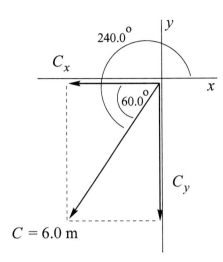

$C_x = C \cos 240.0°$
$C_x = (6.0 \text{ m}) \cos 240.0° = -3.0 \text{ m}$

$C_y = C \sin 240.0°$
$C_y = (6.0 \text{ m}) \sin 240.0° = -5.2 \text{ m}$

(The sketch shows that both C_x and C_y are negative, in agreement with our calculation.)

1-35

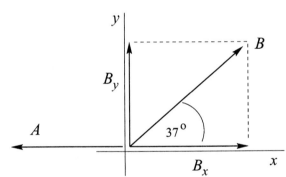

$A_x = -12.0 \text{ m}, \quad A_y = 0$

$B_x = B \cos 37°$
$B_x = (18.0 \text{ m}) \cos 37° = 14.38 \text{ m}$
$B_y = B \sin 37°$
$B_y = (18.0 \text{ m}) \sin 37° = 10.83 \text{ m}$

Note that both B_x and B_y are positive. A_x is negative because because $\vec{A}$ is in the negative x-direction.

a) Let $\vec{R} = \vec{A} + \vec{B}$.
$R_x = A_x + B_x = -12.0 \text{ m} + 14.38 \text{ m} = +2.38 \text{ m}.$
$R_y = A_y + B_y = 0 + 10.83 \text{ m} = +10.83 \text{ m}.$

$R = \sqrt{R_x^2 + R_y^2}$
$= \sqrt{(2.38 \text{ m})^2 + (10.83 \text{ m})^2} = 11.1 \text{ m}$

$\tan \theta = \dfrac{R_y}{R_x} = \dfrac{10.83 \text{ m}}{2.38 \text{ m}} = 4.550$
$\theta = 77.6°$, measured counterclockwise from the $+x$-axis.

b) $\vec{B} + \vec{A} = \vec{A} + \vec{B}$, so the vector sum is the same as in part (a).

c) Now let $\vec{R} = \vec{A} - \vec{B}$.

$R_x = A_x - B_x = -12.0 \text{ m} - 14.38 \text{ m} = -26.38 \text{ m}$.

$R_y = A_y - B_y = 0 - 10.83 \text{ m} = -10.83 \text{ m}$.

Now both R_x and R_y are negative.

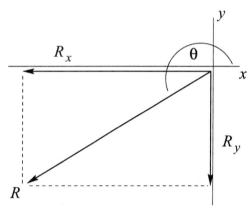

$$R = \sqrt{R_x^2 + R_y^2}$$

$$= \sqrt{(-26.38 \text{ m})^2 + (-10.83 \text{ m})^2} = 28.5 \text{ m}$$

$$\tan \theta = \frac{R_y}{R_x} = \frac{-10.83 \text{ m}}{-26.38 \text{ m}} = 0.4105$$

$\theta = 202°$, measured counterclockwise from the $+x$-axis.

Note: My calculator gives $\arctan(+0.4105) = 22.3°$. But the sketch shows that $\vec{R}$ is in the third quadrant. The angle $22.3° + 180° = 202°$ also has a tangent of $+0.4105$ and the sketch shows this is the correct answer.

d) $\vec{R} = \vec{B} - \vec{A}$. Then $R_x = B_x - A_x = -(A_x - B_x) = +26.38 \text{ m}$ and $R_y = B_y - A_y = -(A_y - B_y) = +10.83 \text{ m}$, using the results of part (b). R_x and R_y now have the same magnitudes but opposite signs from part (c), so now $\vec{R}$ has the same magnitude and opposite direction as in part (c).

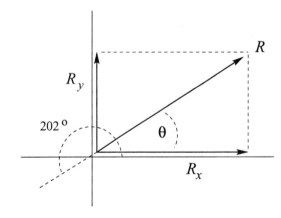

$R = 28.5 \text{ m}$

$$\tan \theta = \frac{R_y}{R_x} = +0.4105$$

$\theta = 22.3°$

1-37

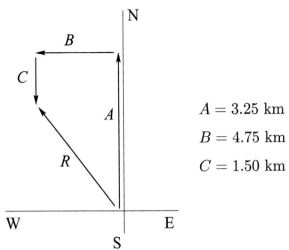

$A = 3.25$ km

$B = 4.75$ km

$C = 1.50$ km

Select a coordinate system where $+x$ is east and $+y$ is north. Let $\vec{A}$, $\vec{B}$ and $\vec{C}$ be the three displacements of the professor. Then the resultant displacement $\vec{R}$ is given by $\vec{R} = \vec{A} + \vec{B} + \vec{C}$. By the method of components, $R_x = A_x + B_x + C_x$ and $R_y = A_y + B_y + C_y$. Find the x and y components of each vector; add them to find the components of the resultant. Then the magnitude and direction of the resultant can be found from its x and y components that we have calculated. As always, it is essential to draw a sketch.

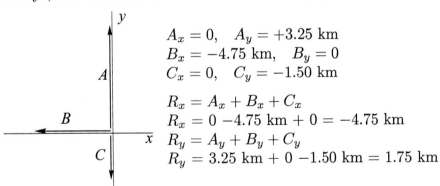

$A_x = 0, \quad A_y = +3.25$ km
$B_x = -4.75$ km, $\quad B_y = 0$
$C_x = 0, \quad C_y = -1.50$ km

$R_x = A_x + B_x + C_x$
$R_x = 0 - 4.75 \text{ km} + 0 = -4.75$ km
$R_y = A_y + B_y + C_y$
$R_y = 3.25 \text{ km} + 0 - 1.50 \text{ km} = 1.75$ km

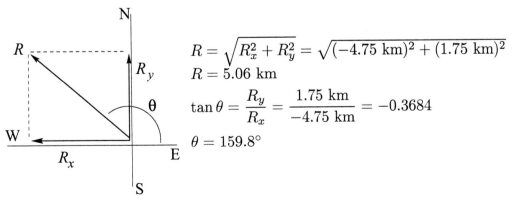

$R = \sqrt{R_x^2 + R_y^2} = \sqrt{(-4.75 \text{ km})^2 + (1.75 \text{ km})^2}$
$R = 5.06$ km

$\tan \theta = \dfrac{R_y}{R_x} = \dfrac{1.75 \text{ km}}{-4.75 \text{ km}} = -0.3684$

$\theta = 159.8°$

The angle θ measured counterclockwise from the $+x$-axis. In terms of compass directions, the resultant displacement is $20.2°$ N of W.

1-39

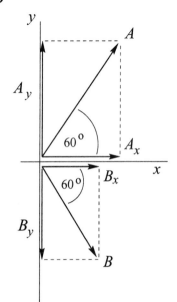

$A_x = A \cos(60.0°)$
$A_x = (2.80 \text{ cm}) \cos(60.0°) = +1.40 \text{ cm}$
$A_y = A \sin(60.0°)$
$A_y = (2.80 \text{ cm}) \sin(60.0°) = +2.425 \text{ cm}$

$B_x = B \cos(-60.0°)$
$B_x = (1.90 \text{ cm}) \cos(-60.0°) = +0.95 \text{ cm}$
$B_y = B \sin(-60.0°)$
$B_y = (1.90 \text{ cm}) \sin(-60.0°) = -1.645 \text{ cm}$

Note that the signs of the components correspond to the directions of the component vectors.

a) Now let $\vec{R} = \vec{A} + \vec{B}$.

$R_x = A_x + B_x = +1.40 \text{ cm} + 0.95 \text{ cm} = +2.35 \text{ m}.$
$R_y = A_y + B_y = +2.425 \text{ cm} - 1.645 \text{ cm} = +0.78 \text{ cm}.$

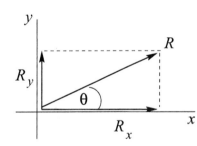

$R = \sqrt{R_x^2 + R_y^2} = \sqrt{(2.35 \text{ cm})^2 + (0.78 \text{ cm})^2}$
$R = 2.48 \text{ km}$

$\tan \theta = \dfrac{R_y}{R_x} = \dfrac{0.78 \text{ cm}}{+2.35 \text{ cm}} = +0.3319$

$\theta = 18.4°$

b) Now let $\vec{R} = \vec{A} - \vec{B}$.

$R_x = A_x - B_x = +1.40 \text{ cm} - 0.95 \text{ cm} = +0.45 \text{ cm}.$
$R_y = A_y - B_y = +2.425 \text{ cm} + 1.645 \text{ cm} = +4.070 \text{ cm}.$

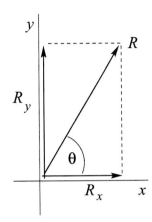

$$R = \sqrt{R_x^2 + R_y^2} = \sqrt{(0.45 \text{ cm})^2 + (4.070 \text{ cm})^2}$$
$$R = 4.09 \text{ cm}$$
$$\tan \theta = \frac{R_y}{R_x} = \frac{4.070 \text{ cm}}{0.45 \text{ cm}} = +9.044$$
$$\theta = 83.7°$$

c)

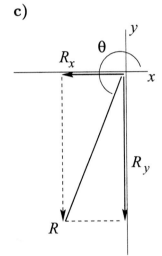

$$\vec{B} - \vec{A} = -(\vec{A} - \vec{B})$$
$\vec{B} - \vec{A}$ and $\vec{A} - \vec{B}$ are equal in magnitude and opposite in direction.

$$R = 4.09 \text{ cm and } \theta = 83.7° + 180° = 264°$$

1-41 We can use the x- and y-components of each vector as calculated in Exercise 1-33.
$$\vec{A} = (+7.22 \text{ m})\hat{i} + (+9.58 \text{ m})\hat{j}$$
$$\vec{B} = (+11.5 \text{ m})\hat{i} + (-9.64 \text{ m})\hat{j}$$
$$\vec{C} = (-3.00 \text{ m})\hat{i} + (-5.2 \text{ m})\hat{j}$$

1-43 **a)** $\vec{A} = 4.00\hat{i} + 3.00\hat{j}$; $A_x = +4.00$; $A_y = +3.00$
$$A = \sqrt{A_x^2 + A_y^2} = \sqrt{(4.00)^2 + (3.00)^2} = 5.00$$

$$\vec{B} = 5.00\hat{i} - 2.00\hat{j}$$; $B_x = +5.00$; $B_y = -2.00$
$$B = \sqrt{B_x^2 + B_y^2} = \sqrt{(5.00)^2 + (-2.00)^2} = 5.39$$

b) $\vec{A} - \vec{B} = 4.00\hat{i} + 3.00\hat{j} - (5.00\hat{i} - 2.00\hat{j}) = (4.00 - 5.00)\hat{i} + (3.00 + 2.00)\hat{j}$
$$\vec{A} - \vec{B} = -1.00\hat{i} + 5.00\hat{j}$$

c) Let $\vec{R} = \vec{A} - \vec{B} = -1.00\hat{i} + 5.00\hat{j}$. Then $R_x = -1.00$, $R_y = 5.00$.

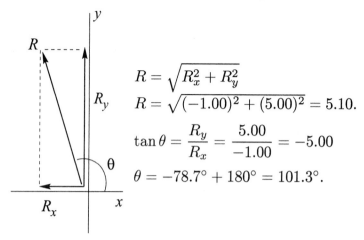

$$R = \sqrt{R_x^2 + R_y^2}$$
$$R = \sqrt{(-1.00)^2 + (5.00)^2} = 5.10.$$

$$\tan\theta = \frac{R_y}{R_x} = \frac{5.00}{-1.00} = -5.00$$

$$\theta = -78.7° + 180° = 101.3°.$$

1-45 a) $\vec{A} = 4.00\hat{i} + 3.00\hat{j}$, $\vec{B} = 5.00\hat{i} - 2.00\hat{j}$

$\vec{A} \cdot \vec{B} = (4.00\hat{i} + 3.00\hat{j}) \cdot (5.00\hat{i} - 2.00\hat{j}) = (4.00)(5.00) + (3.00)(-2.00) = 20.0 - 6.0 = +14.0.$

b) $\cos\phi = \dfrac{\vec{A} \cdot \vec{B}}{AB} = \dfrac{14.0}{(5.00)(5.39)} = 0.519;$ $\phi = 58.7°.$

1-49 $\vec{A} = 4.00\hat{i} + 3.00\hat{j}$, $\vec{B} = 5.00\hat{i} - 2.00\hat{j}$

$\vec{A} \times \vec{B} = (4.00\hat{i} + 3.00\hat{j}) \times (5.00\hat{i} - 2.00\hat{j}) =$
$20.0\ \hat{i} \times \hat{i} - 8.00\ \hat{i} \times \hat{j} + 15.0\ \hat{j} \times \hat{i} - 6.00\ \hat{j} \times \hat{j}$

But $\hat{i} \times \hat{i} = \hat{j} \times \hat{j} = 0$ and $\hat{i} \times \hat{j} = \hat{k}$, $\hat{j} \times \hat{i} = -\hat{k}$, so
$\vec{A} \times \vec{B} = -8.00\ \hat{k} + 15.0\ (-\hat{k}) = -23.0\ \hat{k}.$

The magnitude of $\vec{A} \times \vec{B}$ is 23.0.

Note: Sketch the vectors $\vec{A}$ and $\vec{B}$ in a coordinate system where the xy-plane is in the plane of the paper and the z-axis is directed out toward you.

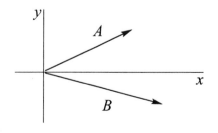

By the right-hand rule $\vec{A} \times \vec{B}$ is directed into the plane of the paper, in the $-z$-direction. This agrees with the above calculation that used unit vectors.

Problems

1-53 a) $f = 1.420 \times 10^9$ cycles/s, so $\dfrac{1}{1.420 \times 10^9}$ s $= 7.04 \times 10^{-10}$ s for one cycle.

b) $\dfrac{3600 \text{ s/h}}{7.04 \times 10^{-10} \text{ s/cycle}} = 5.11 \times 10^{12}$ cycles/h

c) Calculate the number of seconds in 4600 million years $= 4.6 \times 10^9$ y and divide by the time for 1 cycle:

$$\frac{(4.6 \times 10^9)(3.156 \times 10^7 \text{ s/y})}{7.04 \times 10^{-10} \text{ s/cycle}} = 2.1 \times 10^{26} \text{ cycles}$$

d) The clock is off by 1 s in 100,000 y $= 1 \times 10^5$ y, so in 4.60×10^9 y it is off by

$(1 \text{ s}) \left(\dfrac{4.60 \times 10^9}{1 \times 10^5} \right) = 4.6 \times 10^4$ s (about 13 h).

1-57 a) $\vec{A} + \vec{B} = \vec{C}$ (or $\vec{B} + \vec{A} = \vec{C}$)

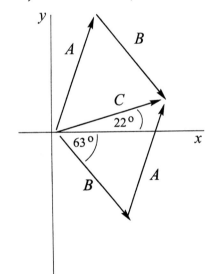

$C_x = A_x + B_x, \quad$ so $A_x = C_x - B_x$
$C_y = A_y + B_y, \quad$ so $A_y = C_y - B_y$

$C_x = C \cos 22.0° = (6.40 \text{ cm}) \cos 22.0°$
$C_x = +5.934$ cm
$C_y = C \sin 22.0° = (6.40 \text{ cm}) \sin 22.0°$
$C_y = +2.397$ cm

$B_x = B \cos(360° - 63.0°) = (6.40 \text{ cm}) \cos 297.0°$
$B_x = +2.906$ cm
$B_y = B \sin 297.0° = (6.40 \text{ cm}) \sin 297.0°$
$B_y = -5.702$ cm

b) $A_x = C_x - B_x = +5.934 \text{ cm} - 2.906 \text{ cm} = +3.03 \text{ cm}$
$A_y = C_y - B_y = +2.397 \text{ cm} - (-5.702) \text{ cm} = +8.10 \text{ cm}$

c)

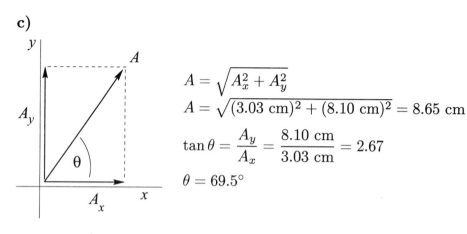

$$A = \sqrt{A_x^2 + A_y^2}$$

$$A = \sqrt{(3.03 \text{ cm})^2 + (8.10 \text{ cm})^2} = 8.65 \text{ cm}$$

$$\tan\theta = \frac{A_y}{A_x} = \frac{8.10 \text{ cm}}{3.03 \text{ cm}} = 2.67$$

$$\theta = 69.5°$$

1-59 Use a coordinate system where east is in the $+x$-direction and north is in the $+y$-direction.

Let $\vec{A}$, $\vec{B}$, and $\vec{C}$ be the three displacements that are given and let $\vec{D}$ be the fourth unmeasured displacement. Then the resultant displacement is $\vec{R} = \vec{A} + \vec{B} + \vec{C} + \vec{D}$. And since she ends up back where she started, $\vec{R} = \mathbf{0}$.

$\mathbf{0} = \vec{A} + \vec{B} + \vec{C} + \vec{D}$, so $\vec{D} = -(\vec{A} + \vec{B} + \vec{C})$

$D_x = -(A_x + B_x + C_x)$ and $D_y = -(A_y + B_y + C_y)$

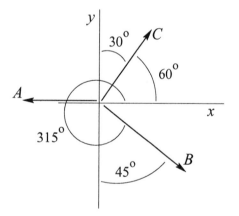

$$A_x = -180 \text{ m}, \ A_y = 0$$

$$B_x = B\cos 315° = (210 \text{ m})\cos 315° = +148.5 \text{ m}$$
$$B_y = B\sin 315° = (210 \text{ m})\sin 315° = -148.5 \text{ m}$$

$$C_x = C\cos 60° = (280 \text{ m})\cos 60° = +140 \text{ m}$$
$$C_y = C\sin 60° = (280 \text{ m})\sin 60° = +242.5 \text{ m}$$

$$D_x = -(A_x + B_x + C_x) = -(-180 \text{ m} + 148.5 \text{ m} + 140 \text{ m}) = -108.5 \text{ m}$$
$$D_y = -(A_y + B_y + C_y) = -(0 - 148.5 \text{ m} + 242.5 \text{ m}) = -94.0 \text{ m}$$

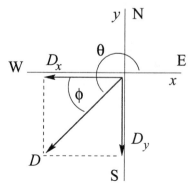

$$D = \sqrt{D_x^2 + D_y^2}$$

$$D = \sqrt{(-108.5 \text{ m})^2 + (-94.0 \text{ m})^2} = 144 \text{ m}$$

$$\tan\theta = \frac{D_y}{D_x} = \frac{-94.0 \text{ m}}{-108.5 \text{ m}} = 0.8664$$

$$\theta = 180° + 40.9° = 220.9°$$
($\vec{D}$ is in the third quadrant since both D_x and D_y are negative.)

The direction of $\vec{D}$ can also be specified in terms of $\phi = \theta - 180° = 40.9°$; $\vec{D}$ is 41° south of west.

-61 a) Let the three displacements that are given in the problem be called $\vec{A}$, $\vec{B}$, and $\vec{C}$.

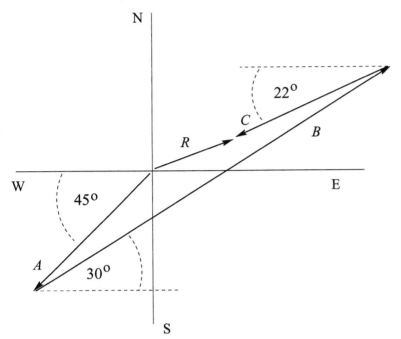

Let the resultant displacement be $\vec{R}$; $\vec{R} = \vec{A} + \vec{B} + \vec{C}$.

Part (b) of the problem is asking for the magnitude. $\vec{R}$.

b) Use a coordinate system where the x-direction is east and the y-direction is north.

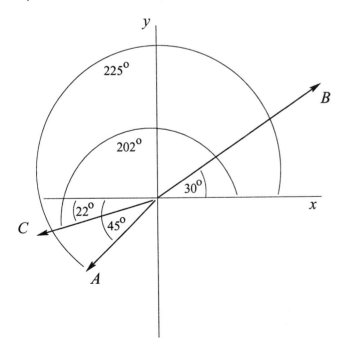

$A_x = A \cos 225.0°$
$A_x = (2.80 \text{ km}) \cos 225.0°$
$A_x - 1.980 \text{ km}$
$A_y = A \sin 225.0°$
$A_y = (2.80 \text{ km}) \sin 225.0°$
$A_y = -1.980 \text{ km}$

$B_x = B \cos 30.0°$
$B_x = (7.40 \text{ km}) \cos 30.0°$
$B_x = 6.409 \text{ km}$
$B_y = B \sin 30.0°$
$B_y = (7.40 \text{ km}) \sin 30.0°$
$B_y = 3.70 \text{ km}$

$C_x = C \cos 202.0° = (3.30 \text{ km}) \cos 202.0° = -3.060 \text{ km}$

$C_y = C \sin 202.0° = (3.30 \text{ km}) \sin 202.0° = -1.236 \text{ km}$

Note that in each case the signs of the components correspond to the directions the component vectors.

$R_x = A_x + B_x + C_x = -1.980 \text{ km} + 6.409 \text{ km} - 3.060 \text{ km} = +1.369 \text{ km}$

$R_y = A_y + B_y + C_y = -1.980 \text{ km} + 3.70 \text{ km} - 1.236 \text{ km} = +0.484 \text{ km}$

$R = \sqrt{R_x^2 + R_y^2} = \sqrt{(+1.369 \text{ km})^2 + (0.484 \text{ km})^2} = 1.45 \text{ km.}$

1-69 $\vec{A} = -2.00\hat{i} + 3.00\hat{j} + 4.00\hat{k}$, $\vec{B} = 3.00\hat{i} + 1.00\hat{j} - 3.00\hat{k}$

a) $A = \sqrt{A_x^2 + A_y^2 + A_z^2} = \sqrt{(-2.00)^2 + (3.00)^2 + (4.00)^2} = 5.38$

$B = \sqrt{B_x^2 + B_y^2 + B_z^2} = \sqrt{(3.00)^2 + (1.00)^2 + (-3.00)^2} = 4.36$

b) $\vec{A} - \vec{B} = (-2.00\hat{i} + 3.00\hat{j} + 4.00\hat{k}) - (3.00\hat{i} + 1.00\hat{j} - 3.00\hat{k})$

$\vec{A} - \vec{B} = (-2.00 - 3.00)\hat{i} + (3.00 - 1.00)\hat{j} + (4.00 - (-3.00))\hat{k} = -5.00\hat{i} + 2.00\hat{j}$ $7.00\hat{k}$.

c) Let $\vec{C} = \vec{A} - \vec{B}$, so $C_x = -5.00$, $C_y = +2.00$, $C_z = +7.00$

$C = \sqrt{C_x^2 + C_y^2 + C_z^2} = \sqrt{(-5.00)^2 + (2.00)^2 + (7.00)^2} = 8.83$

$\vec{B} - \vec{A} = \vec{A} - \vec{B}$, so $\vec{A} - \vec{B}$ and $\vec{B} - \vec{A}$ have the same magnitude but opposi directions.

1-73   Use $\cos\phi = \dfrac{\vec{A} \cdot \vec{B}}{AB}$

a) $\vec{A} = \hat{k}$ (along line ab)

$\vec{B} = \hat{i} + \hat{j} + \hat{k}$ (along line ad)

$A = 1$, $B = \sqrt{1^2 + 1^2 + 1^2} = \sqrt{3}$

$\vec{A} \cdot \vec{B} = \hat{k} \cdot (\hat{i} + \hat{j} + \hat{k}) = 1$

So $\cos\phi = \dfrac{\vec{A} \cdot \vec{B}}{AB} = 1/\sqrt{3}$; $\phi = 54.7°$

b) $\vec{A} = \hat{i} + \hat{j} + \hat{k}$ (along line ad)

$\vec{B} = \hat{j} + \hat{k}$ (along line ac)

$A = \sqrt{1^2 + 1^2 + 1^2} = \sqrt{3}$; $B = \sqrt{1^2 + 1^2} = \sqrt{2}$

$\vec{A} \cdot \vec{B} = (\hat{i} + \hat{j} + \hat{k}) \cdot (\hat{i} + \hat{j}) = 1 + 1 = 2$

So $\cos\phi = \dfrac{\vec{A}\cdot\vec{B}}{AB} = \dfrac{2}{\sqrt{3}\sqrt{2}} = \dfrac{2}{\sqrt{6}};$ $\phi = 35.3°$

-75 a) $\vec{A}\cdot(\vec{B}\times\vec{C}) = A_x(\vec{B}\times\vec{C})_x + A_y(\vec{B}\times\vec{C})_y + A_z(\vec{B}\times\vec{C})_z$

$\vec{A}\cdot(\vec{B}\times\vec{C}) = A_x(B_yC_z - B_zC_y) + A_y(B_zC_x - B_xC_z) + A_z(B_xC_y - B_yC_x)$

$(\vec{A}\times\vec{B})\cdot\vec{C} = (\vec{A}\times\vec{B})_xC_x + (\vec{A}\times\vec{B})_yC_y + (\vec{A}\times\vec{B})_zC_z$

$(\vec{A}\times\vec{B})\cdot\vec{C} = (A_yB_z - A_zB_y)C_x + (A_zB_x - A_xB_z)C_y + (A_xB_y - A_yB_x)C_z$

Comparison of the expressions for $\vec{A}\cdot(\vec{B}\times\vec{C})$ and $(\vec{A}\times\vec{B})\cdot\vec{C}$ shows they contain the same terms, so $\vec{A}\cdot(\vec{B}\times\vec{C}) = (\vec{A}\times\vec{B})\cdot\vec{C}$.

b) $A = 5.00,\ \theta_A = 26.0°;$ $B = 4.00,\ \theta_B = 63.0°$

$|\vec{A}\times\vec{B}| = AB\sin\phi.$

The angle ϕ between $\vec{A}$ and $\vec{B}$ is equal to $\phi = \theta_B - \theta_A = 63.0° - 26.0° = 37.0°$.

So $|\vec{A}\times\vec{B}| = (5.00)(4.00)\sin 37.0° = 12.04$, and by the right hand-rule $\vec{A}\times\vec{B}$ is in the $+z$-direction.

Thus $(\vec{A}\times\vec{B})\cdot\vec{C} = (12.04)(6.00) = 72.2$

CHAPTER 2
MOTION ALONG A STRAIGHT LINE

Exercises 3, 5, 7, 11, 15, 17, 25, 27, 31, 33, 37, 39, 41, 45
Problems 49, 51, 55, 57, 61, 63, 67, 69, 71, 75

Exercises

2-3 $v_{av} = \dfrac{\Delta x}{\Delta t}$ so $\Delta x = v_{av}\,\Delta t$ and $\Delta t = \dfrac{\Delta x}{v_{av}}$.

Use the information given for normal driving conditions to calculate the distan[ce] between the two cities:

$\Delta x = v_{av}\,\Delta t = (105\ \text{km/h})(1\ \text{h}/60\ \text{min})(140\ \text{min}) = 245\ \text{km}.$

Now use v_{av} for a rainy day to calculate Δt; Δx is the same as before:

$\Delta t = \dfrac{\Delta x}{v_{av}} = \dfrac{245\ \text{km}}{70\ \text{km/h}} = 3.50\ \text{h} = 3\ \text{h and 30 min}.$

The trip takes an additional 1 hour and 10 minutes.

2-5 a) $v_{av} = \dfrac{\Delta x}{\Delta t}$, where Δx is the total displacement for the entire 120 s.

$\Delta x = (8.0\ \text{m/s})(60\ \text{s}) + (20.0\ \text{m/s})(60.0\ \text{s}) = 1680\ \text{m}$

$v_{av} = \dfrac{1680\ \text{m}}{120\ \text{s}} = 14\ \text{m/s}$

b) $v_{av} = \dfrac{\Delta x}{\Delta t}$

We know that the total displacement is $240\ \text{m} + 240\ \text{m} = 480\ \text{m}$, but we mu[st] calculate the total elapsed time by calculating the elapsed time for each segment.

$\Delta t = \dfrac{\Delta x}{v_{av}}$, so $\Delta t = \dfrac{240\ \text{m}}{8.0\ \text{m/s}} + \dfrac{240\ \text{m}}{20.0\ \text{m/s}} = 30\ \text{s} + 12\ \text{s} = 42\ \text{s}.$

Then $v_{av} = \dfrac{480\ \text{m}}{42\ \text{m}} = 11.4\ \text{m/s}.$

c) In part (a) the numerical average of the two speeds is $\dfrac{(8.0\ \text{m/s} + 20.0\ \text{m/s})}{2} =$
14 m/s, which does equal v_{av}.

In part (b) the numerical average of the two speeds again is 14 m/s, but this do[es] not equal v_{av}.

The numerical average of the two speeds equals v_{av} in part (a) since there the tw[o] speeds are maintained for equal amounts of time; this is not the case in part (b).

2-7 **a)** $v_{av} = \dfrac{\Delta x}{\Delta t}$ so use $x(t)$ to find the displacement Δx for this time interval.

$t = 0$: $x = 0$

$t = 10.0$ s: $x = (2.40$ m/s$^2)(10.0$ s$)^2 - (0.120$ m/s$^3)(10.0$ s$)^3 = 240$ m $- 120$ m $=$ 120 m.

Then $v_{av} = \dfrac{\Delta x}{\Delta t} = \dfrac{120 \text{ m}}{10.0 \text{ s}} = 12.0$ m/s.

b) $v = \dfrac{dx}{dt} = 2bt - 3ct^2$.

(i) $t = 0$: $v = 0$

(ii) $t = 5.0$ s: $v = 2(2.40$ m/s$^2)(5.0$ s$) - 3(0.120$ m/s$^3)(5.0$ s$)^2 =$ 24.0 m/s $- 9.0$ m/s $= 15.0$ m/s.

(iii) $t = 10.0$ s: $v = 2(2.40$ m/s$^2)(10.0$ s$) - 3(0.120$ m/s$^3)(10.0$ s$)^2 =$ 48.0 m/s $- 36.0$ m/s $= 12.0$ m/s.

c) $v = 2bt - 3ct^2$

$v = 0$ at $t = 0$.

$v = 0$ next when $2bt - 3ct^2 = 0$

$2b = 3ct$ so $t = \dfrac{2b}{3c} = \dfrac{2(2.40 \text{ m/s}^2)}{3(0.120 \text{ m/s}^3)} = 13.3$ s

2-11 **a)** The acceleration a is the slope of the v versus t curve. a has its most positive value when the curve has its largest positive slope; this occurs between approximately 4 s and 7 s.

b) a has its most negative value when the curve has its most negative slope; this ocurs between approximately 30 s to 40 s.

c) At $t = 20$ s, the v versus t curve is a horizontal straight line with zero slope, so $a = 0$.

d) Between 30 s and 40 s the v versus t curve is a horizontal straight line with slope

$\dfrac{0 - 60 \text{ km/h}}{40 \text{ s} - 30 \text{ s}} = \dfrac{(-60 \text{ km/h})(10^3 \text{ m/1 km})(1 \text{ h/3600 s})}{10 \text{ s}} = -1.7$ m/s^2.

The acceleration is constant with this value in this time interval, so at $t = 35$ s it is $a = -1.7$ m/s^2.

e) $\underline{t = 5 \text{ s}}$

The average velocity for $t = 0$ to $t = 5$ s is approximately 15 km/h or 4.1 m/s, so $\Delta x = v_{av}\Delta t = 20$ m; if the car is at $x_0 = 0$ at $t = 0$ then it is at $x = 20$ m at $t = 5$ s.

The velocity at $t = 5$ s is 30 km/h = 8 m/s.

The acceleration at $t = 5$ s is approximately

$$\left(\frac{60 \text{ km/h} - 30 \text{ km/h}}{10 \text{ s} - 5 \text{ s}}\right)\left(\frac{10^3 \text{ m}}{1 \text{ km}}\right)\left(\frac{1 \text{ h}}{3600 \text{ s}}\right) = +1.7 \text{ m/s}^2.$$

$\underline{t = 15 \text{ s}}$

In the first 10 s the car has displacement $\Delta x \approx (8 \text{ m/s})(10 \text{ s}) = 80$ m. From 10 s t
15 s the car has constant speed 60 km/h = 17 m/s, so $\Delta x = (17 \text{ m/s})(5 \text{ s}) = 85$ m
At $t = 15$ s the car is at $x = 80$ m + 85 m = 165 m. The velocity at $t = 15$ s is 1
m/s. The acceleration is zero.

$\underline{t = 25 \text{ s}}$

For the time interval 10 s to 25 s the car has constant speed 17 m/s, so $\Delta x = (1$
m/s$)(15 \text{ s}) = 255$ m. At $t = 25$ s the car is at $x = 80$ m + 255 m = 335 m. Th
velocity is 17 m/s. The acceleration is zero.

$\underline{t = 35 \text{ s}}$

For the time interval 10 s to 30 s, $\Delta x = (17 \text{ m/s})(20 \text{ s}) = 340$ m. From $t = 30$ s t
35 s the average velocity is 45 km/h = 12 m/s, so $\Delta x = (12 \text{ m/s})(5 \text{ s}) = 60$ m. A
$t = 35$ s the car is at $x = 80$ m + 340 m + 60 m = 480 m. The velocity at $t = 35$
is $v = 30$ km/h = 8 m/s. From part (d), $a = -1.7$ m/s^2.

These results allow construction of the motion diagrams.

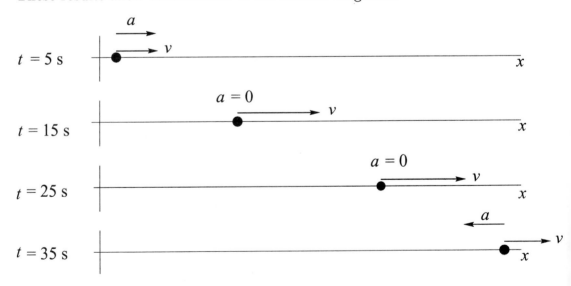

2-15 $\underline{t = 0 \text{ to } t = 5 \text{ s}}$: x versus t is a parabola so a is a constant. The curvature is positiv
so a is positive. v versus t is a straight line with positive slope. $v_0 = 0$.

$\underline{t = 5 \text{ s to } t = 15 \text{ s}}$: x versus t is a straight line so v is constant and $a = 0$. The slop of x versus t is positive so v is positive.

$\underline{t = 15 \text{ s to } t = 25 \text{ s}}$: x versus t is a parabola with negative curvature, so a is constant and negative. v versus t is a straight line with negative slope. The velocity is zero at 20 s, positive for 15 s to 20 s, and negative for 20 s to 25 s.

$\underline{t = 25 \text{ s to } t = 35 \text{ s}}$: x versus t is a parabola with positive curvature, so a is constant and positive. v versus t is a straight line with positive slope. The velocity reaches zero at $t = 40$ s.

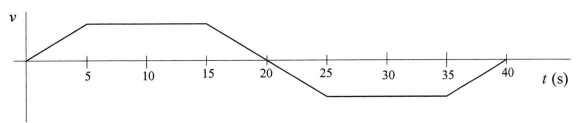

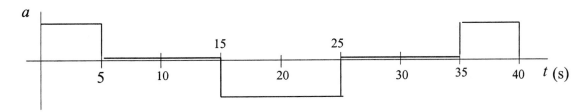

b) The motions diagrams are

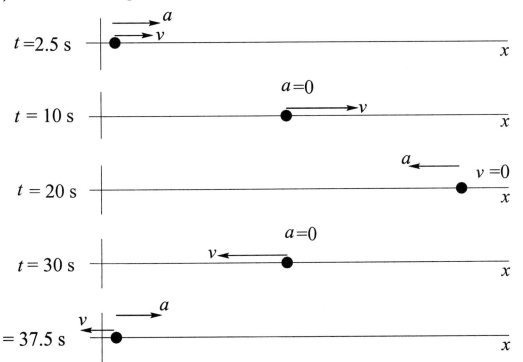

2-17 a)

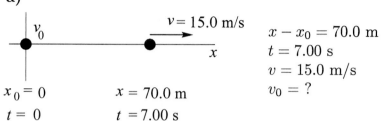

$x - x_0 = 70.0$ m
$t = 7.00$ s
$v = 15.0$ m/s
$v_0 = ?$

Use $x - x_0 = \left(\dfrac{v_0 + v}{2}\right) t$, so $v_0 = \dfrac{2(x - x_0)}{t} - v = \dfrac{2(70.0 \text{ m})}{7.00 \text{ s}} - 15.0 \text{ m/s} = 5.0$ m/

b) Use $v = v_0 + at$, so $a = \dfrac{v - v_0}{t} = \dfrac{15.0 \text{ m/s} - 5.0 \text{ m/s}}{7.00 \text{ s}} = 1.43 \text{ m/s}^2.$

2-25   a) The acceleration a at time t is the slope of the tangent to the v versus t curve at time t.

At $t = 3$ s, the v versus t curve is a horizontal straight line, with zero slope. Th·
$a = 0$.

At $t = 7$ s, the v versus t curve is a straight-line segment with slope $\dfrac{45 \text{ m/s} - 20 \text{ m/}}{9 \text{ s} - 5 \text{ s}}$
6.3 m/s^2. Thus $a = 6.3 \text{ m/s}^2$.

At $t = 11$ s the curve is again a straight-line segment, now with slope $\dfrac{-0 - 45 \text{ m/s}}{13 \text{ s} - 9 \text{ s}}$
-11.2 m/s^2. Thus $a = -11.2 \text{ m/s}^2$.

b) For the time interval $t = 0$ to $t = 5$ s the acceleration is constant and equal ·
zero. For the time interval $t = 5$ s to $t = 9$ s the acceleration is constant and equ·
to 6.25 m/s^2. For the interval $t = 0$ s to $t = 13$ s the acceleration is constant a·
equal to -11.2 m/s^2.

During the first 5 seconds the acceleration is constant, so the constant acceleratic·
kinematic formuals can be used.
$v_0 = 20$ m/s $a = 0$ $t = 5$ s $x - x_0 = ?$
$x - x_0 = v_0 t$ ($a = 0$ so no $\frac{1}{2}at^2$ term)
$x - x_0 = (20 \text{ m/s})(5 \text{ s}) = 100$ m; this is the distance the officer travels in the first
seconds.

During the interval $t = 5$ s to 9 s the acceleration is again constant. The consta·
acceleration formulas can be applied to this 4 second interval. It is convenient ·
restart our clock so the interval starts at time $t = 0$ and ends at time $t = 5$ s. (No·
that the acceleration is <u>not</u> constant over the entire $t = 0$ to $t = 9$ s interval.)
$v_0 = 20$ m/s $a = 6.25 \text{ m/s}^2$ $t = 4$ s $x_0 = 100$ m $x - x_0 = ?$

$x - x_0 = v_0t + \frac{1}{2}at^2$

$x - x_0 = (20 \text{ m/s})(4 \text{ s}) + \frac{1}{2}(6.25 \text{ m/s}^2)(4 \text{ s})^2 = 80 \text{ m} + 50 \text{ m} = 130 \text{ m}.$

Thus $x = x_0 + 130 \text{ m} = 100 \text{ m} + 130 \text{ m} = 230 \text{ m}.$

At $t = 9$ s the officer is at $x = 230$ m, so she has traveled 230 m in the first 9 seconds.

During the interval $t = 9$ s to $t = 13$ s the acceleration is again constant. The constant acceleration formulas can be applied for this 4 second interval but <u>not</u> for the whole $t = 0$ to $t = 13$ s interval. To use the equations restart our clock so this interval begins at time $t = 0$ and ends at time $t = 4$ s.

$v_0 = 45$ m/s (at the start of this time interval)

$a = -11.2 \text{ m/s}^2 \qquad t = 4 \text{ s} \qquad x_0 = 230 \text{ m} \qquad x - x_0 = ?$

$x - x_0 = v_0t + \frac{1}{2}at^2$

$x - x_0 = (45 \text{ m/s})(4 \text{ s}) + \frac{1}{2}(-11.2 \text{ m/s}^2)(4 \text{ s})^2 = 180 \text{ m} - 89.6 \text{ m} = 90.4 \text{ m}.$

Thus $x = x_0 + 90.4 \text{ m} = 230 \text{ m} + 90.4 \text{ m} = 320 \text{ m}.$

At $t = 13$ s the officer is at $x = 320$ m, so she has traveled 320 m in the first 13 seconds.

-27 **a)** The maximum speed occurs at the end of the initial acceleration period.

$a = 20.0 \text{ m/s}^2 \qquad t = 15.0 \text{ min} = 900 \text{ s} \qquad v_0 = 0 \qquad v = ?$

$v = v_0 + at$

$v = 0 + (20.0 \text{ m/s}^2)(900 \text{ s}) = 1.80 \times 10^4 \text{ m/s}$

b) The motion consists of three constant acceleration intervals. In the middle segment of the trip $a = 0$ and $v = 1.80 \times 10^4$ m/s, but we can't directly find the distance traveled during this part of the trip because we don't know the time. Instead, find the distance traveled in the first part of the trip (where $a = +20.0 \text{ m/s}^2$) and in the last part of the trip (where $a = -20.0 \text{ m/s}^2$). Subtract these two distances from the total distance of 3.84×10^8 m to find the distance traveled in the middle part of the trip (where $a = 0$).

first segment

$x - x_0 = ? \qquad t = 15.0 \text{ min} = 900 \text{ s} \qquad a = +20.0 \text{ m/s}^2 \qquad v_0 = 0$

$x - x_0 = v_0t + \frac{1}{2}at^2$

$x - x_0 = 0 + \frac{1}{2}(20.0 \text{ m/s}^2)(900 \text{ s})^2 = 8.10 \times 10^6 \text{ m} = 8.10 \times 10^3 \text{ km}$

second segment

$x - x_0 = ? \qquad t = 15.0 \text{ min} = 900 \text{ s} \qquad a = -20.0 \text{ m/s}^2 \qquad v_0 = 1.80 \times 10^4 \text{ s}$

$x - x_0 = v_0t + \frac{1}{2}at^2$

$x - x_0 = (1.80 \times 10^4 \text{ s})(900 \text{ s}) + \frac{1}{2}(-20.0 \text{ m/s}^2)(900 \text{ s})^2 = 8.10 \times 10^6 \text{ m} = 8.10 \times 10^3$ km (The same distance as traveled as in the first segment.)

Therefore, the distance traveled at constant speed is 3.84×10^8 m $- 8.10 \times 10^6$
$- 8.10 \times 10^6$ m $= 3.678 \times 10^8$ m $= 3.678 \times 10^5$ km.

The fraction this is of the total distance is $\dfrac{3.678 \times 10^8 \text{ m}}{3.84 \times 10^8 \text{ m}} = 0.958$.

c) Find the time for the constant speed segment:

$x - x_0 = 3.678 \times 10^8$ m $v = 1.80 \times 10^4$ m/s $a = 0$ $t = ?$

$x - x_0 = v_0 t + \frac{1}{2} a t^2$

$t = \dfrac{x - x_0}{v_0} = \dfrac{3.678 \times 10^8 \text{ m}}{1.80 \times 10^4 \text{ m/s}} = 2.043 \times 10^4 \text{ s} = 340.5$ min.

The total time for the whole trip is thus 15.0 min + 340.5 min + 15.0 min = 3̃
min.

2-31 **a)** For the car v is constant and $a = 0$, so x versus t is a straight line with positi̇
slope. For the motorcycle $v_0 = 0$ and a is constant and positive, so x versus t is
parabola with positive curvature and zero slope at $t = 0$. Let both vehicles be
$x_0 = 0$ at $t = 0$.

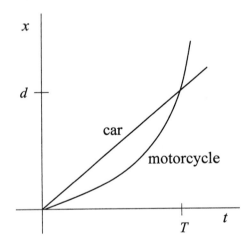

Let T be the time t when the
motorcycle overtakes the car and let
d be the displacement of the car
(and motorcycle) when this occurs

<u>car</u>

$v_0 = v_C$ $a = 0$ $t = T$ $x - x_0 = d$

Putting these values into $x - x_0 = v_0 t + \frac{1}{2} a t^2$ gives $\underline{d = v_C T}$.

<u>motorcycle</u>

$v_0 = 0$ $a = a_M$ $t = T$ $x - x_0 = d$

Putting these values into $x - x_0 = v_0 t + \frac{1}{2} a t^2$ gives $\underline{d = a_M T^2 / 2}$.

Combine these two equations to eliminate d:

$v_C T = \frac{1}{2} a_M T^2$

$T = \dfrac{2 v_C}{a_M}$

Then $v = v_0 + at$ for the motorcycle gives $v_M = 0 + a_M \left(\dfrac{2v_C}{a_M} \right) = 2v_C$, when the motorcycle has overtaken the car.

b) From part (a), $d = v_C T = v_C \left(\dfrac{2v_C}{a_M} \right) = 2v_C^2/a_M$ and $v_C^2 = \frac{1}{2}a_M d$.

Apply $v^2 = v_0^2 + 2a(x - x_0)$ to the motorcycle. This gives $v_M^2 = 2a_M(x - x_0)$. Set $v_M = v_C$; then $x - x_0$ is the distance the motorcycle has traveled when its velocity equals the velocity of the car. $x - x_0 = \dfrac{v_M^2}{2a_M} = \dfrac{v_C^2}{2a_M} = \dfrac{1}{2a_M} \left(\dfrac{a_M d}{2} \right) = \dfrac{d}{4}$.

The motorcycle has traveled a distance $d/4$ when its speed equals v_C.

2-33 Take the origin at the ground and the positive direction to be upward.

a) At the maximum height $v = 0$.

$v = 0 \qquad y - y_0 = 0.440 \text{ m} \qquad a = -9.80 \text{ m/s}^2 \qquad v_0 = ?$

$v^2 = v_0^2 + 2a(y - y_0)$

$v_0 = \sqrt{-2a(y - y_0)} = \sqrt{-2(-9.80 \text{ m/s}^2)(0.440 \text{ m})} = 2.94 \text{ m}$

b) When the flea has returned to the ground $y - y_0 = 0$.

$y - y_0 = 0 \qquad v_0 = +2.94 \text{ m/s} \qquad a = -9.80 \text{ m/s}^2 \qquad t = ?$

$x - x_0 = v_0 t + \frac{1}{2}at^2$

With $y - y_0 = 0$ this gives $t = -\dfrac{2v_0}{a} = -\dfrac{2(2.94 \text{ m/s})}{-9.80 \text{ m/s}^2} = 0.600 \text{ s}$.

2-37 **a)** $v_{av} = \dfrac{v_0 + v}{2}$

Need to use the constant acceleration formulas to solve for the velocity v of the ring just before it strikes the ground. Take the origin of coordinates at the roof and take the positive y-direction to be upward.

$v = ? \qquad v_0 = +5.00 \text{ m/s} \qquad a = -9.80 \text{ m/s}^2 \qquad y - y_0 = -12.0 \text{ m}$ (When the ring is at the ground its displacement is 12.0 m downward.)

$v^2 = v_0^2 + 2a(y - y_0)$

$v = -\sqrt{(5.00 \text{ m/s})^2 + 2(-9.80 \text{ m/s}^2)(-12.0 \text{ m})} = -16.13 \text{ m/s}$

Then $v^2 = v_0^2 + 2a(y - y_0) = \dfrac{+5.00 \text{ m/s} - 16.13 \text{ m/s}}{2} = -5.56 \text{ m/s}$. The minus sign indicates that the average velocity is downward.

b) The acceleration is constant and equal to -9.80 m/s^2, so the average value must equal this value: $a = -9.80$ m/s^2.

c) $t = ?$ $v = -16.13$ m/s $a = -9.80$ m/s^2, $v_0 = +5.00$ m/s

$$v = v_0 + at \text{ so } t = \frac{v - v_0}{a} = \frac{-16.13 \text{ m/s} - 5.00 \text{ m/s}}{-9.80 \text{ m/s}^2} = 2.16 \text{ s}$$

d) We found in part (a) that $v = -16.1$ m/s just before the ring strikes the groun

e)

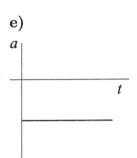
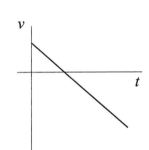
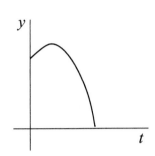

2-39 **a)** Take the $+y$ direction to be upward.

$t = 2.00$ s, $v_0 = -6.00$ m/s, $a = -9.80$ m/s^2, $v = ?$

$v = v_0 + at = -6.00$ m/s $+ (-9.80$ m/s$^2)(2.00$ s$) = -25.5$ m/s

b) $y - y_0 = ?$

$y - y_0 = v_0 t + \frac{1}{2}at^2 = (-6.00 \text{ m/s})(2.00 \text{ s}) + \frac{1}{2}(-9.80 \text{ m/s}^2)(2.00 \text{ s})^2 = -31.6$ m

c) $y - y_0 = -10.0$ m, $v_0 = -6.00$ m/s, $a = -9.80$ m/s^2, $v = ?$

$v^2 = v_0^2 + 2a(y - y_0)$

$v = -\sqrt{2a(y - y_0)} = -\sqrt{(-6.00 \text{ m/s})^2 + 2(-9.80 \text{ m/s}^2)(-10.0 \text{ m})} = -15.2$ m/s

d)

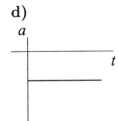
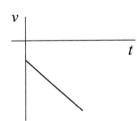
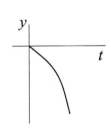

-41 a) $v = 224$ m/s, $v_0 = 0$, $t = 0.900$ s, $a = ?$

$v = v_0 + at$

$$a = \frac{v - v_0}{t} = \frac{224 \text{ m/s} - 0}{0.900 \text{ s}} = 249 \text{ m/s}^2$$

b) $a/g = (249 \text{ m/s}^2)/(9.80 \text{ m/s}^2) = 25.4$

c) $x - x_0 = v_0 t + \frac{1}{2}at^2 = 0 + \frac{1}{2}(249 \text{ m/s}^2)(0.900 \text{ s})^2 = 101$ m

d) Calculate the acceleration, assuming it is constant:

$t = 1.40$ s, $v_0 = 283$ m/s, $v = 0$ (stops), $a = ?$

$v = v_0 + at$

$$a = \frac{v - v_0}{t} = \frac{0 - 283 \text{ m/s}}{1.40 \text{ s}} = -202 \text{ m/s}^2$$

$a/g = (-202 \text{ m/s}^2)/(9.80 \text{ m/s}^2) = -20.6;$ $a = -20.6g$

If the acceleration while the sled is stopping is constant then the magnitude of the acceleration is only $20.6g$. But if the acceleration is not constant it is certainly possible that at some point the instantaneous acceleration could be as large as $40g$.

-45 $a = At - Bt^2$ with $A = 1.50$ m/s^3 and $B = 0.120$ m/s^4

a) $v = v_0 + \int_0^t a\, dt = v_0 + \int_0^t (At - Bt^2)\, dt = v_0 + \frac{1}{2}At^2 - \frac{1}{3}Bt^3$

At rest at $t = 0$ says that $v_0 = 0$, so

$v = \frac{1}{2}At^2 - \frac{1}{3}Bt^3 = \frac{1}{2}(1.50 \text{ m/s}^3)t^2 - \frac{1}{3}(0.120 \text{ m/s}^4)t^3$

$v = (0.75 \text{ m/s}^3)t^2 - (0.040 \text{ m/s}^4)t^3$

$x = x_0 + \int_0^t v\, dt = x_0 + \int_0^t (\frac{1}{2}At^2 - \frac{1}{3}Bt^3)\, dt = x_0 + \frac{1}{6}At^3 - \frac{1}{12}Bt^4$

At the origin at $t = 0$ says that $x_0 = 0$, so

$x = \frac{1}{6}At^3 - \frac{1}{12}Bt^4 = \frac{1}{6}(1.50 \text{ m/s}^3)t^3 - \frac{1}{12}(0.120 \text{ m/s}^4)t^4$

$x = (0.25 \text{ m/s}^3)t^3 - (0.010 \text{ m/s}^4)t^4$

b) At time t, when v is a maximum, $\dfrac{dv}{dt} = 0$. (Since $a = \dfrac{dv}{dt}$, the maximum velocity is when $a = 0$. For earlier times a is positive so v is still increasing. For later times a is negative and v is decreasing.)

$$a = \frac{dv}{dt} = 0 \text{ so } At - Bt^2 = 0$$

One root is $t = 0$, but there $v = 0$ and not a maximum.

The other root is $t = \dfrac{A}{B} = \dfrac{1.50 \text{ m/s}^3}{0.120 \text{ m/s}^4} = 12.5 \text{ s}$

Then $v = (0.75 \text{ m/s}^3)t^2 - (0.040 \text{ m/s}^4)t^3$ gives
$v = (0.75 \text{ m/s}^3)(12.5 \text{ s})^2 - (0.040 \text{ m/s}^4)(12.5 \text{ s})^3 = 117.2 \text{ m/s} - 78.1 \text{ m/s} = 39.1 \text{ m/s}.$

Problems

2-49 Take the origin to be at Seward and the positive direction to be west.

a) average speed $= \dfrac{\text{distance traveled}}{\text{time}}$

The distance traveled (different from the net displacement $(x - x_0)$) is
76 km + 34 km = 110 km.

Find the total elapsed time by using $v_{av} = \dfrac{\Delta x}{\Delta t} = \dfrac{x - x_0}{t}$ to find t for each leg the journey.

Seward to Auora: $t = \dfrac{x - x_0}{v_{av}} = \dfrac{76 \text{ km}}{88 \text{ km/h}} = 0.8636 \text{ h}$

Auora to York: $t = \dfrac{x - x_0}{v_{av}} = \dfrac{-34 \text{ km}}{-72 \text{ km/h}} = 0.4722 \text{ h}$

Total $t = 0.8636 \text{ h} + 0.4722 \text{ h} = 1.336 \text{ h}.$

Then average speed $= \dfrac{110 \text{ km}}{1.336 \text{ h}} = 82 \text{ km/h}.$

b) $v_{av} = \dfrac{\Delta x}{\Delta t}$, where Δx is the displacement, not the total distance traveled.
For the whole trip he ends up 76 km − 34 km = 42 km west of his starting point

$v_{av} = \dfrac{42 \text{ km}}{1.336 \text{ h}} = 31 \text{ km/h}.$

2-51 a) $a_{av} = \dfrac{\Delta v}{\Delta t} = \dfrac{v - v_0}{t}$

$v_0 = 0$ since the runner starts from rest.

$t = 4.0$ s, but we need to calculate v, the speed of the runner at the end of t acceleration period.

For the last $9.1 \text{ s} - 4.0 \text{ s} = 5.1 \text{ s}$ the acceleration is zero and the runner travels a distance of $d_1 = (5.1 \text{ s})v$ (obtained using $x - x_0 = v_0 t + \frac{1}{2}at^2$)

During the acceleration phase of 4.0 s, where the velocity goes from 0 to v, the runner travels a distance

$$d_2 = \left(\frac{v_0 + v}{2}\right)t = \frac{v}{2}(4.0 \text{ s}) = (2.0 \text{ s})v$$

The total distance traveled is 100 m, so $d_1 + d_2 = 100$ m. This gives $(5.1 \text{ s})v + (2.0 \text{ s})v = 100$ m.

$$v = \frac{100 \text{ m}}{7.1 \text{ s}} = 14.08 \text{ m/s}.$$

Now we can calculate a_{av}: $a_{av} = \dfrac{v - v_0}{t} = \dfrac{14.08 \text{ s} - 0}{4.0 \text{ s}} = 3.5 \text{ m/s}^2.$

b) For this time interval the velocity is constant, so $a_{av} = 0$.

c) $a_{av} = \dfrac{v - v_0}{t}$

We have calculated the final speed to be 14.08 m/s, so

$$a_{av} = \frac{14.08 \text{ m/s}}{9.1 \text{ s}} = 1.5 \text{ m/s}^2.$$

d) The runner spends different times moving with the average accelerations of parts (a) and (b).

2-55 Let T be the time when you catch up with the cockroach.

Take $x = 0$ to be at the $t = 0$ location of the roach and positive x to be in the direction of motion of the two objects.

roach:

$v_0 = 1.50 \text{ m/s}, \quad a = 0, \quad x_0 = 0, \quad x = 1.20 \text{ m}, \quad t = T$

you:

$v_0 = 0.80 \text{ m/s}, \quad x_0 = -0.90 \text{ m}, \quad x = 1.20 \text{ m}, \quad t = T, \quad a = ?$

Apply $x - x_0 = v_0 t + \frac{1}{2}at^2$ to both objects:

roach: $1.20 \text{ m} = (1.50 \text{ m/s})T, \quad$ so $T = 0.800$ s.
you: $1.20 \text{ m} - (-0.90 \text{ m}) = (0.80 \text{ m/s})T + \frac{1}{2}aT^2$
$2.10 \text{ m} = (0.80 \text{ m/s})(0.800 \text{ s}) + \frac{1}{2}a(0.800 \text{ s})^2$
$2.10 \text{ m} = 0.64 \text{ m} + (0.320 \text{ s}^2)a$
$a = 4.6 \text{ m/s}^2.$

2-57 Take the origin of coordinates to be at the initial position of the truck.

Let d be the distance that the auto initially is behind the truck, so $x_0(\text{auto}) = -$ and $x_0(\text{truck}) = 0$. Let T be the time it takes the auto to catch the truck. Thus time T the truck has undergone a displacement $x - x_0 = 40.0$ m, so is at $x = x_0 + 40$ m $= 40.0$ m. The auto has caught the truck so at time T it is also at $x = 40.0$ m

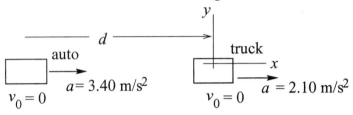

a) Use the motion of the truck to calculate T:

$x - x_0 = 40.0$ m, $v_0 = 0$ (starts from rest), $a = 2.10$ m/s^2, $t = T$

$x - x_0 = v_0 t + \frac{1}{2} a t^2$

Since $v_0 = 0$, this gives $t = \sqrt{\dfrac{2(x - x_0)}{a}}$

$$T = \sqrt{\frac{2(40.0 \text{ m})}{2.10 \text{ m/s}^2}} = 6.17 \text{ s}$$

b) Use the motion of the auto to calculate d:

$x - x_0 = 40.0$ m $+ d$, $v_0 = 0$, $a = 3.40$ m/s^2, $t = 6.17$ s

$x - x_0 = v_0 t + \frac{1}{2} a t^2$

$d + 40.0$ m $= \frac{1}{2}(3.40 \text{ m/s}^2)(6.17 \text{ s})^2$

$d = 64.8$ m $- 40.0$ m $= 24.8$ m

c) auto: $v = v_0 + at = 0 + (3.40 \text{ m/s}^2)(6.17 \text{ s}) = 21.0$ m/s

truck: $v = v_0 + at = 0 + (2.10 \text{ m/s}^2)(6.17 \text{ s}) = 13.0$ m/s

d)

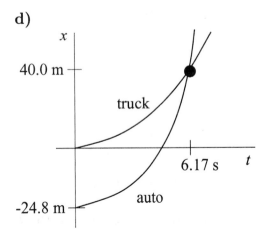

-61 **a)** It is very convenient to work in coordinates attached to the truck.

Note that these coordinates move at constant velocity relative to the earth. in these coordinates the truck is at rest, and the initial velocity of the car is $v_0 = 0$. Also, the car's acceleration in these coordinates is the same as in coordinates fixed to the earth.

First, let's calculate how far the car must travel relative to the truck:

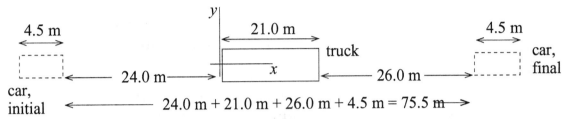

The car goes from $x_0 = -24.0$ m to $x = 51.5$ m. So $x - x_0 = 75.5$ m for the car. Calculate the time it takes the car to travel this distance:

$a = 0.600$ m/s^2, $v_0 = 0$, $x - x_0 = 75.5$ m, $t = ?$

$x - x_0 = v_0 t + \frac{1}{2}at^2$

$t = \sqrt{\dfrac{2(x - x_0)}{a}} = \sqrt{\dfrac{2(75.5 \text{ m})}{0.600 \text{ m/s}^2}} = 15.86$ s

It takes the car 15.9 s to pass the truck.

b) Need how far the car travels relative to the earth, so go now to coordinates fixed to the earth. In these coordinates $v_0 = 20.0$ m/s for the car. Take the origin to be at the initial position of the car.

$v_0 = 20.0$ m/s, $a = 0.600$ m/s^2, $t = 15.86$ s, $x - x_0 = ?$

$x - x_0 = v_0 t + \frac{1}{2}at^2 = (20.0 \text{ m/s})(15.86 \text{ s}) + \frac{1}{2}(0.60 \text{ m/s}^2)(15.86 \text{ s})^2$

$x - x_0 = 317.2$ m $+ 75.5$ m $= 393$ m.

c) In coordinates fixed to the earth:

$v = v_0 + at = 20.0$ m/s $+ (0.600 \text{ m/s}^2)(15.86 \text{ s}) = 29.5$ m/s

2-63 $a(t) = \alpha + \beta t$, with $\alpha = -2.00$ m/s^2 and $\beta = 3.00$ m/s^3

a) Find $v(t)$ and $x(t)$:

$v = v_0 + \int_0^t a \, dt = v_0 + \int_0^t (\alpha + \beta t) \, dt = v_0 + \alpha t + \frac{1}{2}\beta t^2$

$x = x_0 + \int_0^t v \, dt = x_0 + \int_0^t (v_0 + \alpha t + \frac{1}{2}\beta t^2) \, dt = x_0 + v_0 t + \frac{1}{2}\alpha t^2 + \frac{1}{6}\beta t^3$

At $t = 0$, $x = x_0$.

To have $x = x_0$ at $t_1 = 4.00$ s requires that $v_0 t_1 + \frac{1}{2}\alpha t_1^2 + \frac{1}{6}\beta t_1^3 = 0$.

Thus $v_0 = -\frac{1}{6}\beta t_1^2 - \frac{1}{2}\alpha t_1 = -\frac{1}{6}(3.00 \text{ m/s}^3)(4.00 \text{ s})^2 - \frac{1}{2}(-2.00 \text{ m/s}^2)(4.00 \text{ s}) = -4.00 \text{ m/s}$.

b) With v_0 as calculated in part (a) and $t = 4.00$ s,

$v = v_0 + \alpha t + \frac{1}{2}\beta t^2 = -4.00 \text{ s} + (-2.00 \text{ m/s}^2)(4.00 \text{ s}) + \frac{1}{2}(3.00 \text{ m/s}^3)(4.00 \text{ s})^2 = +12.0 \text{ m/s}$.

2-67 Calculate the speed of the diver when she reaches the water.

Take the origin of coordinates to be at the platform, and take the $+y$-direction to be downward.

$y - y_0 = +21.3$ m, $a = +9.80$ m/s^2, $v_0 = 0$ (since diver just steps off), $v = ?$

$v^2 = v_0^2 + 2a(y - y_0)$

$v = +\sqrt{2a(y - y_0)} = +\sqrt{2(9.80 \text{ m/s}^2)(21.3 \text{ m})} = +20.4 \text{ m/s}$.

We know that v is positive because the diver is traveling downward when she reaches the water.

The announcer has exaggerated the speed of the diver.

b) Use the same coordinates as in part (a). Calculate the initial upward velocity needed to give the diver a speed of 25.0 m/s when she reaches the water.

$v_0 = ?$, $v = +25.0$ m/s, $a = +9.80$ m/s^2, $y - y_0 = +21.3$ m

$v^2 = v_0^2 + 2a(y - y_0)$

$v_0 = -\sqrt{v^2 - 2a(y - y_0)} = -\sqrt{(25.0 \text{ m/s})^2 - 2(9.80 \text{ m/s}^2)(21.3 \text{ m})} = -14.4 \text{ m/s}$.

(v_0 is negative since the direction of the initial velocity is upward.)

One way to decide if this speed is reasonable is to calculate the maximum height above the platform it would produce:

$v_0 = -14.4$ m/s, $v = 0$ (at maximum height), $a = +9.80$ m/s^2, $y - y_0 = ?$

$v^2 = v_0^2 + 2a(y - y_0)$

$y - y_0 = \frac{v^2 - v_0^2}{2a} = \frac{0 - (-14.4 \text{ s})^2}{2(+9.80 \text{ m/s})} = -10.6 \text{ m}$

This is not physically attainable; a vertical leap of 10.6 m upward is not possible.

2-69 Let $y = 0$ at the ground and let positive y be upward.

a) Let $t = 0$ when the football is at the window.

$v_0 = 5.00$ m/s, $v = 0$ (at maximum height), $a = -9.80$ m/s^2, $y_0 = +12.0$ m

$y = ?$

$$v^2 = v_0^2 + 2a(y - y_0)$$

$$y - y_0 = \frac{v^2 - v_0^2}{2a} = \frac{0 - (5.00 \text{ s})^2}{2(-9.80 \text{ m/s}^2)} = 1.28 \text{ m}.$$

$$y = y_0 + 1.28 \text{ m} = 12.0 \text{ m} + 1.28 \text{ m} = 13.3 \text{ m} \text{ (maximum height above ground)}.$$

b) Now let $t = 0$ be when the football is on the ground. Use the motion from the ground to the window to find v_0, the speed of the football as it leaves the ground.

$$v_0 = ?, \quad a = -9.80 \text{ m/s}^2, \quad v = +5.00 \text{ s}, \quad y - y_0 = +12.0 \text{ m}$$

$$v^2 = v_0^2 + 2a(y - y_0)$$

$$v_0 = +\sqrt{v^2 - 2a(y - y_0)} = \sqrt{(5.00 \text{ m/s})^2 - 2(-9.80 \text{ m/s}^2)(+12.0 \text{ m})} =$$
$$+16.13 \text{ m/s}$$

Now consider the motion from the ground to the maximum height.

$$v_0 = +16.13 \text{ m/s}, \quad t = ?, \quad v = 0 \text{ (at maximum height)}, \quad a = -9.80 \text{ m/s}^2$$

$$v = v_0 + at$$

$$t = \frac{v - v_0}{a} = \frac{0 - 16.13 \text{ s}}{-9.80 \text{ m/s}} = +1.65 \text{ s}$$

2-71 Take positive y to be upward.

a) Consider the motion from when he applies the acceleration to when the shot leaves his hand.

$$v_0 = 0, \quad v = ?, \quad a = 45.0 \text{ m/s}^2, \quad y - y_0 = 0.640 \text{ m}$$

$$v^2 = v_0^2 + 2a(y - y_0)$$

$$v = \sqrt{2a(y - y_0)} = \sqrt{2(45.0 \text{ m/s}^2)(0.640 \text{ m})} = 7.59 \text{ m/s}$$

b) Consider the motion of the shot from the point where he releases it to its maximum height, where $v = 0$. Take $y = 0$ at the ground.

$$y_0 = 2.20 \text{ m}, \quad y = ?, \quad a = -9.80 \text{ m/s}^2 \text{ (free fall)}, \quad v_0 = 7.59 \text{ m/s (from part (a)}, \quad v = 0 \text{ (at maximum height)}$$

$$v^2 = v_0^2 + 2a(y - y_0)$$

$$y - y_0 = \frac{v^2 - v_0^2}{2a} = \frac{0 - (7.59 \text{ m/s})^2}{2(-9.80 \text{ m/s}^2)} = 2.94 \text{ m}$$

$$y = 2.20 \text{ m} + 2.94 \text{ m} = 5.14 \text{ m}.$$

c) Consider the motion of the shot from the point where he releases it to when it returns to the height of his head. Take $y = 0$ at the ground.

$y_0 = 2.20$ m, $y = 1.83$ m, $a = -9.80$ m/s^2, $v_0 = +7.59$ m/s, $t = ?$

$y - y_0 = v_0 t + \frac{1}{2}at^2$

1.83 m $- 2.20$ m $= (7.59$ m/s$)t + \frac{1}{2}(-9.80$ m/s$^2)t^2$

-0.37 m $= (7.59$ m/s$)t - (4.90$ m/s$^2)t^2$

$4.90t^2 - 7.59t - 0.37 = 0$, with t in seconds.

Use the quadratic formula to solve for t:

$t = \frac{1}{9.80}(7.59 \pm \sqrt{(7.59)^2 - 4(4.90)(-0.37)}) = 0.774 \pm 0.822$

t must be positive, so $t = 0.774$ s $+ 0.822$ s $= 1.60$ s

2-75 First calculate the velocity and acceleration of each car as functions of time.

$$x_A = \alpha t + \beta t^2, \quad v_A = \frac{dx_A}{dt} = \alpha + 2\beta t, \quad a_A = \frac{dv_A}{dt} = 2\beta$$

$$x_B = \gamma t^2 - \delta t^3, \quad v_B = \frac{dx_B}{dt} = 2\gamma t - 3\delta t^2, \quad a_B = \frac{dv_B}{dt} = 2\gamma - 6\delta t$$

a) At $t = 0$, $v_A = \alpha$ and $v_B = 0$. So initially car A moves ahead.

b) Cars at the same point implies $x_A = x_B$.

$\alpha t + \beta t^2 = \gamma t^2 - \delta t^3$

One solution is $t = 0$, which says that they start from the same point.

To find the other solutions, divide by t: $\alpha + \beta t = \gamma t - \delta t^2$

$\delta t^2 + (\beta - \gamma)t + \alpha = 0$

$t = \frac{1}{2\delta}(-(\beta - \gamma) \pm \sqrt{(\beta - \gamma)^2 - 4\delta\alpha}) = \frac{1}{0.40}(+1.60 \pm \sqrt{(1.60)^2 - 4(0.20)(2.60)}) =$

4.00 s ± 1.73 s

So $x_A = x_B$ for $t = 0$, $t = 2.27$ s and $t = 5.73$ s.

c) The distance from A to B is $x_B - x_A$. The rate of change of this distance is $\frac{d(x_B - x_A)}{dt}$. If this distance is not changing, $\frac{d(x_B - x_A)}{dt} = 0$. But this says $v_B - v_A = 0$. (The distance between A and B is neither decreasing nor increasing at the instant when they have the same velocity.)

$v_A = v_B$ requires $\alpha + 2\beta t = 2\gamma t - 3\delta t^2$

$3\delta t^2 + 2(\beta - \gamma)t + \alpha = 0$

$t = \frac{1}{6\delta}(-2(\beta-\gamma) \pm \sqrt{4(\beta - \gamma)^2 - 12\delta\alpha}) = \frac{1}{1.20}(-3.20 \pm \sqrt{4(-1.60)^2 - 12(0.20)(2.60)})$

$t = 2.667$ s ± 1.667 s, so $v_A = v_B$ for $t = 1.00$ s and $t = 4.33$ s.

d) $a_A = a_B$ requires $2\beta = 2\gamma - 6\delta t$

$$t = \frac{\gamma - \beta}{3\delta} = \frac{2.80 \text{ m/s}^2 - 1.20 \text{ m/s}^2}{3(0.20 \text{ m/s}^3)} = 2.67 \text{ s.}$$

CHAPTER 3
MOTION IN TWO OR THREE DIMENSIONS

Exercises

3-1 $(v_{av})_x = \dfrac{\Delta x}{\Delta t} = \dfrac{x_2 - x_1}{t_2 - t_1} = \dfrac{5.3 \text{ m} - 1.1 \text{ m}}{3.0 \text{ s} - 0} = 1.4 \text{ m/s}$

$(v_{av})_y = \dfrac{\Delta y}{\Delta t} = \dfrac{y_2 - y_1}{t_2 - t_1} = \dfrac{-0.5 \text{ m} - 3.4 \text{ m}}{3.0 \text{ s} - 0} = -1.3 \text{ m/s}$

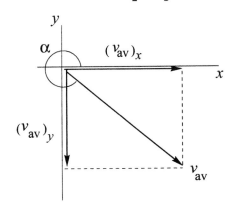

$\tan \alpha = \dfrac{(v_{av})_y}{(v_{av})_x} = \dfrac{-1.3 \text{ m/s}}{1.4 \text{ m/s}} = -0.9286$

$\alpha = 360° - 42.9° = 317°$

$v_{av} = \sqrt{(v_{av})_x^2 + (v_{av})_y^2}$

$v_{av} = \sqrt{(1.4 \text{ m/s})^2 + (-1.3 \text{ m/s})^2} = 1.9 \text{ m/s}$

3-3 **a)** $\vec{r} = [4.0 \text{ cm} + (2.5 \text{ cm/s}^2)t^2]\hat{i} + (5.0 \text{ cm/s})t\hat{j}$

At $t = 0$, $\vec{r} = (4.0 \text{ cm})\hat{i}$.

At $t = 2.0$ s, $\vec{r} = (14.0 \text{ cm})\hat{i} + (10.0 \text{ cm})\hat{j}$.

$(v_{av})_x = \dfrac{\Delta x}{\Delta t} = \dfrac{10.0 \text{ cm}}{2.0 \text{ s}} = 5.0 \text{ cm/s}.$

$(v_{av})_y = \dfrac{\Delta y}{\Delta t} = \dfrac{10.0 \text{ cm}}{2.0 \text{ s}} = 5.0 \text{ cm/s}.$

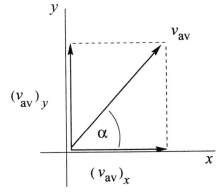

$v_{av} = \sqrt{(v_{av})_x^2 + (v_{av})_y^2} = 7.1 \text{ cm/s}$

$\tan \alpha = \dfrac{(v_{av})_y}{(v_{av})_x} = 1.00$

$\theta = 45°.$

b) $\vec{v} = \dfrac{d\vec{r}}{dt} = ([5.0 \text{ cm/s}^2]t)\hat{i} + (5.0 \text{ cm/s})\hat{j}$

$\underline{t = 0}$: $v_x = 0$, $v_y = 5.0 \text{ cm/s}$; $v = 5.0 \text{ cm/s}$ and $\theta = 90°$

$\underline{t = 1.0 \text{ s}}$: $v_x = 5.0 \text{ cm/s}$, $v_y = 5.0 \text{ cm/s}$; $v = 7.1 \text{ cm/s}$ and $\theta = 45°$

$\underline{t = 2.0 \text{ s}}$: $v_x = 10.0 \text{ cm/s}$, $v_y = 5.0 \text{ cm/s}$; $v = 11 \text{ cm/s}$ and $\theta = 27°$

c) The trajectory is a graph of y versus x.

$x = 4.0 \text{ cm} + (2.5 \text{ cm/s}^2)t^2$, $y = (5.0 \text{ cm/s})t$

For values of t between 0 and 2.0 s, calculate x and y and plot y versus x.

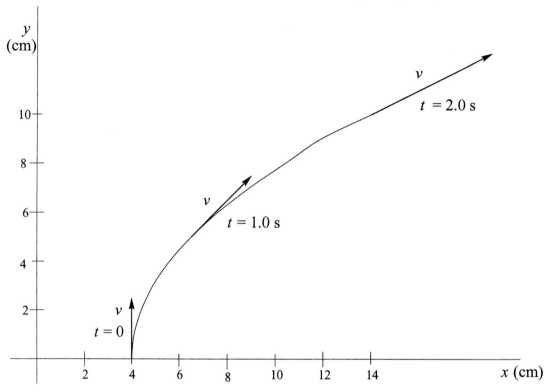

The sketch shows that the instantaneous velocity at any t is tangent to the trajectory.

3-5   **a)** The velocity vectors at $t_1 = 0$ and $t_2 = 30.0$ s are

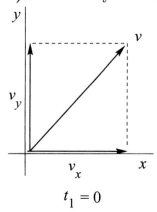

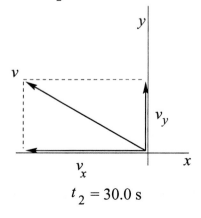

b) $(a_{av})_x = \dfrac{\Delta v_x}{\Delta t} = \dfrac{v_{2x} - v_{1x}}{t_2 - t_1} = \dfrac{-170 \text{ m/s} - 90 \text{ m/s}}{30.0 \text{ s}} = -8.67 \text{ m/s}^2$

$(a_{av})_y = \dfrac{\Delta v_y}{\Delta t} = \dfrac{v_{2y} - v_{1y}}{t_2 - t_1} = \dfrac{40 \text{ m/s} - 110 \text{ m/s}}{30.0 \text{ s}} = -2.33 \text{ m/s}^2$

c)

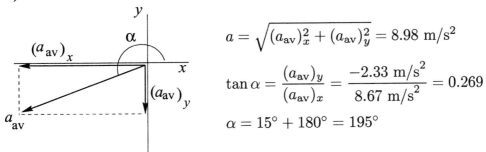

$a = \sqrt{(a_{av})_x^2 + (a_{av})_y^2} = 8.98 \text{ m/s}^2$

$\tan \alpha = \dfrac{(a_{av})_y}{(a_{av})_x} = \dfrac{-2.33 \text{ m/s}^2}{8.67 \text{ m/s}^2} = 0.269$

$\alpha = 15° + 180° = 195°$

3-7 **a)** Calculate x and y for t values in the range 0 to 2.0 s and plot y versus x

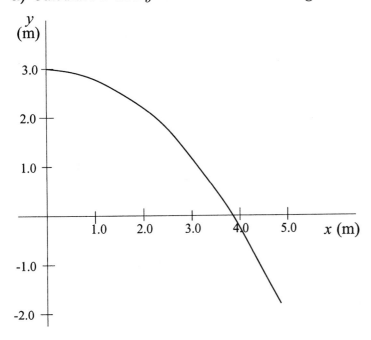

b) $v_x = \dfrac{dx}{dt} = \alpha \qquad v_y = \dfrac{dy}{dt} = -2\beta t$

$a_y = \dfrac{dv_x}{dt} = 0 \qquad a_y = \dfrac{dv_y}{dt} = -2\beta$

Thus $\vec{v} = \alpha \hat{i} - 2\beta t \hat{j} \qquad \vec{a} = -2\beta \hat{j}$

c) <u>velocity:</u> At $t = 2.0$ s, $v_x = 2.4$ m/s, $v_y = -2(1.2 \text{ m/s}^2)(2.0 \text{ s}) = -4.8$ m/s

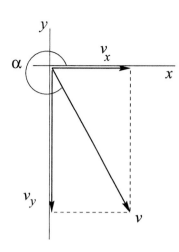

$$v = \sqrt{v_x^2 + v_y^2} = 5.4 \text{ m/s}$$

$$\tan \alpha = \frac{v_y}{v_x} = \frac{-4.8 \text{ m/s}}{2.4 \text{ m/s}} = -2.00$$

$$\alpha = -63.4° + 360° = 297°$$

<u>acceleration:</u> At $t = 2.0$ s, $a_x = 0$, $a_y = -2(1.2 \text{ m/s}^2) = -2.4 \text{ m/s}^2$

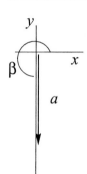

$$a = \sqrt{a_x^2 + a_y^2} = 2.4 \text{ m/s}^2$$

$$\tan \beta = \frac{a_y}{a_x} = \frac{-2.4 \text{ m/s}^2}{0} = -\infty$$

$$\beta = 270°$$

d)

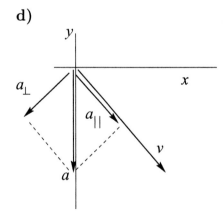

$\vec{a}$ has a component $a_{\parallel}$ in the same direction as $\vec{v}$, so we know that v is increasing (the bird is speeding up.)

$\vec{a}$ also has a component $a_{\perp}$ perpendicular to $\vec{v}$, so that the direction of $\vec{v}$ is changing; the bird is turning toward the $-y$-direction (toward the right)

3-9   Take the positive y-direction to be upward.

Take the origin of coordinates at the initial position of the book, at the point where it leaves the table top.

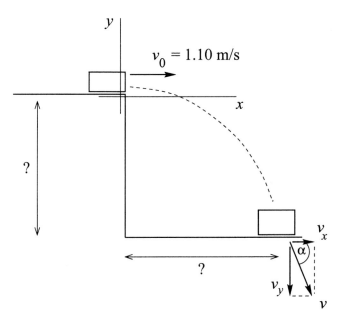

_x_-component:
$$a_x = 0, \quad v_{0x} = 1.10 \text{ m/s},$$
$$t = 0.350 \text{ s}$$

_y_-component:
$$a_y = -9.80 \text{ m/s}^2, \quad v_{0y} = 0,$$
$$t = 0.350 \text{ s}$$

a) $y - y_0 = ?$

$y - y_0 = v_{0y}t + \frac{1}{2}a_yt^2 = 0 + \frac{1}{2}(-9.80 \text{ m/s}^2)(0.350 \text{ s})^2 = -0.600 \text{ m}.$ The table top is 0.600 m above the floor.

b) $x - x_0 = ?$

$x - x_0 = v_{0x}t + \frac{1}{2}a_xt^2 = (1.10 \text{ m/s})(0.350 \text{ s}) + 0 = 0.385 \text{ m}.$

c) $v_x = v_{0x} + a_xt = 1.10 \text{ m/s}$ (The _x_-component of the velocity is constant, since $a_x = 0$.)

$v_y = v_{0y} + a_yt = 0 + (-9.80 \text{ m/s}^2)(0.350 \text{ s}) = -3.43 \text{ m/s}$

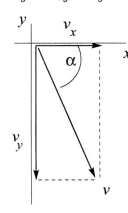

$$v = \sqrt{v_x^2 + v_y^2} = 3.60 \text{ m/s}$$

$$\tan\alpha = \frac{v_y}{v_x} = \frac{-3.43 \text{ m/s}}{1.10 \text{ m/s}} = -3.118$$

$$\alpha = -72.2°$$
Direction of $\vec{v}$ is 72.2° below the horizontal.

d)

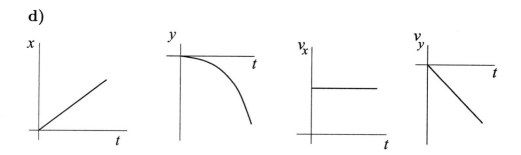

3-15 First find the x- and y-components of the initial velocity.

Use coordiantes where the $+y$-direction is upward, the $+x$-direction is to the right and the origin is at the point where the baseball leaves the bat.

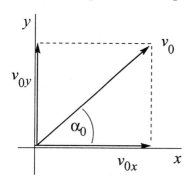

$v_{0x} = v_0 \cos \alpha_0 = (30.0 \text{ m/s}) \cos 36.9° = 24.0 \text{ m/s}$
$v_{0y} = v_0 \sin \alpha_0 = (30.0 \text{ m/s}) \sin 36.9° = 18.0 \text{ m/s}$

a) y-component (vertical motion):

$y - y_0 = +10.0 \text{ m/s}, \quad v_{0y} = 18.0 \text{ m/s}, \quad a_y = -9.80 \text{ m/s}^2, \quad t = ?$
$y - y_0 = v_{0y} + \frac{1}{2} a_y t^2$
$10.0 \text{ m} = (18.0 \text{ m/s})t - (4.90 \text{ m/s}^2)t^2$
$(4.90 \text{ m/s}^2)t^2 - (18.0 \text{ m/s})t + 10.0 \text{ m} = 0$

Apply the quadratic formula: $t = \frac{1}{9.80}[18.0 \pm \sqrt{(-18.0)^2 - 4(4.90)(10.0)}]$ s $=$ (1.837 ± 1.154) s

The ball is at a height of 10.0 above the point where it left the bat at $t_1 = 0.683$ s and at $t_2 = 2.99$ s. At the earlier time the ball passes through a height of 10.0 m as its way up and at the later time it passes through 10.0 m on its way down.

b) $v_x = v_{0x} = +24.0 \text{ m/s}$, at all times since $a_x = 0$.
$v_y = v_{0y} + a_y t$
$\underline{t_1 = 0.683 \text{ s}}$: $v_y = +18.0 \text{ m/s} + (-9.80 \text{ m/s}^2)(0.683 \text{ s}) = +11.3 \text{ m/s}$. ($v_y$ is positive means that the ball is traveling upward at this point.)
$\underline{t_2 = 2.99 \text{ s}}$: $v_y = +18.0 \text{ m/s} + (-9.80 \text{ m/s}^2)(2.99 \text{ s}) = -11.3 \text{ m/s}$. ($v_y$ is negative means that the ball is traveling downward at this point.)

c) $v_x = v_{0x} = 24.0 \text{ m/s}$

Solve for v_y:

$v_y = ?, \quad y - y_0 = 0$ (when ball returns to height where motion started),

$a_y = -9.80 \text{ m/s}^2, \quad v_{0y} = +18.0 \text{ m/s}$

$v^2 = v_{0y}^2 + 2a_y(y - y_0)$

$v_y = -v_{0y} = -18.0 \text{ m/s}$ (negative, since the baseball must be traveling downward at this point)

Now that have the components can solve for the magnitude and direction of $\vec{v}$.

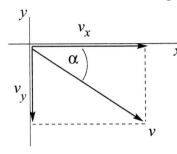

$v = \sqrt{v_x^2 + v_y^2}$

$v = \sqrt{(24.0 \text{ m/s})^2 + (-18.0 \text{ m/s})^2} = 30.0 \text{ m/s}$

$\tan \alpha = \dfrac{v_y}{v_x} = \dfrac{-18.0 \text{ m/s}}{24.0 \text{ m/s}}$

$\alpha = -36.9°, \quad 36.9°$ below the horizontal

The velocity of the ball when it returns to the level where it left the bat has magnitude 30.0 m/s and is directed at an angle of 36.9° below the horizontal.

3-17 Take the origin of coordiates at the point where the quarter leaves your hand and take positive y to be upward.

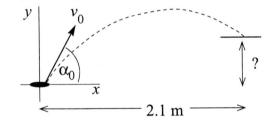

$v_{0x} = v_0 \cos \alpha_0 = (6.4 \text{ m/s}) \cos 60°$
$v_{0x} = 3.20 \text{ m/s}$

$v_{0y} = v_0 \sin \alpha_0 = (6.4 \text{ m/s}) \sin 60°$
$v_{0y} = 5.54 \text{ m/s}$

a) Use the horizontal (x-component) of motion to solve for t, the time the quarter travels through the air:

$t = ?, \quad x - x_0 = 2.1 \text{ m}, \quad v_{0x} = 3.2 \text{ m/s}, \quad a_x = 0$

$x - x_0 = v_{0x}t + \frac{1}{2}a_x t^2 = v_{0x}t$, since $a_x = 0$

$t = \dfrac{x - x_0}{v_{0x}} = \dfrac{2.1 \text{ m}}{3.2 \text{ m/s}} = 0.656 \text{ s}$

Now find the vertical displacement of the quarter after this time:

$y - y_0 = ?, \quad a_y = -9.80 \text{ m/s}^2, \quad v_{0y} = +5.54 \text{ m/s}, \quad t = 0.656 \text{ s}$

$y - y_0 + v_{0y}t + \frac{1}{2}a_y t^2$

$y - y_0 = (5.54 \text{ m/s})(0.656 \text{ s}) + \frac{1}{2}(-9.80 \text{ m/s}^2)(0.656 \text{ s})^2 = 3.63 \text{ m} - 2.11 \text{ m} = 1.5 \text{ m}.$

b) $v_y =$, $t = 0.656$ s, $a_y = -9.80$ m/s^2, $v_{0y} = +5.54$ m/s

$v_y = v_{0y} + a_y t$

$v_y = 5.54$ m/s $+ (-9.80$ m/s$^2)(0.656$ s$) = -0.89$ m/s. The minus sign indicates tha
the y-component of $\vec{v}$ is downward. At this point the quarter has passed throug
the highest point in its path and is on its way down.

3-19 Take the origin of coordinates at the roof and let the $+y$-direction be upward.

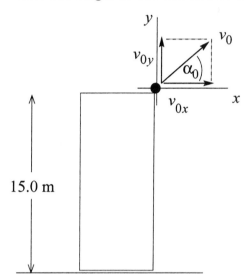

$$v_{0x} = v_0 \cos \alpha_0 = 25.2 \text{ m/s}$$

$$v_{0y} = v_0 \sin \alpha_0 = 16.3 \text{ m/s}$$

a) At the maximum height $v_y = 0$.

$a_y = -9.80$ m/s^2, $v_y = 0$, $v_{oy} = +16.3$ m/s, $y - y_0 = ?$

$v_y^2 = v_{0y}^2 + 2a_y(y - y_0)$

$$y - y_0 = \frac{v_y^2 - v_{0y}^2}{2a_y} = \frac{0 - (16.3 \text{ m/s})^2}{2(-9.80 \text{ m/s}^2)} = +13.6 \text{ m}$$

b) $v_x = v_{0x} = 25.2$ m/s (since $a_x = 0$)

$v_y = ?$, $a_y = -9.80$ m/s^2, $y - y_0 = -15.0$ m (negative because at the ground the
rock is below its initial position), $v_{0y} = 16.3$ m/s

$v_y^2 = v_{0y}^2 + 2a_y(y - y_0)$

$v_y = -\sqrt{v_{0y}^2 + 2a_y(y - y_0)}$ (v_y is negative because at the ground the rock is traveling
downward.)

$$v_y = -\sqrt{(16.3 \text{ m/s})^2 + 2(-9.80 \text{ m/s}^2)(-15.0 \text{ m})} = -23.7 \text{ m/s}$$

Then $v = \sqrt{v_x^2 + v_y^2} = \sqrt{(25.2 \text{ m/s})^2 + (-23.7 \text{ m/s})^2} = 34.6$ m/s.

c) Use the vertical motion (y-component) to find the time the rock is in the air:

$t = ?, \quad v_y = -23.7$ m/s (from part (b)), $\quad a_y = -9.80$ m/s^2, $\quad v_{0y} = +16.3$ m/s

$t = \dfrac{v_y^2 - v_{0y}^2}{a_y} = \dfrac{-23.7 \text{ m/s} - 16.3 \text{ m/s}}{-9.80 \text{ m/s}^2} = +4.08$ s

Can use this t to calculate the horizontal range:

$t = 4.08$ s, $\quad v_{0x} = 25.2$ m/s, $\quad a_x = 0$, $\quad x - x_0 = ?$

$x - x_0 = v_{0x}t + \frac{1}{2}a_x t^2 = (25.2 \text{ m/s})(4.08 \text{ s}) + 0 = 103$ m

d) Graphs of x versus t, y versus t, v_x versus t, and v_y versus t:

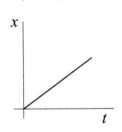 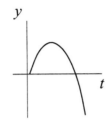 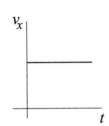 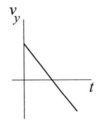

3-25 Since the magnitude of $\vec{v}$ is constant, $v_{\text{tan}} = \dfrac{d\,|\vec{v}|}{dt} = 0$

and the resultant acceleration is equal to the radial component. At each point in the motion the radial component of the acceleration is directed in toward the center of the circular path and its magnitude is given by v^2/R.

a) $a_{\text{rad}} = \dfrac{v^2}{R} = \dfrac{(7.00 \text{ m/s})^2}{14.0 \text{ m}} = 3.50$ m/s^2, upward.

b) The radial acceleration has the same magnitude as in part (a), but now the direction toward the center of the circle is downward. The acceleration at this point in the motion is 3.50 m/s^2, downward.

c) The time to make one rotation is the period T.

$v = \dfrac{2\pi R}{T}$ so $T = \dfrac{2\pi R}{v} = \dfrac{2\pi (14.0 \text{ m})}{7.00 \text{ m/s}} = 12.6$ s.

3-29 Let W stand for the woman, G for the ground, and S for the sidewalk.

Take the positive direction to be the direction in which the sidewalk is moving.

The velocities are $v_{\text{W/G}}$ (woman relative to the ground), $v_{\text{W/S}}$ (woman relative to the sidewalk), and $v_{\text{S/G}}$ (sidewalk relative to the ground).

Eq.(2-24) becomes $v_{\text{W/G}} = v_{\text{W/S}} + v_{\text{S/G}}$.

The time to reach the other end is given by $t = \dfrac{\text{distance traveled relative to ground}}{v_{W/G}}$

a) $v_{S/G} = 1.0$ m/s

$v_{W/S} = +1.5$ m/s

$v_{W/G} = v_{W/S} + v_{S/G} = 1.5$ m/s $+ 1.0$ m/s $= 2.5$ m/s.

$t = \dfrac{35.0 \text{ m}}{v_{W/G}} = \dfrac{35.0 \text{ m}}{2.5 \text{ m}} = 14$ s.

b) $v_{S/G} = 1.0$ m/s

$v_{W/S} = -1.5$ m/s

$v_{W/G} = v_{W/S} + v_{S/G} = -1.5$ m/s $+ 1.0$ m/s $= -0.5$ m/s. (Since $v_{W/G}$ now i
negative, she must get on the moving sidewalk at the opposite end from in part (a).

$t = \dfrac{-35.0 \text{ m}}{v_{W/G}} = \dfrac{-35.0 \text{ m}}{-0.5 \text{ m}} = 70$ s.

3-33 View the motion from above:

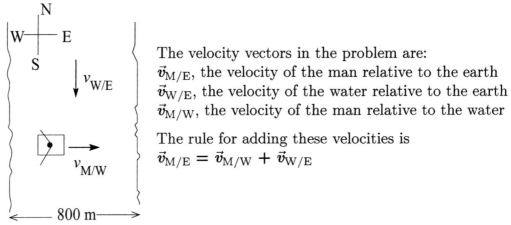

The velocity vectors in the problem are:
$\vec{v}_{M/E}$, the velocity of the man relative to the earth
$\vec{v}_{W/E}$, the velocity of the water relative to the earth
$\vec{v}_{M/W}$, the velocity of the man relative to the water

The rule for adding these velocities is
$\vec{v}_{M/E} = \vec{v}_{M/W} + \vec{v}_{W/E}$

The problem tells us that $\vec{v}_{W/E}$ has magnitude 2.0 m/s and direction due south. I
also tells us that $\vec{v}_{M/W}$ has magnitude 4.2 m/s and direction due east.

The vector addition diagram is then

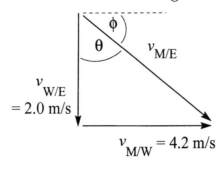

This diagram shows the vector addition
$\vec{v}_{M/E} = \vec{v}_{M/W} + \vec{v}_{W/E}$
and also has $\vec{v}_{M/W}$ and $\vec{v}_{W/E}$ in their
specified directions. Note that the vector
diagram forms a right triangle.

The Pythagorean theorem applied to the vector addition diagram gives
$v_{M/E}^2 = v_{M/W}^2 + v_{W/E}^2.$

$$v_{M/E} = \sqrt{v_{M/W}^2 + v_{W/E}^2} = \sqrt{(4.2 \text{ m/s})^2 + (2.0 \text{ m/s})^2} = 4.7 \text{ m/s}$$

$$\tan\theta = \frac{v_{M/E}}{v_{W/E}} = \frac{4.2 \text{ m/s}}{2.0 \text{ m/s}} = 2.10; \quad \theta = 25°; \text{ or } \phi = 90° - \theta = 65°.$$

The velocity of the man relative to the earth has magnitude 4.7 m/s and direction 25° S of E.

b) This requires careful thought. To cross the river the man must travel 800 m due east relative to the earth. The man's velocity relative to the earth is $\vec{v}_{M/E}$. But, from the vector addition diagram the eastward component of $v_{M/E}$ equals $v_{M/W} = 4.2$ m/s.

$$\text{Thus } t = \frac{x - x_0}{v_x} = \frac{800 \text{ m}}{4.2 \text{ m/s}} = 190 \text{ s}.$$

c) The southward component of $\vec{v}_{M/E}$ equals $v_{W/E} = 2.0$ m/s. Therefore, in the 190 s it takes him to cross the river the distance south the man travels relative to the earth is

$$y - y_0 = v_y t = (2.0 \text{ m/s})(190 \text{ s}) = 380 \text{ m}.$$

3-35 a) $\vec{v}_{P/A}$ is the velocity of the plane relative to the air.

The problem states that $\vec{v}_{P/A}$ has magnitude 35 m/s and direction south.

$\vec{v}_{A/E}$ is the velocity of the air relative to the earth. The problem states that $\vec{v}_{A/E}$ is to the southwest (45° S of W) and has magnitude 10 m/s.

The relative velocity equation is $\vec{v}_{P/E} = \vec{v}_{P/A} + \vec{v}_{A/E}.$

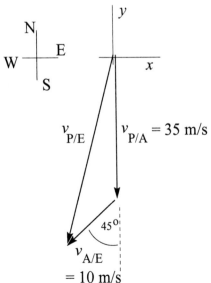

b) $(v_{P/A})_x = 0, \quad (v_{P/A})_y = -35 \text{ m/s}$
$(v_{A/E})_x = -(10 \text{ m/s}) \cos 45° = -7.07 \text{ m/s},$
$(v_{A/E})_y = -(10 \text{ m/s}) \sin 45° = -7.07 \text{ m/s}$

$(v_{P/E})_x = (v_{P/A})_x + (v_{A/E})_x = 0 - 7.07 \text{ m/s} = -7.1 \text{ m/s}$
$(v_{P/E})_y = (v_{P/A})_y + (v_{A/E})_y = -35 \text{ m/s} - 7.07 \text{ m/s} = -42 \text{ m/s}$

c)

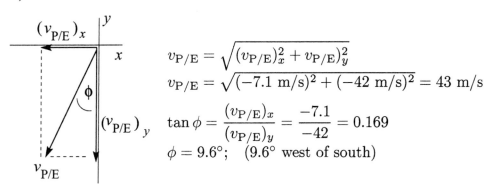

$v_{P/E} = \sqrt{(v_{P/E})_x^2 + v_{P/E})_y^2}$

$v_{P/E} = \sqrt{(-7.1 \text{ m/s})^2 + (-42 \text{ m/s})^2} = 43 \text{ m/s}$

$\tan \phi = \dfrac{(v_{P/E})_x}{(v_{P/E})_y} = \dfrac{-7.1}{-42} = 0.169$

$\phi = 9.6°; \quad (9.6° \text{ west of south})$

Problems

3-37 $x = \alpha t \qquad y = 15.0 \text{ m} - \beta t^2$

   a) $v_x = \alpha \quad v_y = -2\beta t$

   $a_x = 0 \quad a_y = -2\beta$

   $\vec{v}$ perpendicular to $\vec{a}$ means that $\vec{v} \cdot \vec{a} = 0$

   $\vec{v} \cdot \vec{a} = v_x a_x + v_y a_y = +4\beta^2 t$

   Thus $\vec{v} \cdot \vec{a} = 0$ only for $t = 0$.

b) The speed is instantaneously not changing when there is no component of accel
eration in the direction of the velocity, that is, when $\vec{v} \cdot \vec{a} = 0$. From part (a) thi
happens only at $t = 0$.

c) $\vec{v}$ perpendicular to $\vec{r}$ means that $\vec{v} \cdot \vec{r} = 0$

$\vec{v} \cdot \vec{r} = v_x x + v_y y = \alpha^2 t - 2\beta t(15.0 \text{ m} - \beta t^2) = 0$

$t = 0$ is one root; at $t = 0$, $x = 0$ and $y = 15.0$ m.

The other roots are given by $\alpha^2 - 2\beta(15.0 \text{ m} - \beta t^2) = 0$.

With the numerical values for α and β this equation becomes $0.5t^2 = 13.56 \text{ s}^2$

The positive, physical root is $t = 5.21$ s. At this t, $x = 6.25$ m and $y = 1.44$ m.

d) The distance from the origin is given by $r = \sqrt{x^2 + y^2}$. The minumum value of r occurs when

$$\frac{dr}{dt} \frac{d\sqrt{x^2 + y^2}}{dt} = (x^2 + y^2)^{-1/2} \left[x \frac{dx}{dt} + y \frac{dy}{dt} \right] = 0.$$

But this says $xv_x + yv_y = 0$, which is the same condition as in part (c). Thus the minimum distance occurs at $t = 5.21$ s and at this t the distance r equals 6.41 m. (Note: For the other solution to this equation, $t = 0$, the distance is 15.0 m, which is larger.)

e)

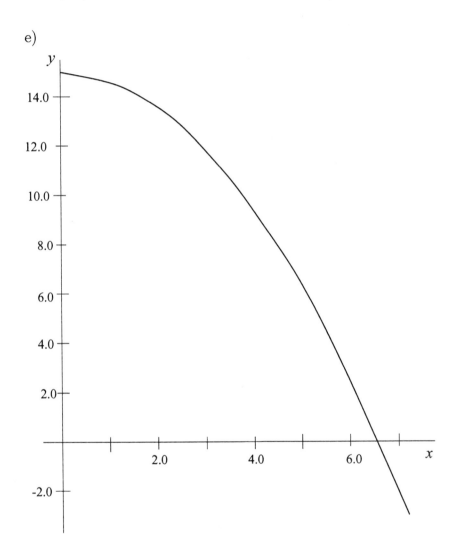

3-43

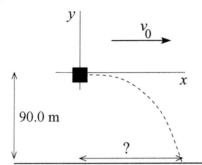

Take the origin of coordinates at the point where the canister is released. Take $+y$ to be upward. The initial velocity of the canister is the velocity of the plane, 64.0 m/s in the $+x$-direction.

Use the vertical motion to find the time of fall:

$t = ?$, $v_{0y} = 0$, $a_y = -9.80$ m/s^2, $y - y_0 = -90.0$ m (When the canister reaches the ground it is 90.0 m <u>below</u> the origin.)

$y - y_0 = v_{0y}t + \frac{1}{2}a_y t^2$

Since $v_{0y} = 0$, $t = \sqrt{\dfrac{2(y - y_0)}{a_y}} = \sqrt{\dfrac{2(-90.0 \text{ m})}{-9.80 \text{ m/s}^2}} = 4.286$ s.

Then use the horizontal component of the motion to calculate how far the canister falls in this time:

$x - x_0 = ?$, $a_x = 0$, $v_{0x} = 64.0$ m/s,

$x - x_0 = v_0 t + \frac{1}{2}at^2 = (64.0 \text{ m/s})(4.286 \text{ s}) + 0 = 274$ m.

3-45

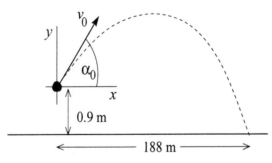

Take the origin of coordinates at the point where the ball leaves the bat, and take $+y$ to be upward.

$v_{0x} = v_0 \cos \alpha_0$

$v_{0y} = v_0 \sin \alpha_0$,

but we don't know v_0.

Write down the equation for the horizontal displacement when the ball hits the ground and the corresponding equation for the vertical displacement. The time t is the same for both components, so this will give us two equations in two unknowns $(v_0$ and $t)$.

<u>y-component</u>:

$a_y = -9.80$ m/s^2, $y - y_0 = -0.9$ m, $v_{0y} = v_0 \sin 45°$

$y - y_0 = v_{0y}t + \frac{1}{2}a_y t^2$

-0.9 m $= (v_0 \sin 45°)t + \frac{1}{2}(-9.80 \text{ m/s}^2)t^2$

x-component:

$$a_x = 0, \quad x - x_0 = 188 \text{ m}, \quad v_{0x} = v_0 \cos 45°$$

$$x - x_0 = v_{0x}t + \tfrac{1}{2}a_x t^2$$

$$t = \frac{x - x_0}{v_{0x}} = \frac{188 \text{ m}}{v_0 \cos 45°}$$

Put the expression for t from the x-component motion into the y-component equation and solve for v_0. (Note that $\sin 45° = \cos 45°$.)

$$-0.9 \text{ m} = (v_0 \sin 45°) \left(\frac{188 \text{ m}}{v_0 \cos 45°} \right) - (4.90 \text{ m/s}^2) \left(\frac{188 \text{ m}}{v_0 \cos 45°} \right)^2$$

$$4.90 \text{ m/s}^2 \left(\frac{188 \text{ m}}{v_0 \cos 45°} \right)^2 = 188 \text{ m} + 0.9 \text{ m} = 188.9 \text{ m}$$

$$\left(\frac{v_0 \cos 45°}{188 \text{ m}} \right)^2 = \frac{4.90 \text{ m/s}^2}{188.9 \text{ m}}, \quad v_0 = \left(\frac{188 \text{ m}}{\cos 45°} \right) \sqrt{\frac{4.90 \text{ m/s}^2}{188.9 \text{ m}}} = 42.8 \text{ m/s}$$

b) Use the horizontal motion to find the time it takes the ball to reach the fence:

x-component:

$$x - x_0 = 116 \text{ m}, \quad a_x = 0, \quad v_{0x} = v_0 \cos 45° = (42.8 \text{ m/s}) \cos 45° = 30.3 \text{ m/s},$$
$$t = ?$$

$$x - x_0 = v_{0x}t + \tfrac{1}{2}a_x t^2$$

$$t = \frac{x - x_0}{v_{0x}} = \frac{116 \text{ m}}{30.3 \text{ m/s}} = 3.83 \text{ s}$$

Find the vertical displacement of the ball at this t:

y-component:

$$y - y_0 = ?, \quad a_y = -9.80 \text{ m/s}^2, \quad v_{0y} = v_0 \sin 45° = 30.3 \text{ m/s}, \quad t = 3.83 \text{ s}$$
$$y - y_0 = v_{0y}t + \tfrac{1}{2}a_y t^2$$
$$y - y_0 = (30.3 \text{ s})(3.83 \text{ s}) + \tfrac{1}{2}(-9.80 \text{ m/s}^2)(3.83 \text{ s})^2$$
$y - y_0 = 116.0 \text{ m} - 71.9 \text{ m} = +44.1 \text{ m}$, above the point where the ball was hit. The height of the ball above the ground is $44.1 \text{ m} + 0.90 \text{ m} = 45.0 \text{ m}$. It's height then above the top of the fence is $45.0 \text{ m} - 3.0 \text{ m} = 42.0 \text{ m}$.

3-47 Take $+y$ to be upward and the origin of coordinates at the floor.

Neither the launch angle α_0 nor the initial speed v_0 are specified. We want the maximum height to be D.

First, use the vertical motion to solve for α_0:

$y - y_0 = D$, $\quad v_{0y} = v_0 \sin \alpha_0 = \sqrt{6gD} \sin \alpha_0$, $\quad v_y = 0$, $\quad a_y = -g$

$v_y^2 = v_{0y}^2 + 2a_y(y - y_0)$

$0 = 6gD \sin^2 \alpha_0 - 2gD$

This gives $\sin \alpha_0 = \sqrt{1/3}$. Note: $\cos^2 \alpha_0 = 1 - \sin^2 \alpha_0 = 2/3$, so $\cos \alpha_0 = \sqrt{2/3}$.

Now use the vertical motion to express the time in the air, t, in terms of v_0. Aft time t the ball has returned to the floor and $y - y_0 = 0$.

$y - y_0 = v_{0y}t + \frac{1}{2}a_y t^2$

$0 = v_0 \sin \alpha_0 - \frac{1}{2}gt$, $\quad$ so $t = \dfrac{2v_0 \sin \alpha_0}{g} = \dfrac{2v_0}{\sqrt{3g}}$.

In the horizontal motion $a_x = 0$ and $v_{0x} = v_0 \cos \alpha_0$.

$x - x_0 = v_{0x}t + \frac{1}{2}a_x t^2 = (v_0 \cos \alpha_0)t = (v_0 \cos \alpha_0)\left(\dfrac{2v_0 \sin \alpha_0}{g}\right)$

$x - x_0 = \left(\dfrac{2 \sin \alpha_0 \cos \alpha_0}{g}\right)(6gD) = (12 \sin \alpha_0 \cos \alpha_0)D = 12(\sqrt{1/3})(\sqrt{2/3})D$

$4\sqrt{2}D$.

3-51 **a)** Use the equation derived in Example 3-9:

$R = (v_0 \cos \alpha_0)\left(\dfrac{2v_0 \sin \alpha_0}{g}\right)$

Call the range R_1 when the angle is α_0 and R_2 when the angle is $90° - \alpha$.

$R_1 = (v_0 \cos \alpha_0)\left(\dfrac{2v_0 \sin \alpha_0}{g}\right)$

$R_2 = (v_0 \cos(90° - \alpha_0))\left(\dfrac{2v_0 \sin(90° - \alpha_0)}{g}\right)$

The problem asks us to show that $R_1 = R_2$.

We can use the trig identities in appendix B to show:

$\cos(90° - \alpha_0) = \cos(\alpha_0 - 90°) = \sin \alpha_0$

$\sin(90° - \alpha_0) = -\sin(\alpha_0 - 90°) = -(-\cos \alpha_0) = +\cos \alpha_0$

Thus $R_2 = (v_0 \sin \alpha_0)\left(\dfrac{2v_0 \cos \alpha_0}{g}\right) = (v_0 \cos \alpha_0)\left(\dfrac{2v_0 \sin \alpha_0}{g}\right) = R_1$.

b) $R = \dfrac{v_0^2 \sin 2\alpha_0}{g}$ $\quad$ so $\sin 2\alpha_0 = \dfrac{Rg}{v_0^2} = \dfrac{(0.25 \text{ m})(9.80 \text{ m/s}^2)}{(2.2 \text{ m/s})^2}$.

This gives $\alpha = 15°$ or $75°$.

-53 a)

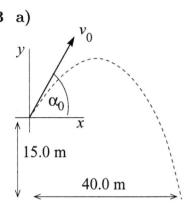

Take the origin of coordinates at the top of the ramp and take $+y$ to be upward. The problem specifies that the object is displaced 40.0 m to the right when it is 15.0 m below the origin.

We don't know t, the time in the air, and we don't know v_0. Write down the equations for the horizontal and vertical displacements. Combine these two equations to eliminate one unknown.

y-component:

$y - y_0 = -15.0$ m, $\quad a_y = -9.80$ m/s^2, $\quad v_{0y} = v_0 \sin 53.0°$

$y - y_0 = v_{0y}t + \frac{1}{2}a_y t^2$

-15.0 m $= (v_0 \sin 53.0°)t - (4.90$ m/s$^2)t^2$

y-component:

$x - x_0 = 40.0$ m, $\quad a_x = 0$, $\quad v_{0x} = v_0 \cos 53.0°$

$x - x_0 = v_{0x}t + \frac{1}{2}a_x t^2$

40.0 m $= (v_0 t)\cos 53.0°$

The second equation says $v_0 t = \dfrac{40.0 \text{ m}}{\cos 53.0°} = 66.47$ m.

Use this to replace $v_0 t$ in the first equation:

-15.0 m $= (66.47$ m$)\sin 53° - (4.90$ m/s$^2)t^2$

$$t = \sqrt{\frac{(66.46 \text{ m})\sin 53° + 15.0 \text{ m}}{4.90 \text{ m/s}^2}} = \sqrt{\frac{68.08 \text{ m}}{4.90 \text{ m/s}^2}} = 3.727 \text{ s}.$$

Now that we have t we can use the x-component equation to solve for v_0:

$$v_0 = \frac{40.0 \text{ m}}{t\cos 53.0°} = \frac{40.0 \text{ m}}{(3.727 \text{ s})\cos 53.0°} = 17.8 \text{ m/s}.$$

b) $v_0 = (17.8 \text{ m/s})/2 = 8.9$ m/s

This is less than the speed required to make it to the other side, so he lands in the river.

Use the vertical motion to find the time it takes him to reach the water:

$y - y_0 = -100$ m; $v_{0y} = +v_0 \sin 53.0° = 7.11$ m/s; $a_y = -9.80$ m/s^2

$y - y_0 = v_{0y}t + \frac{1}{2}a_y t^2$ gives $-100 = 7.11t - 4.90t^2$

$4.90t^2 - 7.11t - 100 = 0$ and $t = \frac{1}{9.80}(7.11 \pm \sqrt{(7.11)^2 - 4(4.90)(-100)})$

$t = 0.726$ s ± 4.57 s so $t = 5.30$ s.

The horizontal distance he travels in this time is

$x - x_0 = v_{0x}t = (v_0 \cos 53.0°)t = (5.36$ m/s$)(5.30$ s$) = 28.4$ m.

He lands in the river a horizontal distance of 28.4 m from his launch point.

3-55

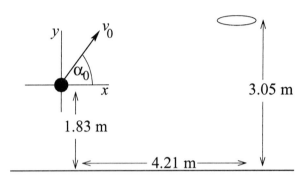

Take the origin of coordinates at the point where the player releases the ball.

$v_{0x} = v_0 \cos \alpha_0$

$v_{0y} = v_0 \sin \alpha_0$

3.05 m

1.83 m

4.21 m

a) y-component:

$v_y = 0$ (at the maximum height), $v_{0y} = v_0 \sin \alpha_0 = (4.88$ m/s$) \sin 35° = 2.80$ m/s

$a_y = -9.80$ m/s^2, $y - y_0 = ?$

$v_y^2 = v_{0y}^2 + 2a_y(y - y_0)$

$y - y_0 = \dfrac{v_y^2 - v_{0y}^2}{2a_y} = \dfrac{0 - (2.80 \text{ m/s})^2}{2(-9.80 \text{ m/s}^2)} = +0.400$ m

The height above the floor then is 0.400 m $+ 1.83$ m $= 2.23$ m.

b) Use the vertical motion to find the time the ball is in the air:

y-component:

$y - y_0 = -1.83$ m (vertical displacement when the ball reaches the floor),

$v_{0y} = 2.80$ m/s, $a_y = -9.80$ m/s^2, $t = ?$

$y - y_0 = v_{0y}t + \frac{1}{2}a_y t^2$

-1.83 m $= (2.80$ m/s$)t - (4.90$ m/s$^2)t^2$

$4.90t^2 - 2.80t - 1.83 = 0$, with t in seconds.

The quadratic formula gives

$t = \dfrac{1}{9.80}(2.80 \pm \sqrt{(2.80)^2 + 4(4.90)(1.83)})$ s $= 0.286$ s ± 0.675 s.

t must be positive, so $t = 0.286$ s $+ 0.675$ s $= 0.961$ s

Find the horizontal displacement for this t:

$x - x_0 = ?$, $\quad v_{0x} = v_0 \cos \alpha_0 = (4.88 \text{ m/s}) \cos 35° = 4.00 \text{ m/s}$, $\quad a_x = 0$,
$t = 0.961$ s

$x - x_0 = v_{0x}t + \frac{1}{2}a_x t^2 = (4.00 \text{ m/s})(0.961 \text{ s}) = 3.84$ m

c) We don't know either v_0 or the time t for the ball to reach the basket. Write down the equations for the horizontal and vertical displacements to get two equations for these two unknowns.

y-component:

$a_y = -9.80 \text{ m/s}^2$, $\quad v_{0y} = v_0 \sin 35°$, $\quad y - y_0 = 3.05 \text{ m} - 1.83 \text{ m} = 1.22$ m,
$y - y_0 = v_{0y}t + \frac{1}{2}a_y t^2$
$1.22 \text{ m} = (v_0 \sin 35°)t - (4.90 \text{ m/s}^2)t^2$

x-component:

$a_x = 0$, $\quad v_{0x} = v_0 \cos 35°$, $\quad x - x_0 = 4.21$ m
$x - x_0 = v_{0x}t + \frac{1}{2}a_x t^2$

Since $a_x = 0$ this equation gives $v_0 t = \dfrac{x - x_0}{\cos 35°} = \dfrac{4.21 \text{ m}}{\cos 35°} = 5.139$ m

Use this result in the y-component equation:
$1.22 \text{ m} = (5.139 \text{ m}) \sin 35° - (4.90 \text{ m/s}^2)t^2$
$(4.90 \text{ m/s}^2)t^2 = 1.728$ m

$t = \sqrt{\dfrac{1.728 \text{ m}}{4.90 \text{ m/s}^2}} = 0.594$ s

Then $v_0 = \dfrac{5.139 \text{ m}}{t} = \dfrac{5.139 \text{ m}}{0.594 \text{ s}} = 8.65 \text{ m/s}$.

d) We know from part (c) that $v_0 = 8.65 \text{ m/s}$.

Use the y-component motion to find the maximum height above the floor:
$v_y = 0$ (at maximum height), $\quad a_y = -9.80 \text{ m/s}^2$,
$v_{0y} = v_0 \sin 35° = (8.65 \text{ m/s}) \sin 35° = 4.96 \text{ m/s}$, $\quad y - y_0 = ?$
$v_y^2 = v_{0y}^2 + 2a_y(y - y_0)$

$y - y_0 = \dfrac{v_y^2 - v_{0y}^2}{2a_y} = \dfrac{0 - (4.96 \text{ m/s})^2}{2(-9.80 \text{ m/s}^2)} = +1.26$ m.
The maximum height above the floor is given by $1.83 \text{ m} + 1.26 \text{ m} = 3.09$ m.

Also use the y-component to find the time t to reach the maximum height. Can then use this t in the x-component equations to find the horizontal displacement at this point.

$$v_y = v_{0y} + a_y t, \text{ so } t = \frac{v_y - v_{0y}}{a_y} = \frac{0 - 4.96 \text{ m/s}}{-9.80 \text{ m/s}^2} = +0.506 \text{ s}.$$

$x - x_0 = ?, \quad a_x = 0, \quad v_{0x} = v_0 \cos 35° = (8.65 \text{ m/s}) \cos 35° = 7.09 \text{ m/s}, \quad t = 0.506$

$x - x_0 = v_{0x}t + \frac{1}{2}a_x t^2 = (7.09 \text{ m/s})(0.506 \text{ s}) + 0 = 3.59 \text{ m}$

This is the distance of the ball from the point where it was released. Its distanc
from the basket is 4.21 m − 3.59 m = 0.62 m.

3-57 Take $+y$ to be upward. The trajectory of the rocket is:

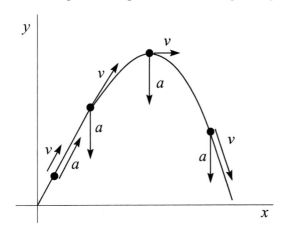

a) Find $v_x(t)$ and $v_y(t)$. Note that $\sin \alpha_0 = 4/5$ and $\cos \alpha_0 = 3/5$.

time 0 to time T:

$a_x = g \cos \alpha_0 = (3/5)g \qquad v_{0x} = 0, \quad v_x = v_{0x} = a_x t = (3/5)gt$

$a_y = g \sin \alpha_0 = (4/5)g \qquad v_{0y} = 0, \quad v_y = v_{0y} = a_y t = (4/5)gt$

at time T: $v_x = (3/5)gT \qquad v_y = (4/5)gT$

time T to end:

$a_x = 0 \qquad v_x = (3/5)gT$ (constant during this time interval)

$a_y = -g \qquad v_y = v_{0y} + a_y t = (4/5)gT - g(t - T)$ (Note that t is the time measure
from when the rocket is first fired, so $t - T$ is the elapsed time for this part of th
motion.)

The graphs of $v_x(t)$ and $v_y(t)$ are thus

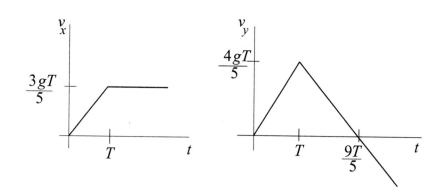

c) The maximum altitude is reached after the engines shut off, so occurs at a time t that is greater than T.

Set $v_y = 0$ and solve for t: $(4/5)gT - g(t - T) = 0$, $\quad t = (4/5)T + T = (9/5)T$.

The acceleration is constant in the time interval $t = 0$ to $t = T$ and again in the interval $t = T$ to the end. But the acceleration changes at $t = T$ so the constant acceleration equations don't apply for the entire time period of the motion.

First use $y - y_0 = v_{0y}t + \frac{1}{2}a_y t^2$ in the $t = 0$ to $t = T$ interval to solve for y_T, the altitude at time $t = T$: $y_T = 0 + \frac{1}{2}a_y T^2 = \frac{1}{2}(4/5)gT^2 = (2/5)gT^2$.

For the second part of the motion the initial time is $t = T$ and the constant acceleration equation is $y - y_T = v_{Ty}t + \frac{1}{2}a_y(t-T)^2$. At the maximum height $t-T = (4/5)T$.
$y - (2/5)gT^2 = [(4/5)gT][(4/5)T] - \frac{1}{2}g[(4/5)T]^2$
This gives $y = (18/25)gT^2$.

d) We know that $t = (9/5)T$ when the rocket is at the maximum height and that this maximum height is $y = (18/25)gT^2$. Apply the constant acceleration equations for the y-component of the motion from the maximum height to the ground to find the time t when the rocket reaches the ground.

$y - y_0 = -(18/25)gT^2$, $\quad t = ?$, $\quad a_y = -g$, $\quad v_{0y} = 0$

Since $v_{0y} = 0$ (starts at maximum height), the equation $y - y_0 = v_{0y}t + \frac{1}{2}a_y t^2$ gives

$$t = \sqrt{\frac{2(y - y_0)}{a_y}} = \sqrt{\frac{36T^2}{25}} = (6/5)T.$$

$(6/5)T$ is the time from the maximum height to the ground and $(9/5)T$ is the time from the start of the motion to the maximum height, so the total time in the air is $t = 3T$. The rocket travels for time T in the first constant acceleration segment and for time $2T$ in the second constant acceleration segment.

Find x at $t = T$, at the end of the first segment:

$x - x_0 = v_{0x}t + \frac{1}{2}a_x t^2$ gives $x = 0 + \frac{1}{2}(3/5)gT^2 = (3/10)gT^2$

The horizontal displacement during the second segment, from $t = T$ to $t = 3T$ and where $a_x = 0$, is

$x - x_0 = v_{Tx}t = [(3/5)gT][3T - T] = (6/5)gT^2.$

The total horizontal displacement is the sum, $(3/10)gT^2 + (6/5)gT^2 = (3/2)gT^2.$

3-59 $x = R\cos\omega t, \quad y = R\sin\omega t$

a) $r = \sqrt{x^2 + y^2} = \sqrt{R^2\cos^2\omega t + R^2\sin^2\omega t} = \sqrt{R^2(\sin^2\omega t + \cos^2\omega t)} = \sqrt{R^2}$ R, since $\sin^2\omega t + \cos^2\omega t = 1.$

b) $v_x = \dfrac{dx}{dt} = -R\omega\sin\omega t, \quad v_y = \dfrac{dy}{dt} = R\omega\cos\omega t$

$\vec{v}\cdot\vec{r} = v_x x + v_y y = (-R\omega\sin\omega t)(R\cos\omega t) + (R\omega\cos\omega t)(R\sin\omega t)$

$\vec{v}\cdot\vec{r} = R^2\omega(-\sin\omega t\cos\omega t + \sin\omega t\cos\omega t) = 0,$ so $\vec{v}$ is perpendicular to $\vec{r}$.

c) $a_x = \dfrac{dv_x}{dt} = -R\omega^2\cos\omega t = -\omega^2 x$

$a_y = \dfrac{dv_y}{dt} = -R\omega^2\sin\omega t = -\omega^2 y$

$a = \sqrt{a_x^2 + a_y^2} = \sqrt{\omega^4 x^2 + \omega^4 y^2} = \omega^2\sqrt{x^2 + y^2} = R\omega^2.$

$\vec{a} = a_x\hat{i} + a_y\hat{j} = -\omega^2(x\hat{i} + y\hat{j}) = -\omega^2\vec{r}.$

Since ω^2 is positive this means that the direction of $\vec{a}$ is opposite to the directio of $\vec{r}$.

d) $v = \sqrt{v_x^2 + v_y^2} = \sqrt{R^2\omega^2\sin^2\omega t + R^2\omega^2\cos^2\omega t} = \sqrt{R^2\omega^2(\sin^2\omega t + \cos^2\omega t)}$

$v = \sqrt{R^2\omega^2} = R\omega.$

e) $a = R\omega^2, \quad \omega = v/R, \quad$ so $a = R(v^2/R^2) = v^2/R.$

3-63   The trick is to find the time it takes for the twins to meet.

Since the pigeon flies at constant speed, the total distance the pigeon flies is jus this speed times the time it flies.

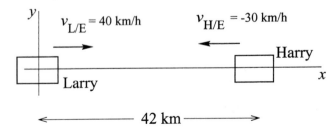

Work in a coordinate systend the velocity of Harry is $v_{H/L}$, the velocity of Harry relative to Larry. Let $v_{L/E} = +40$ km/h be the velocity of Larry relative to the earth and let $v_{H/E} = -30$ km/h be the velocity of Harry relative to the earth (in the opposite direction). Then the relative velocity formula (Eq.(3-36) gives $v_{H/E} = v_{H/L} + v_{L/E}$, or

$$v_{H/L} = v_{H/E} - v_{L/E} = -30 \text{ km/h} - 40 \text{ km/h} = -70 \text{ km/h}.$$

Then $x - x_0 = vt$ gives $t = \dfrac{x - x_0}{v} = \dfrac{-42 \text{ km}}{-70 \text{ km/h}} = +0.600$ h.

The distance the pigeon flies in this time is

distance = (speed)(time) = (50 km/h)(0.600 h) = 30 km.

-65 Select a coordinate system where $+y$ is north and $+x$ is east.

The velocity vectors in the problem are:

$\vec{v}_{P/E}$, the velocity of the plane relative to the earth.

$\vec{v}_{P/A}$, the velocity of the plane relative to the air (the magnitude $v_{P/A}$ is the air speed of the plane and the direction of $\vec{v}_{P/A}$ is the compass course set by the pilot).

$\vec{v}_{A/E}$, the velocity of the air relative to the earth (the wind velocity).

The rule for combining relative velocities gives $\vec{v}_{P/E} = \vec{v}_{P/A} + \vec{v}_{A/E}$.

a) We are given the following information about the relative velocities:

$\vec{v}_{P/A}$ has magnitude 220 km/h and its direction is west. In our coordinates it has components $(v_{P/A})_x = -220$ km/h and $(v_{P/A})_y = 0$.

From the displacement of the plane relative to the earth after 0.500 h, we find that $\vec{v}_{P/E}$ has components in our coordinate system of

$$(v_{P/E})_x = -\frac{120 \text{ km}}{0.500 \text{ h}} = -240 \text{ km/h (west)}$$

$$(v_{P/E})_y = -\frac{20 \text{ km}}{0.500 \text{ h}} = -40 \text{ km/h (south)}$$

With this information the diagram corresponding to the velocity addition equation is

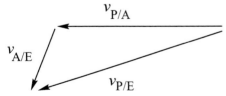

We are asked to find $\vec{v}_{A/E}$, so solve for this vector:

$\vec{v}_{P/E} = \vec{v}_{P/A} + \vec{v}_{A/E}$ gives $\vec{v}_{A/E} = \vec{v}_{P/E} - \vec{v}_{P/A}$.

The x-component of this equation gives

$(v_{A/E})_x = (v_{P/E})_x - (v_{P/A})_x = -240 \text{ km/h} - (-220 \text{ km/h}) = -20 \text{ km/h}.$

The y-component of this equation gives

$(v_{A/E})_y = (v_{P/E})_y - (v_{P/A})_y = -40 \text{ km/h}.$

Now that we have the components of $\vec{v}_{A/E}$ we can find its magnitude and directio

$v_{A/E} = \sqrt{(v_{A/E})_x^2 + (v_{A/E})_y^2}$

$v_{A/E} = \sqrt{(-20 \text{ km/h})^2 + (-40 \text{ km/h})^2} = 44.7 \text{ km/h}$

$\tan \phi = \dfrac{40 \text{ km/h}}{20 \text{ km/h}} = 2.00; \quad \phi = 63.4°$

The direction of the wind velocity is
63.4° S of W, or 26.6° W of S.

b) The rule for combining the relative velocities is still $\vec{v}_{P/E} = \vec{v}_{P/A} + \vec{v}_{A/E}$, b
some of these velocities have different values than in part (a).

$\vec{v}_{P/A}$ has magnitude 220 km/h but its direction is to be found.

$\vec{v}_{A/E}$ has magnitude 40 km/h and its direction is due south.

The direction of $\vec{v}_{P/E}$ is west; its magnitude is not given.

The vector diagram for $\vec{v}_{P/E} = \vec{v}_{P/A} + \vec{v}_{A/E}$ and the specified directions for t
vectors is

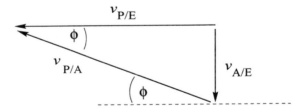

The vector addition diagram
forms a right triangle.

$\sin \phi = \dfrac{v_{A/E}}{v_{P/A}} = \dfrac{40 \text{ km/h}}{220 \text{ km/h}} = 0.1818; \quad \phi = 10.5°.$

The pilot should set her course 10.5° north of west.

Exercises 5, 11, 13, 15, 17, 21, 23, 25, 27, 29
Problems 33, 35, 37, 39, 41, 43, 47

Exercises

4-5   Use a coordinate system where the $+x$-axis is in the
       direction of $\vec{F}_A$, the force applied by dog A.

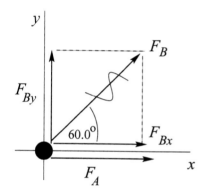

$$F_{Ax} = +270 \text{ N}, \quad F_{Ay} = 0$$

$$F_{Bx} = F_B \cos 60.0° = (300 \text{ N}) \cos 60.0° = +150 \text{ N}$$
$$F_{By} = F_B \sin 60.0° = (300 \text{ N}) \sin 60.0° = +260 \text{ N}$$

$$\vec{R} = \vec{F}_A + \vec{F}_B$$
$$R_x = F_{Ax} + F_{Bx} = +270 \text{ N} + 150 \text{ N} = +420 \text{ N}$$
$$R_y = F_{Ay} + F_{By} = 0 + 260 \text{ N} = +260 \text{ N}$$

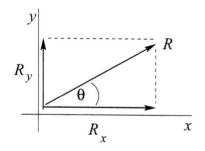

$$R = \sqrt{R_x^2 + R_y^2}$$
$$R = \sqrt{(420 \text{ N})^2 + (260 \text{ N})^2} = 494 \text{ N}$$

$$\tan \theta = \frac{R_y}{R_x} = 0.619$$

$$\theta = 31.8°$$

4-11   a) During this time interval the acceleration is constant and equal to

$$a_x = \frac{F_x}{m} = \frac{0.250 \text{ N}}{0.160 \text{ kg}} = 1.562 \text{ m/s}^2$$

We can use the constant acceleration kinematic equations from Chapter 2.
$$x - x_0 = v_{0x}t + \tfrac{1}{2}a_x t^2 = 0 + \tfrac{1}{2}(1.562 \text{ m/s}^2)(2.00 \text{ s})^2,$$

so the puck is at $x = 3.12$ m.

$v_x = v_{0x} + a_x t = 0 + (1.562 \text{ m/s}^2)(2.00 \text{ s}) = 3.12 \text{ m/s}.$

b) In the time interval from $t = 2.00 \text{ s} + 5.00 \text{ s}$ the force has been removed so th acceleration is zero. The speed stays constant at $v_x = 3.12 \text{ m/s}$. The distance th puck travels is $x - x_0 = v_{0x} t = (3.12 \text{ s})(5.00 \text{ s} - 2.00 \text{ s}) = 9.36 \text{ m}$. At the end of th interval it is at $x = x_0 + 9.36 \text{ m} = 12.5 \text{ m}$.

In the time interval from $t = 5.00 \text{ s}$ to 7.00 s the acceleration is again $a_x = 1.5$ m/s^2. At the start of this interval $v_{0x} = 3.12 \text{ m/s}$ and $x_0 = 12.5 \text{ m}$.

$x - x_0 = v_{0x} t + \frac{1}{2} a_x t^2 = (3.12 \text{ m/s})(2.00 \text{ s}) + \frac{1}{2}(1.562 \text{ m/s}^2)(2.00 \text{ s})^2.$

$x - x_0 = 6.24 \text{ m} + 3.12 \text{ m} = 9.36 \text{ m}.$

Therefore, at $t = 7.00 \text{ s}$ the puck is at $x = x_0 + 9.36 \text{ m} = 12.5 \text{ m} + 9.36 \text{ m} = 21.9 \text{ m}$.

$v_x = v_{0x} + a_x t = 3.12 \text{ m/s} + (1.562 \text{ m/s}^2)(2.00 \text{ s}) = 6.24 \text{ m/s}$

4-13   **a)** Constant velocity implies $\vec{a} = 0$, so $\sum \vec{F} = 0$;

the resultant force is zero; the vector sum of all forces on the puck is zero.

b)

Constant $\vec{v}$ says the puck travels in a straight line.

c) The new force produces a constant acceleration that is in the direction perpe dicular to the line connecting points A and B. The path is a parabola, just as projectile motion.

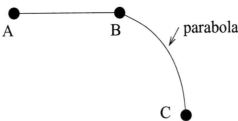

d) Now the acceleration is always perpendicular to the instantaneous velocity $\vec{v}$, the puck moves in a circular path with constant speed.

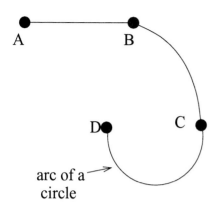

arc of a
circle

-15 $F = ma$

We must use $w = mg$ to find the mass of the boulder:

$$m = \frac{w}{g} = \frac{2400 \text{ N}}{9.80 \text{ m/s}^2} = 244.9 \text{ kg}$$

Then $F = ma = (244.9 \text{ kg})(12.0 \text{ m/s}^2) = 2940 \text{ N}$.

-17 $w = mg$

The mass of the watermelon is constant, independent of its location. Its weight differs on earth and Jupiter's moon. Use the information about the watermelon's weight on earth to calculate its mass:

$$w = mg \text{ gives that } m = \frac{w}{g} = \frac{44.0 \text{ N}}{9.80 \text{ m/s}^2} = 4.49 \text{ kg}.$$

On Jupiter's moon, $m = 4.49$ kg, the same as on earth. Thus the weight on Jupiter's moon is $w = mg = (4.49 \text{ kg})(1.81 \text{ m/s}^2) = 8.13 \text{ N}$.

-21 a) The free-body diagram for the bottle is

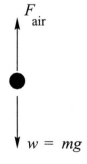

The only forces on the bottle are
gravity (downward) and air resistance (upward).

F_{air}

$w = mg$

b)

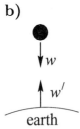

w is the force of gravity that the earth exerts on the bottle. The reaction to this force is w', force that the bottle exerts on the earth

Note that these two equal and opposite forces produce very different acceleratio because the bottle and the earth have very different masses.

F_{air} is the force that the air exerts on the bottle and is upward. The reaction to th force is a downward force F'_{air} that the bottle exerts on the air. These two forc have equal magnitudes and opposite directions.

4-23 The force of gravity that the earth exerts on her is her weight,

$w = mg = (45 \text{ kg})(9.8 \text{ m/s}^2) = 441 \text{ N}$. By Newton's 3rd law, she exerts an equ and opposite force on the earth.

Apply $\sum \vec{F} = m\vec{a}$ to the earth, with $|\sum \vec{F}| = w = 441 \text{ N}$, but must use the ma of the earth for m.

$$a = \frac{w}{m} = \frac{441 \text{ N}}{6.0 \times 10^{24} \text{ kg}} = 7.4 \times 10^{-23} \text{ m/s}^2.$$

(This is <u>much</u> smaller than her acceleration of 9.8 m/s^2.)

4-25 The free-body diagram for the bucket is

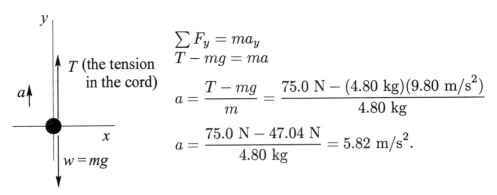

$$\sum F_y = ma_y$$
$$T - mg = ma$$

$$a = \frac{T - mg}{m} = \frac{75.0 \text{ N} - (4.80 \text{ kg})(9.80 \text{ m/s}^2)}{4.80 \text{ kg}}$$

$$a = \frac{75.0 \text{ N} - 47.04 \text{ N}}{4.80 \text{ kg}} = 5.82 \text{ m/s}^2.$$

4-27 Since the crates are connected by a rope, they both have the same acceleration.

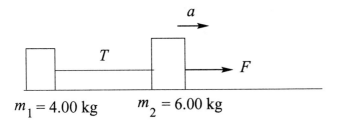

$m_1 = 4.00$ kg $m_2 = 6.00$ kg

a) Consider the two crates and the rope connecting them as a single object of mass $m = m_1 + m_2 = 10.0$ kg. The free-body diagram is

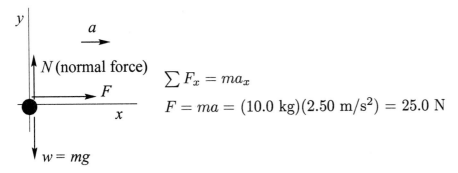

$$\sum F_x = ma_x$$

$$F = ma = (10.0 \text{ kg})(2.50 \text{ m/s}^2) = 25.0 \text{ N}$$

b) Consider the forces on the 4.00 kg crate:

$$\sum F_x = ma_x$$

$$T = m_1 a = (4.00 \text{ kg})(2.50 \text{ m/s}^2) = 10.0 \text{ N}.$$

As a check, can also consider the forces on the 6.00 kg crate:

$$\sum F_x = ma_x$$

$$F - T = m_2 a$$

$$T = F - m_2 a = 25.0 \text{ N} - (6.00 \text{ kg})(2.50 \text{ m/s}^2)$$

$$T = 25.0 \text{ N} - 15.0 \text{ N} = 10.0 \text{ N}$$

4-29 Take the $+y$-direction to be upward.

The free-body diagam for the person is

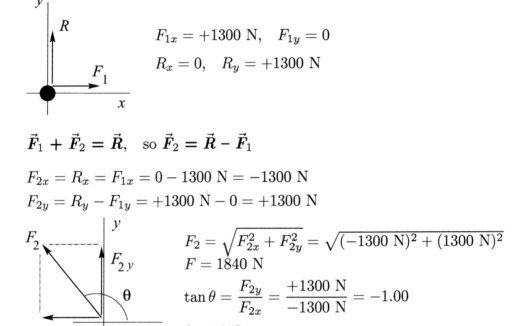

$$\sum F_y = ma_y$$
$$F_D - mg = ma$$

$$a = \frac{F_D - mg}{m} = \frac{620 \text{ N} - (55.0 \text{ kg})(9.80 \text{ m/s}^2)}{55.0 \text{ kg}}$$

$$a = 1.47 \text{ m/s}^2$$

$w = mg$

The acceleration is positive, so its direction is upward.

Problems

4-33 Use coordinates with the $+x$-axis along $\vec{F}_1$ and the $+y$-axis along $\vec{R}$.

$$F_{1x} = +1300 \text{ N}, \quad F_{1y} = 0$$

$$R_x = 0, \quad R_y = +1300 \text{ N}$$

$$\vec{F}_1 + \vec{F}_2 = \vec{R}, \quad \text{so } \vec{F}_2 = \vec{R} - \vec{F}_1$$

$$F_{2x} = R_x = F_{1x} = 0 - 1300 \text{ N} = -1300 \text{ N}$$
$$F_{2y} = R_y - F_{1y} = +1300 \text{ N} - 0 = +1300 \text{ N}$$

$$F_2 = \sqrt{F_{2x}^2 + F_{2y}^2} = \sqrt{(-1300 \text{ N})^2 + (1300 \text{ N})^2}$$
$$F = 1840 \text{ N}$$

$$\tan \theta = \frac{F_{2y}}{F_{2x}} = \frac{+1300 \text{ N}}{-1300 \text{ N}} = -1.00$$

$$\theta = 135°$$

The magnitude of $\vec{F}_2$ is 1840 N and its direction is 135° counterclockwise from the direction of $\vec{F}_1$.

4-35 If the box moves in the $+x$-direction it must have $a_y = 0$, so $\sum F_y = 0$.

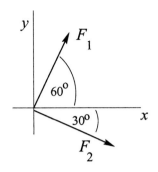

The smallest force the child can exert and still produce such motion is a force that makes the y-components of all three forces sum to zero, but that doesn't have any x-component.

Let $\vec{F}_3$ be the force exerted by the child.

$\sum F_y = ma_y$ implies $F_{1y} + F_{2y} + F_{3y} = 0$,   so $F_{3y} = -(F_{1y} + F_{2y})$.

$F_{1y} = +F_1 \sin 60° = (100\text{ N}) \sin 60° = 86.6\text{ N}$
$F_{2y} = +F_2 \sin(-30°) = -F_2 \sin 30° = -(140\text{ N}) \sin 30° = -70.0\text{ N}$

Then $F_{3y} = -(F_{1y} + F_{2y}) = -(86.6\text{ N} - 70.0\text{ N}) = -16.6\text{ N};\quad F_{3x} = 0$

The smallest force the child can exert has magnitude 17 N and is directed at 90° clockwise from the $+x$-axis shown in the figure.

b) Apply $\sum F_x = ma_x$. The force exerted by the child is in the $-y$-direction and has no x-component.

$F_{1x} = F_1 \cos 60° = 50\text{ N}$
$F_{2x} = F_2 \cos 30° = 121.2\text{ N}$
$\sum F_x = F_{1x} + F_{2x} = 50\text{ N} + 121.2\text{ N} = 171.2\text{ N}$

$$m = \frac{\sum F_x}{a_x} = \frac{171.2\text{ N}}{2.00\text{ m/s}^2} = 85.6\text{ kg}$$

Then $w = mg = 840\text{ N}$.

4-37 First use the information given about the height of the jump to calculate the speed he has at the instant his feet leave the ground. Use a coordinate system with the $+y$-axis upward and the origin at the position when his feet leave the ground.

$v_y = 0$ (at the maximum height), $v_{0y} = ?$, $a_y = -9.80\text{ m/s}^2$, $y - y_0 = +1.2\text{ m}$
$v_y^2 = v_{0y}^2 + 2a_y(y - y_0)$

$v_{0y} = \sqrt{-2a_y(y - y_0)} = \sqrt{-2(-9.80\text{ m/s}^2)(1.2\text{ m})} = 4.85\text{ m/s}$

Now consider the acceleration phase, from when he starts to jump until when his feet leave the ground. Use a coordinate system where the $+y$-axis is upward and the origin is at his position when he starts his jump.

Calculate the average acceleration:

$$(a_{av})_y = \frac{v_y - v_{0y}}{t} = \frac{4.85 \text{ m/s} - 0}{0.300 \text{ s}} = 16.17 \text{ m/s}^2$$

Finally, find the average upward force that the ground must exert on him to produce this average upward acceleration. (Don't forget about the downward force o gravity.)

$$m = w/g = \frac{890 \text{ N}}{9.80 \text{ m/s}^2} = 90.8 \text{ kg}$$

$(a_{av})_y \uparrow$ F_{av} (the average force the ground exerts on him)

mg

$$\sum F_y = ma_y$$
$$F_{av} - mg = m(a_{av})_y$$

$$F_{av} = m(g + (a_{av})_y)$$
$$F_{av} = 90.8 \text{ kg}(9.80 \text{ m/s}^2 + 16.17 \text{ m/s}^2$$
$$F_{av} = 2360 \text{ N}$$

This is the average force exerted on him by the ground. But by Newton's 3rd law, the average force he exerts on the ground is equal and opposite, so is 2360 N downward.

4-39 a) $x = (9.0 \times 10^3 \text{ m/s}^2)t^2 - (8.0 \times 10^4 \text{ m/s}^3)t^3$

$x = 0$ at $t = 0$

When $t = 0.025$ s, $x = (9.0 \times 10^3 \text{ m/s}^2)(0.025 \text{ s})^2 - (8.0 \times 10^4 \text{ m/s}^3)(0.025\text{s})^3 = 4.4$ m.

The length of the barrel must be 4.4 m.

b) $v = \dfrac{dx}{dt} = (18.0 \times 10^3 \text{ m/s}^2)t - (24.0 \times 10^4 \text{ m/s}^3)t^2$

At $t = 0$, $v = 0$ (object starts from rest).

At $t = 0.025$ s, when the object reaches the end of the barrel,

$v = (18.0 \times 10^3 \text{ m/s}^2)(0.025 \text{ s}) - (24.0 \times 10^4 \text{ m/s}^3)(0.025 \text{ s})^2 = 300$ m/s

c) $\sum F_x = ma_x$, so must find a.

$$a = \frac{dv}{dt} = 18.0 \times 10^3 \text{ m/s}^2 - (48.0 \times 10^4 \text{ m/s}^3)t$$

(i) At $t = 0$, $a = 18.0 \times 10^3 \text{ m/s}^2$ and $\sum F_x = (1.50 \text{ kg})(18.0 \times 10^3 \text{ m/s}^2) = 2.7 \times 10^4$ N.

(ii) At $t = 0.025$ s, $a = 18.0 \times 10^3$ m/s$^2 - (48.0 \times 10^4$ m/s$^3)(0.025$ s$) = 6.0 \times 10^3$ m/s^2
and $\sum F_x = (1.50$ kg$)(6.0 \times 10^3$ m/s$^2) = 9.0 \times 10^3$ N.

4-41 a)   Consider all four cars together as one object.
The horizontal force on this combined object is the force of the engine on the first car. The mass of all four cars together is $4m$.

$$\sum F_x = ma_x$$
$$F = 4ma$$

b) Treat the last three cars together as one object. The horizontal force on this combined object is the force of the first car on the second car. The mass of these three cars together is $3m$.

$$\sum F_x = ma_x$$
$$F = 3ma$$

c) Treat the last two cars together. Their total mass is $2m$.

$$\sum F_x = ma_x$$
$$F = 2ma$$

d) By Newton's third law the force of the fourth car on the third car is equal and opposite to the force of the third car on the fourth. Free-body diagram for the fourth car:

$$\sum F_x = ma_x$$
$$F = ma$$

e) The forces would all be the same magnitude but would be in the opposite direction.

4-43 a) Take the $+y$-direction to be upward since that is the direction
of the acceleration. The maximum upward acceleration is obtained from the maxi-

mum possible tension in the cables.

$$\sum F_y = ma_y$$
$$T - mg = ma$$

$$a = \frac{T - mg}{m}$$

$$a = \frac{28,000 \text{ N} - (2200 \text{ kg})(9.80 \text{ m/s}^2)}{2200 \text{ kg}} = 2.93 \text{ m/s}^2.$$

b) What changes is the weight mg of the elevator.

$$a = \frac{T - mg}{m} = \frac{28,000 \text{ N} - (2200 \text{ kg})(1.62 \text{ m/s}^2)}{2200 \text{ kg}} = 11.1 \text{ m/s}^2.$$

4-47 Note that in this problem the mass of the rope is given, and that it is not negligible compared to the other masses.

a)

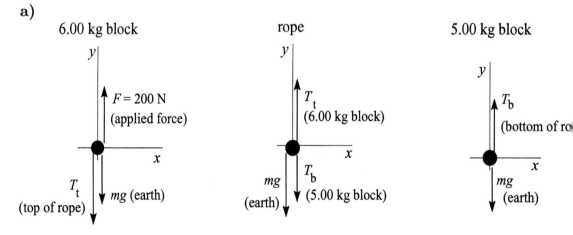

6.00 kg block   rope   5.00 kg block

$F = 200$ N (applied force)

T_t (top of rope)

mg (earth)

T_t (6.00 kg block)

mg (earth)

T_b (5.00 kg block)

T_b (bottom of ro

mg (earth)

T_t is the tension at the top of the rope and T_b is the tension at the bottom of th rope.

b) Treat the rope and the two blocks together as a single object, with mass m = 6.00 kg + 4.00 kg + 5.00 kg = 15.0 kg. Take $+y$ upward, since the acceleration upward.

$$\sum F_y = ma_y$$
$$F - mg = ma$$

$$a = \frac{F - mg}{m}$$

$$a = \frac{200 \text{ N} - (15.0 \text{ kg})(9.80 \text{ m/s}^2)}{15.0 \text{ kg}} = 3.53 \text{ m/s}^2$$

c) Consider the forces on the top block ($m = 6.00$ kg), since the tension at the top of the rope (T_t) will be one of these forces.

$$\sum F_y = ma_y$$
$$F - mg - T_t = ma$$
$$T_t = F - m(g + a)$$
$$T = 200 \text{ N} - (6.00 \text{ kg})(9.80 \text{ m/s}^2 + 3.53 \text{ m/s}^2) = 120 \text{ N}$$

Alternatively, can consider the forces on the combined object rope plus bottom block ($m = 9.00$ kg):

$$\sum F_y = ma_y$$
$$T_t - mg = ma$$
$$T_t = m(g + a) = 9.00 \text{ kg}(9.80 \text{ m/s}^2 + 3.53 \text{ m/s}^2) = 120 \text{ N},$$
which checks

d) One way to do this is to consider the forces on the top half of the rope ($m = 2.00$ kg). Let T_m be the tension at the midpoint of the rope.

$$\sum F_y = ma_y$$
$$T_t - T_m - mg = ma$$
$$T_m = T_t - m(g+a) = 120 \text{ N} - 2.00 \text{ kg}(9.80 \text{ m/s}^2 + 3.53 \text{ m/s}^2) = 93.3 \text{ N}$$

To check this answer we can alternatively consider the forces on the bottom half of the rope plus the lower block taken together as a combined object ($m = 2.00$ kg + 5.00 kg $= 7.00$ kg):

$$\sum F_y = ma_y$$
$$T_m - mg = ma$$
$$T_m = m(g + a) = 7.00 \text{ kg}(9.80 \text{ m/s}^2 + 3.53 \text{ m/s}^2) = 93.3 \text{ N},$$
which checks

CHAPTER 5
APPLICATIONS OF NEWTON'S LAWS

Exercises 3, 5, 7, 9, 11, 15, 17, 19, 21, 25, 27, 29, 33, 35, 37, 43, 45, 51, 53
Problems 61, 63, 65, 67, 69, 71, 77, 79, 81, 85, 89, 91, 97, 99, 101, 103

Exercises

5-3 **a)** Force diagram for the person:

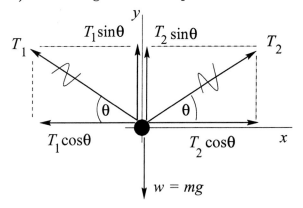

T_1 and T_2 are the tensions in each half of the rope.

$\sum F_x = 0$

$T_2 \cos \theta - T_1 \cos \theta = 0$

This says that $T_1 = T_2 = T$ (The tension is the same on both sides of the person.)

$\sum F_y = 0$

$T_1 \sin \theta + T_2 \sin \theta - mg = 0$

But $T_1 = T_2 = T$, so $2T \sin \theta = mg$

$$T = \frac{mg}{2 \sin \theta} = \frac{(90.0 \text{ kg})(9.80 \text{ m/s}^2)}{2 \sin 10.0°} = 2540 \text{ N}$$

b) The relation $2T \sin \theta = mg$ still applies but now we are given that $T = 2.50 \times 10^4$ N (the breaking strength) and are asked to find θ.

$$\sin \theta = \frac{mg}{2T} = \frac{(90.0 \text{ kg})(9.80 \text{ m/s}^2)}{2(2.50 \times 10^4 \text{ N})} = 0.01764, \quad \theta = 1.01°.$$

5-5   (We use the symbol N for the normal force.)

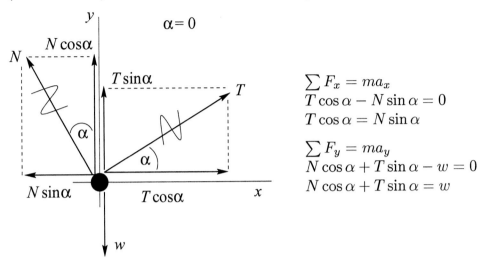

$$\sum F_x = ma_x$$
$$T\cos\alpha - N\sin\alpha = 0$$
$$T\cos\alpha = N\sin\alpha$$

$$\sum F_y = ma_y$$
$$N\cos\alpha + T\sin\alpha - w = 0$$
$$N\cos\alpha + T\sin\alpha = w$$

The first equation gives $N = T\left(\dfrac{\cos\alpha}{\sin\alpha}\right)$.

Use this in the second equation to eliminate N:

$$\left(T\frac{\cos\alpha}{\sin\alpha}\right)\cos\alpha + T\sin\alpha = w$$

Multiply this equation by $\sin\alpha$:

$$T(\cos^2\alpha + \sin^2\alpha) = w\sin\alpha$$
$$T = w\sin\alpha \text{ (since } \cos^2\alpha + \sin^2\alpha = 1).$$

Then $N = T\left(\dfrac{\cos\alpha}{\sin\alpha}\right) = w\sin\alpha\left(\dfrac{\cos\alpha}{\sin\alpha}\right) = w\cos\alpha.$

These results are the same as obtained in Example 5-3.

5-7   Force diagram for the wrecking ball:

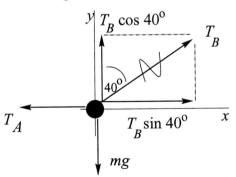

a) $\sum F_y = ma_y$

$T_B \cos 40° - mg = 0$

$$T_B = \frac{mg}{\cos 40°} = \frac{(4090 \text{ kg})(9.80 \text{ m/s}^2)}{\cos 40°} = 5.23 \times 10^4 \text{ N}$$

b) $\sum F_x = ma_x$

$T_B \sin 40° - T_A = 0$

$T_A = T_B \sin 40° = 3.36 \times 10^4 \text{ N}$

5-9   Let $\vec{F}$ be the resistive force.

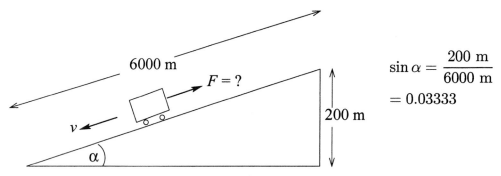

$$\sin \alpha = \frac{200 \text{ m}}{6000 \text{ m}}$$

$$= 0.03333$$

Free-body diagram for the car:

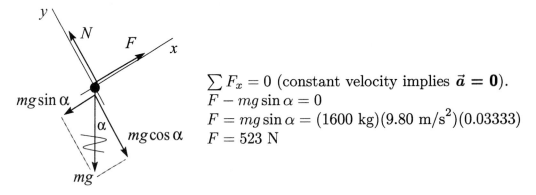

$\sum F_x = 0$ (constant velocity implies $\vec{a} = 0$).

$F - mg \sin \alpha = 0$

$F = mg \sin \alpha = (1600 \text{ kg})(9.80 \text{ m/s}^2)(0.03333)$

$F = 523 \text{ N}$

5-11   **a)** Let the tensions in the three strings be T, T', and T''

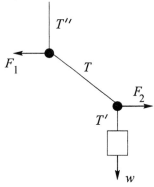

Free-body diagram for the block:

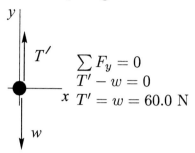

$$\sum F_y = 0$$
$$T' - w = 0$$
$$T' = w = 60.0 \text{ N}$$

Free-body diagram for the lower knot:

$$\sum F_y = 0$$
$$T \sin 45° - T' = 0$$
$$T = \frac{T'}{\sin 45°} = \frac{60.0 \text{ N}}{\sin 45°} = 84.9 \text{ N}$$

b) Apply $\sum F_x = 0$ to the force diagram for the lower knot:
$$\sum F_x = 0$$
$$F_2 = T \cos 45° = (84.9 \text{ N}) \cos 45° = 60.0 \text{ N}$$

Free-body diagram for the upper knot:

$$\sum F_x = 0$$
$$T \cos 45° - F_1 = 0$$
$$F_1 = (84.9 \text{ N}) \cos 45°$$
$$F_1 = 60.0 \text{ N}$$

Note that $F_1 = F_2$.

5-15 **a)** The free-body diagrams for the bricks and counterweight are:

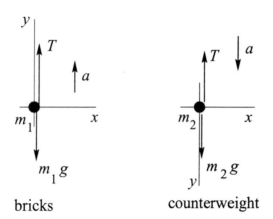

bricks   counterweight

b) Apply $\sum F_y = ma_y$ to each object. The acceleration magnitude is the same for the two objects. For the bricks take $+y$ to be upward since $\vec{a}$ for the bricks is upward. For the counterweight take $+y$ to be downward since $\vec{a}$ is downward.

bricks: $\sum F_y = ma_y$

$T - m_1 g = m_1 a$

counterweight: $\sum F_y = ma_y$

$m_2 g - T = m_2 a$

Add these two equations to eliminate T:

$(m_2 - m_1)g = (m_1 + m_2)a$

$a = \left(\dfrac{m_2 - m_1}{m_1 + m_2} \right) g = \left(\dfrac{28.0 \text{ kg} - 15.0 \text{ kg}}{15.0 \text{ kg} + 28.0 \text{ kg}} \right) (9.80 \text{ m/s}^2) = 2.96 \text{ m/s}^2$

c) $T - m_1 g = m_1 a$ gives $T = m_1(a + g) = (15.0 \text{ kg})(2.96 \text{ m/s}^2 + 9.80 \text{ m/s}^2) = 191$ N

As a check, calculate T using the other equation:

$m_2 g - T = m_2 a$ gives $T = m_2(g - a) = 28.0 \text{ kg}(9.80 \text{ m/s}^2 - 2.96 \text{ m/s}^2) = 191$ N, which checks.

The tension is 1.30 times the weight of the bricks; this causes the bricks to accelerate upward. The tension is 0.696 times the weight of the counterweight; this causes the counterweight to accelerate downward.

5-17   A sketch of the situation is

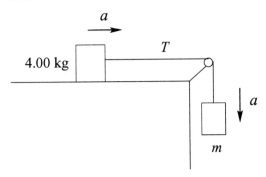

a) Free-body diagrams for each of the two blocks:

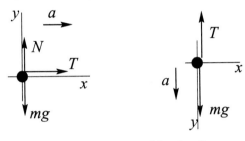

4.00 kg block   block of mass m

Note that for the hanging block we take the $+y$-direction to be downward since that is the direction of the acceleration of that block.

b) Apply $\sum F_x = ma_x$ to the 4.00 kg block:

$T = ma$

$$a = \frac{T}{m} = \frac{10.0 \text{ N}}{4.00 \text{ kg}} = 2.50 \text{ m/s}^2$$

Since the two blocks are connected by the rope, they have the same magnitude of acceleration.

c) Apply $\sum F_y = ma_y$ to the block of mass m:

$mg - T = ma$

$m(g - a) = T$

$$m = \frac{T}{g - a} = \frac{10.0 \text{ N}}{9.80 \text{ m/s}^2 - 2.50 \text{ m/s}^2} = 1.37 \text{ kg}$$

d) The tension is less than the weight of the hanging block. The net force on the hanging block must be downward if that block is to accelerate downward.

5-19   **a)** Consider the forces on the student.

The reading on the scale equals the upward normal force exerted on the student; also, the student is at rest relative to the elevator, so the acceleration of the student equals that of the elevator.

We can calculate the mass of the student from the weight: $m = \dfrac{w}{g} = \dfrac{550 \text{ N}}{9.80 \text{ m/s}^2} =$ 56.1 kg

We know that $\vec{a}$ is downward since $w < N$. Thus take the $+y$-direction to be downward, in the direction of $\vec{a}$.

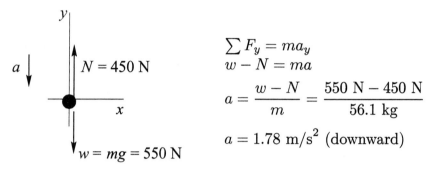

$$\sum F_y = ma_y$$
$$w - N = ma$$
$$a = \frac{w - N}{m} = \frac{550 \text{ N} - 450 \text{ N}}{56.1 \text{ kg}}$$
$$a = 1.78 \text{ m/s}^2 \text{ (downward)}$$

b) Now $N > w$ so the student (and elevator) is accelerating upward. Take the $+y$-direction to be upward, in the direction of $\vec{a}$.

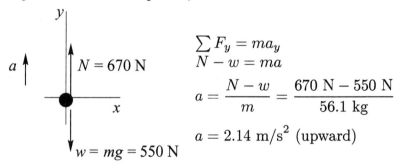

$$\sum F_y = ma_y$$
$$N - w = ma$$
$$a = \frac{N - w}{m} = \frac{670 \text{ N} - 550 \text{ N}}{56.1 \text{ kg}}$$
$$a = 2.14 \text{ m/s}^2 \text{ (upward)}$$

c) Now $N = 0$. $N < w$ so the acceleration is downward.

$$\sum F_y = ma_y$$
$$mg = ma$$
$$a = g \text{ (downward)}$$

The elevator and student are in free-fall; the elevator cable may have snapped.

5-21 **a)** At rest implies that $a = 0$, so the forces on the lead sinker must sum to zero.

The gravity force on the sinker is vertically downward, toward the center of the earth. The only other force on the sinker is the tension in the string. This force must be vertically upward in order for the two forces to sum to zero.

The sinker hangs such that the string is vertical relative to the earth. Thus relative to the car the sinker appears to be deflected down the slope.

b) <u>car moving up</u>

First consider the forces on the car, to calculate its acceleration. Take the x-axis parallel to the incline and the y-axis perpendicular to the incline.

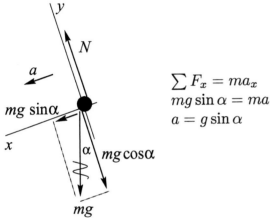

$$\sum F_x = ma_x$$
$$mg \sin \alpha = ma$$
$$a = g \sin \alpha$$

If the hanging sinker is at rest relative to the car then it also has acceleration $a = g \sin \alpha$ directed down the incline. Consider the forces on the sinker. Take the same x and y axes as for the car, so that $a_x = g \sin \alpha$ and $a_y = 0$. Assume that the string makes an angle β with the direction perpendicular to the ceiling of the car (the y-axis).

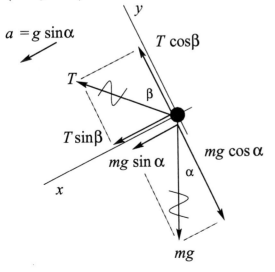

$$\sum F_x = ma_x$$
$$mg \sin \alpha + T \sin \beta = mg \sin \alpha$$

But this says $T \sin \beta = 0$,
so $\beta = 0$ and the string is perpendicular to the ceiling of the car.

car moving down the incline

The force diagrams for the car and for the sinker are the same as before, so agai
the sinker hangs such that the string is perpendicualr to the ceiling.

There is no deflection during each stage of the motion.

5-25   **a)** Constant speed implies $a = 0$

Consider the free-body diagram for the crate. Let $\vec{F}$ be the horizontal force applie
by the worker. The friction is kinetic friction since the crate is sliding along th
surface.

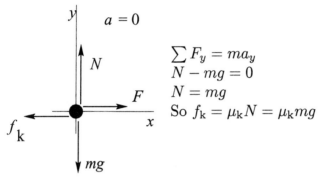

$$\sum F_y = ma_y$$
$$N - mg = 0$$
$$N = mg$$
So $f_k = \mu_k N = \mu_k mg$

$$\sum F_x = ma_x$$
$$F - f_k = 0$$
$$F = f_k = \mu_k mg = (0.20)(11.2 \text{ kg})(9.80 \text{ m/s}^2) = 22 \text{ N}$$

b) Now the only horizontal force on the crate is the kinetic friction force. Calculat
the acceleration force it produces. The friction force is $f_k = \mu_k mg$, just as in par
(a).

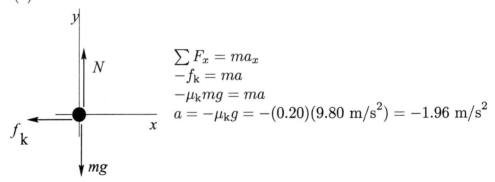

$$\sum F_x = ma_x$$
$$-f_k = ma$$
$$-\mu_k mg = ma$$
$$a = -\mu_k g = -(0.20)(9.80 \text{ m/s}^2) = -1.96 \text{ m/s}^2$$

Use the constant acceleration equations to find the distance the crate travels:
$$v = 0, \quad v_0 = 3.50 \text{ m/s}, \quad a = -1.96 \text{ m/s}^2, \quad x - x_0 = ?$$
$$v^2 = v_0^2 + 2a(x - x_0)$$

$$x - x_0 = \frac{v^2 - v_0^2}{2a} = \frac{0 - (3.50 \text{ m/s})^2}{2(-1.96 \text{ m/s}^2)} = 3.1 \text{ m}$$

5-27 **a)** Constant speed implies $a = 0$.

Consider the free-body diagram for the box:

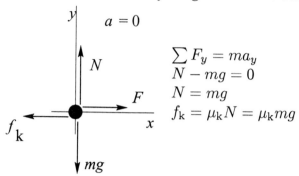

$$\sum F_y = ma_y$$
$$N - mg = 0$$
$$N = mg$$
$$f_k = \mu_k N = \mu_k mg$$

$$\sum F_x = ma_x$$
$$F - f_k = 0$$
$$F = f_k = \mu_k mg = (0.12)(6.00 \text{ kg})(9.80 \text{ m/s}^2) = 7.06 \text{ N}$$

b)

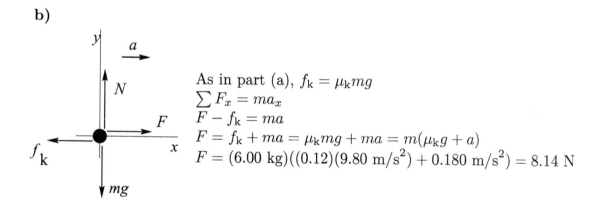

As in part (a), $f_k = \mu_k mg$
$$\sum F_x = ma_x$$
$$F - f_k = ma$$
$$F = f_k + ma = \mu_k mg + ma = m(\mu_k g + a)$$
$$F = (6.00 \text{ kg})((0.12)(9.80 \text{ m/s}^2) + 0.180 \text{ m/s}^2) = 8.14 \text{ N}$$

c) The normal force $N = mg$ is reduced. This in turn reduces the friction force, so the magnitude of the force $\vec{F}$ required to move the box is less.

part (a): $F = \mu_k mg = (0.12)(6.00 \text{ kg})(1.62 \text{ m/s}^2) = 1.17 \text{ N}$

part (b): $F = m(\mu_k + a) = 6.00 \text{ kg}((0.12)(1.62 \text{ m/s}^2) + 0.180 \text{ m/s}^2) = 2.25 \text{ N}$

5-29 Constant speed implies $a = 0$.

The angle α of the incline is given by $\sin \alpha = \dfrac{2.00 \text{ m}}{20.0 \text{ m}}$, so $\alpha = 5.739°$.

a) Consider the force diagram for the safe, with all the forces except for the applied force we are being asked to calculate. We must decide whether this force must be down the incline or up the incline, to make the total resultant force zero. Note that we know that the kinetic friction force f_k is directed up the incline, since the friction force opposes the motion. Use coordinates parallel and perpendicular to the incline.

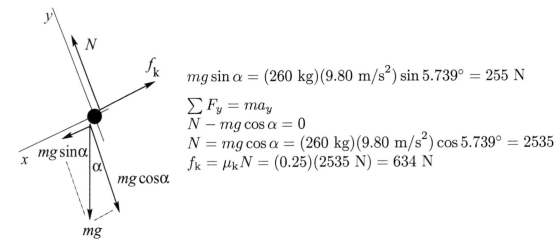

$mg \sin \alpha = (260 \text{ kg})(9.80 \text{ m/s}^2) \sin 5.739° = 255 \text{ N}$

$\sum F_y = ma_y$
$N - mg \cos \alpha = 0$
$N = mg \cos \alpha = (260 \text{ kg})(9.80 \text{ m/s}^2) \cos 5.739° = 2535$
$f_k = \mu_k N = (0.25)(2535 \text{ N}) = 634 \text{ N}$

We see that $mg \sin \alpha < f_k$, so for the forces to balance ($a = 0$) more force directe
down the incline is needed. The safe must be pulled down if it is to travel .
constant speed.

b) Now we can add the applied force $\vec{F}$ to the free-body diagram, since we ha
determined its direction.

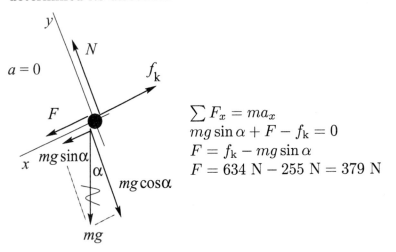

$\sum F_x = ma_x$
$mg \sin \alpha + F - f_k = 0$
$F = f_k - mg \sin \alpha$
$F = 634 \text{ N} - 255 \text{ N} = 379 \text{ N}$

5-33   Constant v implies $a = 0$.

Free-body diagram for A:

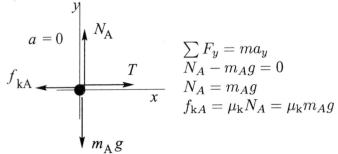

$\sum F_y = ma_y$
$N_A - m_A g = 0$
$N_A = m_A g$
$f_{kA} = \mu_k N_A = \mu_k m_A g$

$\sum F_x = ma_x$

$T - f_{kA} = 0$

$T = \mu_k m_A g$

Free-body diagram for B:

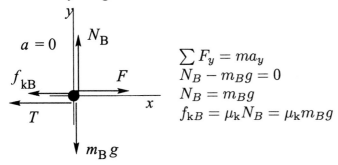

$\sum F_y = ma_y$

$N_B - m_B g = 0$

$N_B = m_B g$

$f_{kB} = \mu_k N_B = \mu_k m_B g$

$\sum F_x = ma_x$

$F - T - f_{kB} = 0$

$F = T + \mu_k m_B g$

Use the first equation to replace T in the second:

$F = \mu_k m_A g + \mu_k m_B g.$

a) $F = \mu_k(m_A + m_B)g$

b) $T = \mu_k m_A g$

5-35   The magnitude of the acceleration is the same for both blocks.

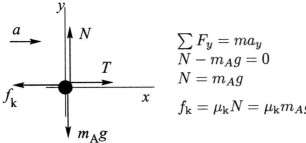

For each block take a positive coordinate direction to be the direction of the block's acceleration.

block on the table:

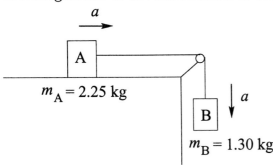

$\sum F_y = ma_y$

$N - m_A g = 0$

$N = m_A g$

$f_k = \mu_k N = \mu_k m_A g$

$$\sum F_x = ma_x$$
$$T - f_k = m_A a$$
$$T - \mu_k m_A g = m_A a$$

hanging block:

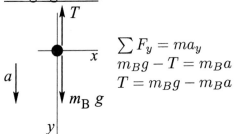

$$\sum F_y = ma_y$$
$$m_B g - T = m_B a$$
$$T = m_B g - m_B a$$

a) Use the second equation in the first

$$m_B g - m_B a - \mu_k m_A g = m_A a$$
$$(m_A + m_B)a = (m_B - \mu_k m_A)g$$
$$a = \frac{(m_B - \mu_k m_A)g}{m_A + m_B} = \frac{(1.30 \text{ kg} - (0.45)(2.25 \text{ kg}))(9.80 \text{ m/s}^2)}{2.25 \text{ kg} + 1.30 \text{ kg}} = 0.7937 \text{ m/s}^2$$

Now use the constant acceleration equations to find the final speed. Note that the blocks have the same speeds.

$$x - x_0 = 0.0300 \text{ m}, \quad a = 0.7937 \text{ m/s}^2, \quad v_0 = 0, \quad v = ?$$
$$v^2 = v_0^2 + 2a(x - x_0)$$
$$v = \sqrt{2a(x - x_0)} = \sqrt{2(0.7937 \text{ m/s}^2)(0.0300 \text{ m})} = 0.218 \text{ m/s} = 21.8 \text{ cm/s}.$$

b) $T = m_B g - m_B a = m_B(g - a) = 1.30 \text{ kg}(9.80 \text{ m/s}^2 - 0.7937 \text{ m/s}^2) = 11.7 \text{ N}$

Or, to check,

$$T - \mu_k m_A g = m_A a$$
$$T = m_A(a + \mu_k g) = 2.25 \text{ kg}(0.7937 \text{ m/s}^2 + (0.45)(9.80 \text{ m/s}^2)) = 11.7 \text{ N, which}$$
checks

5-37 **a)** Constant v implies $a = 0$.

Crate moving says that the friction is kinetic friction.

Free-body diagram for the crate:

$$\sum F_y = ma_y$$
$$N - mg - F\sin\theta = 0$$
$$N = mg + F\sin\theta$$

$$f_k = \mu_k N = \mu_k mg + \mu_k F\sin\theta$$

$$\sum F_x = ma_x$$
$$F\cos\theta - f_k = 0$$
$$F\cos\theta - \mu_k mg - \mu_k F\sin\theta = 0$$
$$F(\cos\theta - \mu_k\sin\theta) = \mu_k mg$$

$$F = \frac{\mu_k mg}{\cos\theta - \mu_k\sin\theta}$$

b) "start the crate moving" means the same force diagram as in part (a), except that μ_k is replaced by μ_s.

Thus $F = \dfrac{\mu_s mg}{\cos\theta - \mu_s\sin\theta}$.

$F \to \infty$ if $\cos\theta - \mu_s\sin\theta = 0$.

This gives $\mu_s = \dfrac{\cos\theta}{\sin\theta} = \dfrac{1}{\tan\theta}$.

-43 $\vec{a}_{\mathrm{rad}}$ is in toward the center of the circular path.

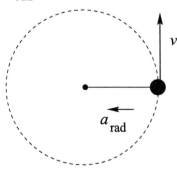

view from above

Free-body diagram:

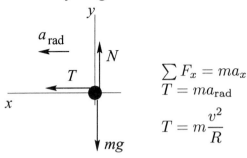

$$\sum F_x = ma_x$$
$$T = ma_{\text{rad}}$$

$$T = m\frac{v^2}{R}$$

view from side

The maximum v produces $T = 600$ N. Thus

$$v = \sqrt{\frac{RT}{m}} = \sqrt{\frac{(0.90 \text{ m})(600 \text{ N})}{0.80 \text{ kg}}} = 26.0 \text{ m/s}$$

5-45   **a)** Free-body diagram for the plane:

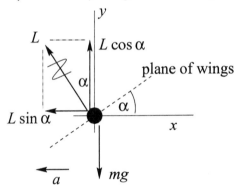

The lift force is perpendicular to the wings. The banking angle α is the angle between the plane of the wings and the horizontal.

For a constant-altitude turn the acceleration is horizontal, in toward the center of the circular path.

Use coordinates where x is horizontal and y is vertical. Then $a_y = 0$.

$$\sum F_y = ma_y$$

$L \cos \alpha - w = 0$

Then $L = 3.8w$ gives $3.8w \cos \alpha = w$

$\cos \alpha = 1/3.8$ and $\alpha = 75°$.

b) The answer to part (a) is determined from the vertical component of $\sum \vec{F} = m\vec{a}$ and the speed of the plane does not enter.

5-51   At each point $\vec{a}_{\text{rad}}$ is directed toward the center of the circular path.

a) "the pilot feels weightless" means that the vertical normal force N exerted on the pilot by the chair on which the pilot sits is zero. Force diagram for the pilot at the top of the path:

$$\sum F_y = ma_y$$
$$mg = ma_{\text{rad}}$$
$$g = \frac{v^2}{R}$$

Thus $v = \sqrt{gR} = \sqrt{(9.80 \text{ m/s}^2)(150 \text{ m})} = 38.34 \text{ m/s}$

$v = (38.34 \text{ m/s}) \left(\dfrac{1 \text{ km}}{10^3 \text{ m}} \right) \left(\dfrac{3600 \text{ s}}{1 \text{ h}} \right) = 138 \text{ km/h}$

b) Force diagram for the pilot at the bottom of the path. Note that the vertical normal force exerted on the pilot by the chair on which the pilot sits is now upward.

$$\sum F_y = ma_y$$
$$N - mg = m\frac{v^2}{R}$$
$$N = mg + m\frac{v^2}{R}$$

This normal force is the pilot's apparent weight.

$w = 700 \text{ N}$, so $m = \dfrac{w}{g} = 71.43 \text{ kg}$

$v = (280 \text{ km/h}) \left(\dfrac{1 \text{ h}}{3600 \text{ s}} \right) \left(\dfrac{10^3 \text{ m}}{1 \text{ km}} \right) = 77.78 \text{ m/s}$

Thus $N = 700 \text{ N} + 71.43 \text{ kg}\dfrac{(77.78 \text{ m/s})^2}{150 \text{ m}} = 3580 \text{ N}.$

5-53 Consider the free-body diagram for the water when the pail is at the top of its circular path.

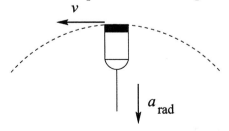

The radial acceleration is in toward the center of the circle so at this point is downward.
N is the downward normal force exerted on the water by the bottom of the pail.

$$\sum F_y = ma_y$$

$$N + mg = m\frac{v^2}{R}$$

At the minimum speed the water is just ready to lose contact with the bottom the pail, so at this speed, $N \to 0$. (Note that the force N cannot be upward.)

With $N \to 0$ the equation becomes $mg = m\dfrac{v^2}{R}$

$$v = \sqrt{gR} = \sqrt{(9.80 \text{ m/s}^2)(0.600 \text{ m})} = 2.42 \text{ m/s}.$$

Problems

5-61   No friction. Constant speed implies $a = 0$.

Since $a = 0$ it is equally convenient to use

(1) coordinate axes parallel and perpendicular to the incline;

(2) coordinate axes that are horizontal and vertical.

We will work the problem both ways.

coordinate axes parallel and perpendicular to the incline

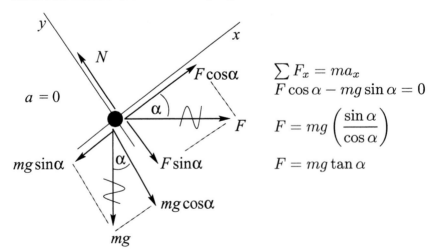

$$\sum F_x = ma_x$$
$$F \cos \alpha - mg \sin \alpha = 0$$

$$F = mg \left(\frac{\sin \alpha}{\cos \alpha} \right)$$

$$F = mg \tan \alpha$$

coordinates that are horizontal and vertical

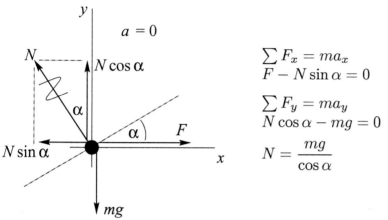

$$\sum F_x = ma_x$$
$$F - N \sin \alpha = 0$$

$$\sum F_y = ma_y$$
$$N \cos \alpha - mg = 0$$

$$N = \frac{mg}{\cos \alpha}$$

Combine these two equations to eliminate N:

$$F = N \sin \alpha = (mg/\cos \alpha) \sin \alpha$$

$F = mg \tan \alpha$, the same as before.

-63 First note the limiting values of the tension.

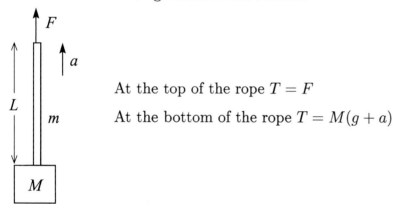

At the top of the rope $T = F$

At the bottom of the rope $T = M(g + a)$

Consider the rope and block as one combined object, in order to calculate the acceleration:

$$\sum F_y = ma_y$$
$$F - (M + m)g = (M + m)a$$

$$a = \frac{F}{M + m} - g$$

Now consider the forces on a section of the rope that extends a distance $x < L$ below the top. The tension at the bottom of this section is $T(x)$ and the mass of this section is $m(x/L)$.

$$\sum F_y = ma_y$$
$$F - T(x) - m(x/L)g = m(x/L)a$$
$$T(x) = F - m(x/L)g - m(x/L)a$$

Using our expression for a and simplifying gives

$$T(x) = F\left(1 - \frac{mx}{L(M+m)}\right)$$

Important to check this result for the limiting cases:

$x = 0$: The expression gives the correct value of $T = F$.

$x = L$: The expression gives $T = F(M/(M+m))$. This should equal $T = M(g+a)$ and when we use the expression for a we see that it does.

5-65 **a)** Free-body diagram for the hanging block:

$a = 0$

$$\sum F_y = ma_y$$
$$T_3 - w = 0$$
$$T_3 = 12.0 \text{ N}$$

Free-body diagram for the knot:

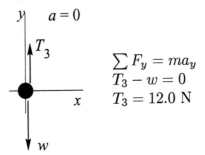

$a = 0$

$45.0°$

$$\sum F_y = ma_y$$
$$T_2 \sin 45.0° - T_3 = 0$$

$$T_2 = \frac{T_3}{\sin 45.0°} = \frac{12.0 \text{ N}}{\sin 45.0°}$$

$$T_2 = 17.0 \text{ N}$$

$$\sum F_x = ma_x$$
$$T_2 \cos 45.0° - T_1 = 0$$
$$T_1 = T_2 \cos 45.0° = 12.0 \text{ N}$$

Free-body diagram for block A:

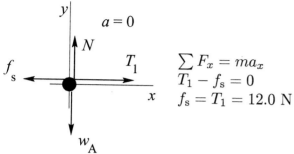

$$\sum F_x = ma_x$$
$$T_1 - f_s = 0$$
$$f_s = T_1 = 12.0 \text{ N}$$

Note: Also can apply $\sum F_y = ma_y$ to this block:

$N - w_A = 0$

$N = w_A = 60.0$ N

Then $\mu_s N = (0.25)(60.0 \text{ N}) = 15.0$ N; this is the maximum possible value for the static friction force.

We see that $f_s < \mu_s N$; for this value of w the static friction force can hold the blocks in place.

b) We have all the same free-body diagrams and force equations as in part (a) but now the static friction force has its largest possbile value, $f_s = \mu_s N = 15.0$ N. Then $T_1 = f_s = 15.0$ N.

From the equations for the forces on the knot:

$$T_2 \cos 45.0° - T_1 = 0 \text{ implies } T_2 = T_1 / \cos 45.0° = \frac{15.0 \text{ N}}{\cos 45.0°} = 21.2 \text{ N}$$

$T_2 \sin 45.0° - T_3 = 0$ implies $T_3 = T_2 \sin 45.0° = (21.2 \text{ N}) \sin 45.0° = 15.0$ N

And finally $T_3 - w = 0$ implies $w = T_3 = 15.0$ N.

5-67 Consider the forces on the scrub brush.

Note that the normal force exerted by the wall is horizontal, since it is perpendicular to the wall. The kinetic friction force exerted by the wall is parallel to the wall and opposes the motion, so it is vertically downward.

$$\sum F_x = ma_x$$
$$N - F \cos 53.1° = 0$$
$$N = F \cos 53.1°$$

$$f_k = \mu_k N = \mu_k F \cos 53.1°$$

$$\sum F_y = ma_y$$

$$F \sin 53.1° - w - f_k = 0$$

$$F \sin 53.1° - w - \mu_k F \cos 53.1° = 0$$

$$F(\sin 53.1° - \mu_k \cos 53.1°) = w$$

$$F = \frac{w}{\sin 53.1° - \mu_k \cos 53.1°}$$

a) $F = \dfrac{w}{\sin 53.1° - \mu_k \cos 53.1°} = \dfrac{12.0 \text{ N}}{\sin 53.1° - (0.15) \cos 53.1°} = 16.9 \text{ N}$

b) $N = F \cos 53.1° = (16.9 \text{ N}) \cos 53.1° = 10.1 \text{ N}$

5-69   Let m_1 be the mass of that part of the rope that is on

the table, and let m_2 be the mass of that part of the rope that is hanging over t
edge. ($m_1 + m_2 = m$, the total mass of the rope).

Since the mass of the rope is not being neglected, the tension in the rope vari
along the length of the rope. Let T be the tension in the rope at that point that
at the edge of the table.

Free-body diagram for the hanging section of the rope:

$$\sum F_y = ma_y$$
$$T - m_2 g = 0$$
$$T = m_2 g$$

Free-body diagram for that part of the rope that is on the table:

$$\sum F_y = ma_y$$
$$N - m_1 g = 0$$
$$N = m_1 g$$

When the maximum amount of rope hangs over the edge the static friction has
maximum value:

$$f_s = \mu_s N = \mu_s m_1 g$$

$$\sum F_x = ma_x$$

$$T - f_s = 0$$

$$T = \mu_s m_1 g$$

Use the first equation to replace T:

$$m_2 g = \mu_s m_1 g$$

$$m_2 = \mu_s m_1$$

The fraction that hangs over is $\dfrac{m_2}{m} = \dfrac{m_2}{m_1 + \mu_s m_1} = \dfrac{\mu_s}{1 + \mu_s}$.

5-71 First calculate the maximum acceleration that the static friction force can give to the case:

The static friction force is to the right in the sketch (northward) since it tries to make the case move with the truck. The maximum value it can have is $f_s = \mu_s N$.

$$\sum F_y = m a_y$$

$$N - mg = 0$$

$$N = mg$$

$$f_s = \mu_s N = \mu_s mg$$

$$\sum F_x = m a_x$$

$$f_s = ma$$

$$\mu_s mg = ma$$

$$a = \mu_s g = (0.30)(9.80 \text{ m/s}^2) = 2.94 \text{ m/s}^2$$

The truck's acceleration is less than this so the case doesn't slip relative to the truck; the case's acceleration is $a = 2.20 \text{ m/s}^2$ (northward).

Then $f_s = ma = (30.0 \text{ kg})(2.20 \text{ m/s}^2) = 66$ N, northward.

b) Now the acceleration of the truck is greater than the acceleration that static friction can give the case. Therefore, the case slips relative to the truck and the friction is kinetic friction. The friction force still tries to keep the case moving with the truck, so the acceleration of the case and the friction force are both southward.

$$\sum F_y = ma_y$$
$$N - mg = 0$$
$$N = mg$$

$$f_k = \mu_k mg = (0.20)(30.0 \text{ kg})(9.80 \text{ m/s}^2)$$
$$f_k = 59 \text{ N, southward}$$

Note: $f_k = ma$ implies $a = \dfrac{f_k}{m} = \dfrac{59 \text{ N}}{30.0 \text{ kg}} = 2.0 \text{ m/s}^2$. The magnitude of t
acceleration of the case is less than that of the truck and the case slides toward t
front of the truck.

5-77   Block B is pulled to the left at constant speed, so block A moves to the right
at constant speed and $a = 0$ for each block.

Free-body diagram for block A:

N_{BA} is the normal force that B exerts on A.

$f_{BA} = \mu_k N_{BA}$ is the kinetic friction force that B exerts on A. Block A moves
the right relative to B, and f_{BA} opposes this motion, so f_{BA} is to the left.

Note also that F acts just on B, not on A.

$$\sum F_y = ma_y$$
$$N_{BA} - w_A = 0$$
$$N_{BA} = 1.40 \text{ N}$$

$$f_{BA} = \mu_k N_{BA} = (0.30)(1.40 \text{ N}) = 0.420 \text{ N}$$

$$\sum F_x = ma_x$$
$$T - f_{BA} = 0$$
$$T = f_{BA} = 0.420 \text{ N}$$

Free-body diagram for block B:

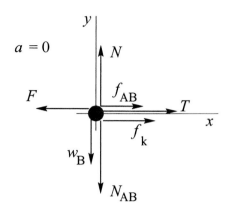

N_{AB} is the normal force that block A exerts on block B. By Newton's third law N_{AB} and N_{BA} are equal in magnitude and opposite in direction, so $N_{AB} = 1.40$ N.

f_{AB} is the kinetic friction force that A exerts on B. Block B moves to the left relative to A and f_{AB} opposes this motion, so f_{AB} is to the right.

$f_{AB} = \mu_k N_{AB} = (0.30)(1.40 \text{ N}) = 0.420$ N.

But also note that f_{AB} and f_{BA} are a third law action-reaction pair, so they must be equal in magnitude and opposite in direction and this is indeed what our calculation gives.

N and f_k are the normal and friction force exerted by the floor on block B; $f_k = \mu_k N$. Note that block B moves to the left relative to the floor and f_k opposes this motion, so f_k is to the right.

$$\sum F_y = ma_y$$
$$N - w_B - N_{AB} = 0$$
$$N = w_B + N_{AB} = 4.20 \text{ N} + 1.40 \text{ N} = 5.60 \text{ N}$$
Then $f_k = \mu_k N = (0.30)(5.60 \text{ N}) = 1.68$ N.

$$\sum F_x = ma_x$$
$$f_{AB} + T + f_k - F = 0$$
$$F = T + f_{AB} + f_k = 0.420 \text{ N} + 0.420 \text{ N} + 1.68 \text{ N} = 2.52 \text{ N}$$

-79   Parts (a) and (b) will be done together.

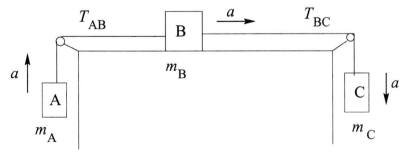

Consider the forces on each block. Note that each block has the same magnitude of acceleration, but in different directions. For each block let the direction of $\vec{a}$ be a positive coordinate direction.

Block A:

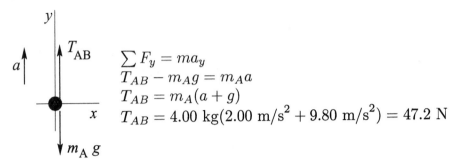

$$\sum F_y = ma_y$$
$$T_{AB} - m_A g = m_A a$$
$$T_{AB} = m_A(a + g)$$
$$T_{AB} = 4.00 \text{ kg}(2.00 \text{ m/s}^2 + 9.80 \text{ m/s}^2) = 47.2 \text{ N}$$

Block B:

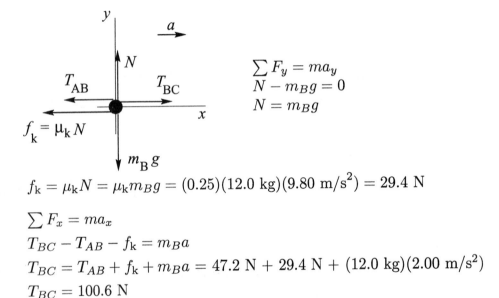

$$\sum F_y = ma_y$$
$$N - m_B g = 0$$
$$N = m_B g$$

$$f_k = \mu_k N = \mu_k m_B g = (0.25)(12.0 \text{ kg})(9.80 \text{ m/s}^2) = 29.4 \text{ N}$$

$$\sum F_x = ma_x$$
$$T_{BC} - T_{AB} - f_k = m_B a$$
$$T_{BC} = T_{AB} + f_k + m_B a = 47.2 \text{ N} + 29.4 \text{ N} + (12.0 \text{ kg})(2.00 \text{ m/s}^2)$$
$$T_{BC} = 100.6 \text{ N}$$

Block C:

$$\sum F_y = ma_y$$
$$m_C g - T_{BC} = m_C a$$
$$m_C(g - a) = T_{BC}$$

$$T = \frac{T_{BC}}{g - a}$$

$$T = \frac{100.6 \text{ N}}{9.80 \text{ m/s}^2 - 2.00 \text{ m/s}^2} = 12.9 \text{ kg}$$

5-81 Let the tensions in the ropes be T_1 and T_2.

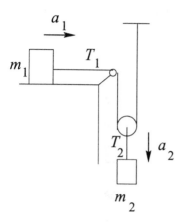

Consider the forces on each block. In each case take a positive coordinate direction in the direction of the acceleration of that block.

Forces on m_1:

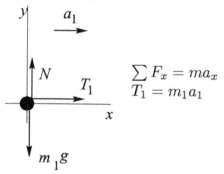

$$\sum F_x = ma_x$$
$$T_1 = m_1 a_1$$

Forces on m_2:

$$\sum F_y = ma_y$$
$$m_2 g - T_2 = m_2 a_2$$

This gives us two equations, but there are 4 unknowns (T_1, T_2, a_1, and a_2) so two more equations are required.

Free-body diagram for the moveable pulley (mass m):

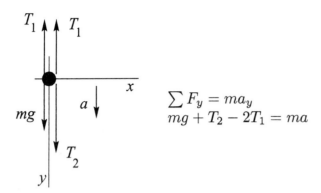

$$\sum F_y = ma_y$$
$$mg + T_2 - 2T_1 = ma$$

But our pulleys have negligible mass, so $mg = ma = 0$ and $T_2 = 2T_1$.

Combine these three equations to eliminate T_1 and T_2:

$m_2 g - T_2 = m_2 a_2$ gives $m_2 g - 2T_1 = m_2 a_2$.

And then with $T_1 = m_1 a_1$ we have $m_2 g - 2m_1 a_1 = m_2 a_2$.

There are still two unknowns, a_1 and a_2. But the accelerations a_1 and a_2 are related. In any time interval, if m_1 moves to the right a distance d, then in the same time m_2 moves downward a distance $d/2$. One of the constant acceleration kinematic equations says $x - x_0 = v_0 t + \frac{1}{2}at^2$, so if m_2 moves half the distance must have half the acceleration of m_1: $a_2 = a_1/2$, or $a_1 = 2a_2$.

This is the additional equation we need. Use it in the previous quation and get

$$m_2 g - 2m_1(2a_2) = m_2 a_2.$$
$$a_2(4m_1 + m_2) = m_2 g$$
$$a_2 = \frac{m_2 g}{4m_1 + m_2} \quad \text{and} \quad a_1 = 2a_2 = \frac{2m_2 g}{4m_1 + m_2}.$$

5-85   The cart and the block have the same acceleration.

The normal force exerted by the cart on the block is perpendicular to the front of the cart, so is horizontal and to the right.

The friction force on the block is directed so as to hold the block up against the downward pull of gravity. We want to calculate the minimum a required, so take static friction to have its maximum value, $f_s = \mu_s N$.

Free-body diagram for the block:

$$\sum F_x = ma_x$$
$$N = ma$$
$$f_s = \mu_s N = \mu_s ma$$

$$\sum F_y = ma_y$$
$$f_s - mg = 0$$
$$\mu_s ma = mg$$
$$a = g/\mu_s$$

An observer on the cart sees the block pinned there, with no reason for a horizontal force on it because the block is at rest relative to the cart. Therefore, such an observer concludes that $N = 0$ and thus $f_s = 0$, and he doesn't understand what holds the block up against the downward force of gravity. The reason for this difficulty is that $\sum \vec{F} = m\vec{a}$ does not apply in a coordinate frame attached to the cart. This reference frame is accelerated, and hence not inertial.

5-89 **a)** Force diagram for the two blocks taken as a single combined object:

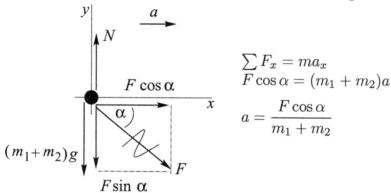

$$\sum F_x = ma_x$$
$$F \cos\alpha = (m_1 + m_2)a$$
$$a = \frac{F \cos\alpha}{m_1 + m_2}$$

$$\sum F_y = ma_y$$
$$N - F\sin\alpha - (m_1 + m_2)g = 0$$
$$N = F\sin\alpha + (m_1 + m_2)g$$

b) We now know the acceleration of each block if they move together. Now apply $\sum \vec{F} = m\vec{a}$ to each block.

bottom block:

N is the normal force applied by the surface, and was calculated in part (a). N' is the normal force applied by the top block. Note that $\vec{F}$ acts only on the top block so doesn't appear in this force diagram.

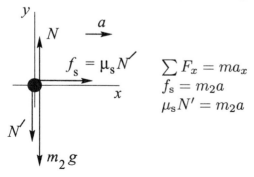

$$\sum F_x = ma_x$$
$$f_s = m_2 a$$
$$\mu_s N' = m_2 a$$

$$\sum F_y = ma_y$$
$$N - N' - m_2 g = 0$$
$$N' = N - m_2 g$$

In this equation use the value of N calculated in part (a);

$$N' = N - m_2 g = F \sin \alpha + m_1 g + m_2 g - m_2 g = F \sin \alpha + m_1 g.$$

Use this result, and also $a = F \cos \alpha / (m_1 + m_2)$ from part (a), in $\mu_s N' = m_2 a$:

$$\mu_s F \sin \alpha + \mu_s m_1 g = m_2 F \cos \alpha / (m_1 + m_2).$$

Solve for F:

$$(\mu_s \sin \alpha (m_1 + m_2) - m_2 \cos \alpha) F = -\mu_s m_1 g (m_1 + m_2)$$

$$F = \frac{\mu_s m_1 (m_1 + m_2) g}{m_2 \cos \alpha - \mu_s \sin \alpha (m_1 + m_2)}, \text{ as was to be shown.}$$

An alternative approach is to apply $\sum \vec{F} = m\vec{a}$ to the top block:

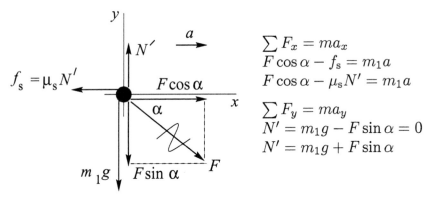

$$\sum F_x = ma_x$$
$$F \cos \alpha - f_s = m_1 a$$
$$F \cos \alpha - \mu_s N' = m_1 a$$

$$\sum F_y = ma_y$$
$$N' = m_1 g - F \sin \alpha = 0$$
$$N' = m_1 g + F \sin \alpha$$

Combine these two equations to eliminate N':

$$F \cos \alpha - \mu_s m_1 g - \mu_s F \sin \alpha = m_1 a$$

Use the value of a calculated in part (a):

$$F \cos \alpha - \mu_s m_1 g - \mu_s F \sin \alpha = \frac{m_1 F \cos \alpha}{m_1 + m_2}$$

Solve for F:

$$F((m_1 + m_2) \cos \alpha - \mu_s (m_1 + m_2) \sin \alpha - m_1 \cos \alpha) = \mu_s m_1 (m_1 + m_2) g$$

$$F = \frac{\mu_s m_1 (m_1 + m_2) g}{m_2 \cos \alpha - \mu_s \sin \alpha (m_1 + m_2)}, \text{ which is the same result.}$$

5-91   **a)** To keep the car from sliding up the banking the static friction force is directed down the incline. At maximum speed the static friction force has its maximum value $f_s = \mu_s N$.

Free-body diagram for the car:

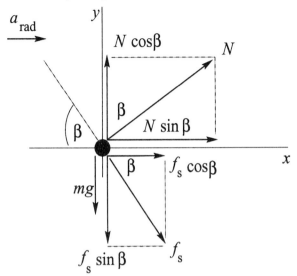

$$\sum F_y = ma_y$$
$$N\cos\beta - f_s\sin\beta - mg = 0$$
But $f_s = \mu_s N$, so
$$N\cos\beta - \mu_s N\sin\beta - mg = 0$$
$$N = \frac{mg}{\cos\beta - \mu_s\sin\beta}$$

$$\sum F_x = ma_x$$
$$N\sin\beta + \mu_s N\cos\beta = ma_{\text{rad}}$$
$$N(\sin\beta + \mu_s\cos\beta) = ma$$

Use the $\sum F_y$ equation to replace N:

$$\left(\frac{mg}{\cos\beta - \mu_s\sin\beta}\right)(\sin\beta + \mu_s\cos\beta) = ma_{\text{rad}}$$

$$a_{\text{rad}} = \left(\frac{\sin\beta + \mu_s\cos\beta}{\cos\beta - \mu_s\sin\beta}\right)g = \left(\frac{\sin 25° + (0.30)\cos 25°}{\cos 25° - (0.30)\sin 25°}\right)(9.80\text{ m/s}^2) = 8.73\text{ m/s}^2$$

$a_{\text{rad}} = v^2/R$ implies $v = \sqrt{a_{\text{rad}}R} = \sqrt{(8.73\text{ m/s}^2)(50\text{ m})} = 21\text{ m/s}.$

b) To keep the car from sliding <u>down</u> the banking the static friction force is directed up the incline. At the minimum speed the static friction force has its maximum value $f_s = \mu_s N$. Free-body diagram for the car:

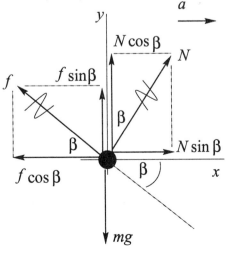

The free-body diagram is identical to that in part (a) except that now the components of f_s have opposite directions. The force equations are all the same except for the opposite sign for terms containing μ_s.

Thus $a_{rad} = \left(\dfrac{\sin\beta - \mu_s \cos\beta}{\cos\beta + \mu_s \sin\beta} \right) g = \left(\dfrac{\sin 25° - (0.30)\cos 25°}{\cos 25° + (0.30)\sin 25°} \right) (9.80 \text{ m/s}^2)$

$= 1.43 \text{ m/s}^2$

$v = \sqrt{a_{rad} R} = \sqrt{(1.43 \text{ m/s}^2)(50 \text{ m})} = 8.5 \text{ m/s}.$

5-97 **a)** Use the information given about Jena to find the time t for one revolution of the merry-go-round. Her acceleration is a_{rad}, directed in toward the axis. Le $\vec{F}_1$ be the horizontal force that keeps her from sliding off. Let her speed be v_1 an let R_1 be her distance from the axis.

$\sum F_x = ma_x$
$F_1 = ma_{rad}$

$F_1 = m\dfrac{v_1^2}{R_1}, \quad v_1 = \sqrt{\dfrac{R_1 F_1}{m}}$

The time for one revolution is $t = \dfrac{2\pi R_1}{v_1} = 2\pi R_1 \sqrt{\dfrac{m}{R_1 F_1}}.$

Jackie goes around once in the same time but her speed (v_2) and the radius of he circular path (R_2) are different.

$v_2 = \dfrac{2\pi R_2}{t} = 2\pi R_2 \left(\dfrac{1}{2\pi R_1} \right) \sqrt{\dfrac{R_1 F_1}{m}} = \dfrac{R_2}{R_1} \sqrt{\dfrac{R_1 F_1}{m}}.$

Free-body diagram for Jackie:

$\sum F_x = ma_x$
$F_2 = ma_{rad}$

$F_2 = m\dfrac{v_2^2}{R_2} = \left(\dfrac{m}{R_2} \right)\left(\dfrac{R_2^2}{R_1^2} \right)\left(\dfrac{R_1 F_1}{m} \right) = \left(\dfrac{R_2}{R_1} \right) F_1 = \left(\dfrac{3.60 \text{ m}}{1.80 \text{ m}} \right)(60.0 \text{ N})$
$= 120.0 \text{ N}$

b) $F_2 = m\dfrac{v_2^2}{R_2}$, so $v_2 = \sqrt{\dfrac{F_2 R_2}{m}} = \sqrt{\dfrac{(120.0 \text{ N})(3.60 \text{ m})}{30.0 \text{ kg}}} = 3.79 \text{ m/s}$

5-99 $v = (0.60 \text{ rev/s}) \left(\dfrac{2\pi R}{1 \text{ rev}} \right) = (0.60 \text{ rev/s}) \left(\dfrac{2\pi (2.5 \text{ m})}{1 \text{ rev}} \right) = 9.425 \text{ m/s}$

a)

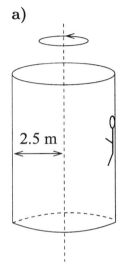

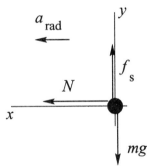

The person is held up against gravity by the static friction force exerted on him by the wall. The acceleration of the person is a_{rad}, directed in towards the axis of rotation.

b) To calculate the minimum μ_s required, take f_s to have its maximum value, $f_s = \mu_s N$.

$\sum F_y = ma_y$

$f_s - mg = 0$

$\mu_s N = mg$

$\sum F_x = ma_x$

$N = mv^2/R$

Combine these two equations to eliminate N:

$\mu_s mv^2/R = mg$

$\mu_s = \dfrac{Rg}{v^2} = \dfrac{(2.5 \text{ m})(9.80 \text{ m/s}^2)}{(9.425 \text{ m/s})^2} = 0.28$

c) No, the mass of the person divided out of the equation for μ_s.

5-101 a) Turn to the right.

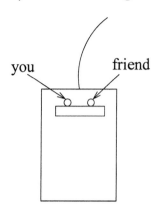

you   friend

As viewed in an inetial frame, in the absence of
sufficient friction your friend doesn't make
the turn completely and you move to the right
toward your friend.

b) The maximum radius of the turn is the one that makes a_{rad} just equal to th
maximum acceleration that static friction can give to your friend, and for thi
situation f_s has its maximum value $f_s = \mu_s N$.

Free-body diagram for your friend, as viewed by someone standing behind the car

$$\sum F_y = ma_y$$
$$N - mg = 0$$
$$N = mg$$

$$\sum F_x = ma_x$$
$$f_s = ma_{rad}$$
$$\mu_s N = mv^2/R$$
$$\mu_s mg = mv^2/R$$

$$R = \frac{v^2}{\mu_s g} = \frac{(20 \text{ m/s}^2)^2}{(0.35)(9.80 \text{ m/s}^2)} = 117 \text{ m}$$

5-103

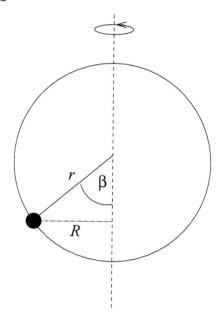

The bead moves in a circle of radius $R = r \sin \beta$.
The normal force exerted on the bead by the hoop is radially inward.

Free-body diagram for the bead:

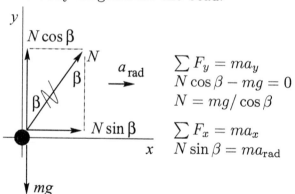

$$\sum F_y = ma_y$$
$$N \cos \beta - mg = 0$$
$$N = mg / \cos \beta$$

$$\sum F_x = ma_x$$
$$N \sin \beta = ma_{\text{rad}}$$

Combine these two equations to eliminate N:

$$\left(\frac{mg}{\cos \beta} \right) \sin \beta = ma_{\text{rad}}$$

$$\frac{\sin \beta}{\cos \beta} = \frac{a_{\text{rad}}}{g}$$

$a_{\text{rad}} = v^2/R$ and $v = 2\pi R/T$, so $a_{\text{rad}} = 4\pi^2 R/T^2$, where T is the time for one revolution.

$R = r \sin \beta$, so $a_{\text{rad}} = \dfrac{4\pi^2 r \sin \beta}{T^2}$

Use this in the above equation: $\dfrac{\sin \beta}{\cos \beta} = \dfrac{4\pi^2 r \sin \beta}{T^2 g}$

This equation is satisfied by $\sin \beta = 0$, so $\beta = 0$, or by

$$\frac{1}{\cos \beta} = \frac{4\pi^2 r}{T^2 g}, \text{ which gives } \cos \beta = \frac{T^2 g}{4\pi^2 r}$$

a) 4.00 rev/s implies $T = (1/4.00)$ s $= 0.250$ s

Then $\cos \beta = \dfrac{(0.250 \text{ s})^2 (9.80 \text{ m/s}^2)}{4\pi^2 (0.100 \text{ m})}$ and $\beta = 81.1°$.

b) This would mean $\beta = 90°$. But $\cos 90° = 0$, so this requires $T \to 0$. So β approaches 90° as the hoop rotates very fast, but $\beta = 90°$ is not possible.

c) 1.00 rev/s implies $T = 1.00$ s

The $\cos \beta = \dfrac{T^2 g}{4\pi^2 r}$ equation then says $\cos \beta = \dfrac{(1.00 \text{ s})^2 (9.80 \text{ m/s}^2)}{4\pi^2 (0.100 \text{ m})} = 2.48$, which is not possible.

The only way to have the $\sum \vec{F} = m\vec{a}$ equations satisfied is for $\sin \beta = 0$. This means $\beta = 0$; the bead sits at the bottom of the hoop.

CHAPTER 6
WORK AND KINETIC ENERGY

Exercises 1, 5, 7, 13, 15, 19, 21, 23, 27, 29, 31, 33, 37, 39, 43, 47
Problems 49, 55, 57, 59, 61, 63, 67, 73, 77, 79, 81, 85

Exercises

6-1 **a)** $W_F = (F\cos\phi)s = (2.40\text{ N})(\cos 0°)(1.50\text{ m}) = +3.60\text{ J}$

The force and displacement vectors are in the same direction and the work done is positive.

b) $W_F = (F\cos\phi)s = (0.600\text{ N})(\cos 180°)(1.50\text{ m}) = -0.900\text{ J}$

The force and displacement vectors are in opposite directions and the work done is negative.

c) $W_{\text{tot}} = W_F + W_f = 3.60\text{ J} - 0.900\text{ J} = +2.70\text{ J}$

6-5 Constant speed implies $a = 0$.

a) Free-body diagram for the crate:

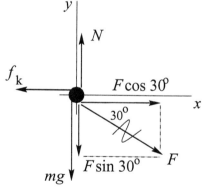

$$\sum F_y = ma_y$$
$$N - mg - F\sin 30° = 0$$
$$N = mg + F\sin 30°$$

$$f_k = \mu_k N = \mu_k mg + F\mu_k \sin 30°$$

$$\sum F_x = ma_x$$
$$F\cos 30° - f_k = 0$$
$$F\cos 30° - \mu_k mg - \mu_k \sin 30° F = 0$$

$$F = \frac{\mu_k mg}{\cos 30° - \mu_k \sin 30°} = \frac{0.25(30.0\text{ kg})(9.80\text{ m/s}^2)}{\cos 30° - (0.25)\sin 30°} = 99.2\text{ N}$$

b) $W_F = (F\cos\phi)s = (99.2\text{ N})(\cos 30°)(4.5\text{ m}) = 387\text{ J}$

($F\cos 30°$ is the horizontal component of $\vec{F}$; the work done by $\vec{F}$ is the displacement times the component of $\vec{F}$ in the direction of the displacement.)

c) We have an expression for f_k from part (a):

$f_k = \mu_k(mg + F\sin 30°) = (0.250)[(30.0 \text{ kg})(9.80 \text{ m/s}^2) + (99.2 \text{ N})(\sin 30°)] = 85.9 \text{ N}$

$\phi = 180°$ since f_k is opposite to the displacement.

Thus $W_f = (f_k\cos\phi)s = (85.9 \text{ N})(\cos 180°)(4.5 \text{ m}) = -387 \text{ J}$

d) The normal force is perpendicular to the displacement so $\phi = 90°$ and $W_N = ($
The gravity force (the weight) is perpendicular to the displacement so $\phi = 90°$ an
$W_w = 0$.

e) $W_{\text{tot}} = W_F + W_f + W_N + W_w = +387 \text{ J} + (-387 \text{ J}) = 0$

6-7 $\quad W_F = (F\cos\phi)s$

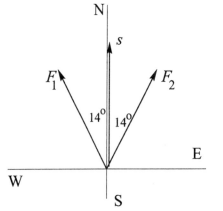

$W_1 = F_1 s\cos\phi_1$
$W_1 = (1.80 \times 10^6 \text{ N})(0.75 \times 10^3 \text{ m})\cos 14°$
$W_1 = 1.31 \times 10^9 \text{ J}$

$W_2 = F_2 s\cos\phi_2 = W_1$

$W_{\text{tot}} = W_1 + W_2 = 2(1.31 \times 10^9 \text{ J}) = 2.62 \times 10^9 \text{ J}$

6-13 a) $\quad W = K_2 - K_1$

$K_1 = \frac{1}{2}mv_1^2, \quad K_2 = \frac{1}{2}mv_2^2$

$v_2 = \frac{1}{4}v_1$ gives that $K_2 = \frac{1}{2}m(\frac{1}{4}v_1)^2 = \frac{1}{16}(\frac{1}{2}mv_1^2) = \frac{1}{16}K_1$

$W = K_2 - K_1 = \frac{1}{16}K_1 - K_1 = -\frac{15}{16}K_1$

b) K depends only on the magnitude of $\vec{v}$ not on its direction, so the answer for W
in part (a) does <u>not</u> depend on the final direction of the electron's motion.

6-15 $\quad W_{\text{tot}} = K_2 - K_1$

$K_1 = \frac{1}{2}mv_1^2 = \frac{1}{2}(0.420 \text{ kg})(2.00 \text{ m/s})^2 = 0.84 \text{ J}$

$K_2 = \frac{1}{2}mv_2^2 = \frac{1}{2}(0.420 \text{ kg})(6.00 \text{ m/s})^2 = 7.56 \text{ J}$

$W_{\text{tot}} = K_2 - K_1 = 7.56 \text{ J} - 0.84 \text{ J} = 6.72 \text{ J}$

The 40.0 N force is the only force doing work on the ball, so it must do 6.72 J o
work.

$$W_F = (F\cos\phi)s \text{ gives that } s = \frac{W}{F\cos\phi} = \frac{6.72\text{ J}}{(40.0\text{ N})(\cos 0)} = 0.168\text{ m}$$

6-19 a) $W_{\text{tot}} = K_2 - K_1$, so $K_2 = W_{\text{tot}} + K_1$

$K_1 = \frac{1}{2}mv_1^2 = \frac{1}{2}(7.00\text{ kg})(4.00\text{ m/s})^2 = 56.0\text{ J}$

The only force that does work on the wagon is the 10.0 N force. This force is in the direction of the displacement so $\phi = 0°$ and the force does positive work:

$W_F = (F\cos\phi)s = (10.0\text{ N})(\cos 0)(3.0\text{ m}) = 30.0\text{ J}$

Then $K_2 = W_{\text{tot}} + K_1 = 30.0\text{ J} + 56.0\text{ J} = 86.0\text{ J}.$

$$K_2 = \frac{1}{2}mv_2^2; \quad v_2 = \sqrt{\frac{2K_2}{m}} = \sqrt{\frac{2(86.0\text{ J})}{7.00\text{ kg}}} = 4.96\text{ m/s}$$

b)

$$\sum F_x = ma_x$$
$$F = ma$$
$$a = \frac{F}{m} = \frac{10.0\text{ N}}{7.00\text{ kg}} = 1.43\text{ m/s}^2$$

$v_2^2 = v_1^2 + 2a(x - x_0)$

$v_2 = \sqrt{v_1^2 + 2a(x - x_0)} = \sqrt{(4.00\text{ m/s})^2 + 2(1.43\text{ m/s}^2)(3.0\text{ m})} = 4.96\text{ m/s}$

This agrees with the result calculated in part (a).

6-21 a) $W_{\text{tot}} = K_2 - K_1$

$f_k = \mu_k N = \mu_k mg$

f_k is the only force that does work, so $W_{\text{tot}} = W_f$

f_k is opposite to the displacement so $\phi = 180°$ and

$W_f = f(\cos\phi)s = \mu_k mg(\cos 180°)s = -\mu_k mgs$

$K_2 = 0$ (car stops)

$K_1 = \frac{1}{2}mv_0^2$

$W_{\text{tot}} = K_2 - K_1$, so $-\mu_k mgs = -\frac{1}{2}mv_0^2$ and $s = \dfrac{v_0^2}{2\mu_k g}$

b) $s = \dfrac{v_0^2}{2\mu_k g}$ says that $\dfrac{v_0^2}{s} = 2\mu_k g = \text{constant, so } \dfrac{s_1}{v_{01}^2} = \dfrac{s_2}{v_{02}^2}$

$$s_2 = s_1 \left(\dfrac{v_{02}}{v_{01}}\right)^2 = 91.2 \text{ m} \left(\dfrac{60.0 \text{ km/h}}{80.0 \text{ km/h}}\right)^2 = 51.3 \text{ m}$$

6-23 Use the information given to calculate the force constant of the spring:

$$F = kx \text{ gives } k = \dfrac{F}{x} = \dfrac{160 \text{ N}}{0.050 \text{ s}} = 3200 \text{ N/m}$$

a) $F = kx = (3200 \text{ N/m})(0.015 \text{ m}) = 48 \text{ N}$

$F = kx = (3200 \text{ N/m})(-0.020 \text{ m}) = -64 \text{ N} \text{ (magnitude 64 N)}$

b) $W = \frac{1}{2}kx^2 = \frac{1}{2}(3200 \text{ N/m})(0.015 \text{ m})^2 = 0.36 \text{ J}$

$W = \frac{1}{2}kx^2 = \frac{1}{2}(3200 \text{ N/m})(-0.020 \text{ m})^2 = 0.64 \text{ J}$

Note that in each case the work done is positive.

6-27 a) Free-body diagram for the glider:

$\sum F_x = ma_x$

$f_s - F_{\text{spring}} = 0$

$\mu_s mg - kd = 0$

$$\mu_s = \dfrac{kd}{mg} = \dfrac{(20.0 \text{ N/m})(0.086 \text{ m})}{(0.100 \text{ kg})(9.80 \text{ m/s}^2)} = 1.76$$

b) Use the results of part (a) to calculate d, the amount the spring is stretched when the glider stops instantaneously:

$\mu_s mg = kd$

$$d = \dfrac{\mu_s mg}{k} = \dfrac{(0.60)(0.100 \text{ kg})(9.80 \text{ m/s}^2)}{20.0 \text{ N/m}} = 0.0294 \text{ m}$$

Now apply the work-energy theorem to the motion of the glider:

$W_{\text{tot}} = K_2 - K_1$

$K_1 = \frac{1}{2}mv_1^2$, $K_2 = 0$ (instantaneously stops)

$W_{\text{tot}} = W_{\text{spring}} + W_{\text{fric}} = -\frac{1}{2}kd^2 - \mu_k mgd$ (as in Example 6-8)

$W_{\text{tot}} = -\frac{1}{2}(20.0 \text{ N/m})(0.0294 \text{ m})^2 - 0.47(0.100 \text{ kg})(9.80 \text{ m/s}^2)(0.0294 \text{ m}) = -0.02218$ J

Then $W_{\text{tot}} = K_2 - K_1$ gives -0.02218 J $= -\frac{1}{2}mv_1^2$.

$$v_1 = \sqrt{\frac{2(0.02218 \text{ J})}{0.100 \text{ kg}}} = 0.67 \text{ m/s}$$

6-29 The magnitude of the work done by F_x equals the area under the F_x versus x curve. The work is positive when F_x and the displacement are in the same direction; it is negative when they are in opposite directions.

a) F_x is positive and the displacement Δx is positive, so $W > 0$.

$W = \frac{1}{2}(2.0 \text{ N})(2.0 \text{ m}) + (2.0 \text{ N})(1.0 \text{ m}) = +4.0$ J

b) During this displacement $F_x = 0$, so $W = 0$.

c) F_x is negative, Δx is positive, so $W < 0$.

$W = -\frac{1}{2}(1.0 \text{ N})(2.0 \text{ m}) = -1.0$ J

d) The work is the sum of the answers to parts (a), (b), and (c), so

$W = 4.0$ J $+ 0 - 1.0$ J $= +3.0$ J

e) The work done for $x = 7.0$ m to $x = 3.0$ m is $+1.0$ J. This work is positive since the displacement and the force are both in the $-x$-direction. The magnitude of the work done for $x = 3.0$ m to $x = 2.0$ m is 2.0 J, the area under F_x versus x. This work is negative since the displacement is in the $-x$-direction and the force is in the $+x$-direction. Thus $W = +1.0$ J $- 2.0$ J $= -1.0$ J

6-31 **a)** $W_{\text{tot}} = K_2 - K_1$ so $K_2 = K_1 + W_{\text{tot}}$

$K_1 = 0$ (released with no initial velocity), $K_2 = \frac{1}{2}mv_2^2$

The only force doing work is the spring force. Eq.(6-10) gives the work done <u>on</u> the spring to move its end from x_1 to x_2. The force the spring exerts on an object attached to it is $F = -kx$, so the work the spring does is

$W_{\text{spr}} = -(\frac{1}{2}kx_2^2 - \frac{1}{2}kx_1^2) = \frac{1}{2}kx_1^2 - \frac{1}{2}kx_2^2$. Here $x_1 = -0.375$ m and $x_2 = 0$.

Thus $W_{\text{spr}} = \frac{1}{2}(4000 \text{ N/m})(-0.375 \text{ m})^2 - 0 = 281$ J.

$K_2 = K_1 + W_{\text{tot}} = 0 + 281$ J $= 281$ J

Then $K_2 = \frac{1}{2}mv_2^2$ implies $v_2 = \sqrt{\dfrac{2K_2}{m}} = \sqrt{\dfrac{2(281 \text{ J})}{70.0 \text{ kg}}} = 2.83$ m/s.

b) $K_2 = K_1 + W_{\text{tot}}$

$K_1 = 0$

$W_{\text{tot}} = W_{\text{spr}} = \frac{1}{2}kx_1^2 - \frac{1}{2}kx_2^2$. Now $x_2 = -0.200$ m, so

$W_{\text{spr}} = \frac{1}{2}(4000 \text{ N/m})(-0.375 \text{ m})^2 - \frac{1}{2}(4000 \text{ N/m})(-0.200 \text{ m})^2 = 281 \text{ J} - 80 \text{ J} = 201$ J

Thus $K_2 = 0 + 201 \text{ J} = 201$ J and $K_2 = \frac{1}{2}mv_2^2$ gives

$$v_2 = \sqrt{\frac{2K_2}{m}} = \sqrt{\frac{2(201 \text{ J})}{70.0 \text{ kg}}} = 2.40 \text{ m/s}.$$

6-33 $W_{\text{tot}} = K_2 - K_1 = 0$

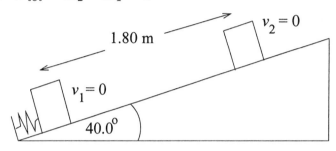

$W_{\text{tot}} = W_{\text{spr}} + W_w = 0$
So $W_{\text{spr}} = -W_w$
(The spring does positive work on the glider since the spring force is directed up the incline, the same as the direction of the displacement.

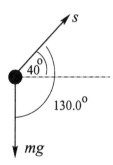

$W_w = (w\cos\phi)s = (mg\cos 130.0°)s$
$W_w = (0.0900 \text{ kg})(9.80 \text{ m/s}^2)(\cos 130.0°)(1.80 \text{ m}) = -1.020$ J
(The component of w parallel to the incline is directed down the incline, opposite to the displacement, so gravity does negative work.)

$W_{\text{spr}} = -W_w = +1.020$ J

$W_{\text{spr}} = \frac{1}{2}kx^2$ so $x = \sqrt{\frac{2W_{\text{spr}}}{k}} = \sqrt{\frac{2(1.020 \text{ J})}{640 \text{ N/m}}} = 0.0565$ m

b) The spring was compressed only 0.0565 m so at this point in the motion the glider is no longer in contact with the spring.

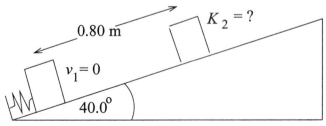

$W_{\text{tot}} = K_2 - K_1$
$K_2 = K_1 + W_{\text{tot}}$
$K_1 = 0$

$W_{\text{tot}} = W_{\text{spr}} + W_w$

From part (a), $W_{\text{spr}} = 1.020$ J and

$W_w = (mg\cos 130.0°)s = (0.0900 \text{ kg})(9.80 \text{ m/s}^2)(\cos 130.0°)(0.80 \text{ m}) = -0.454$ J

Then $K_2 = W_{\text{spr}} + W_w = +1.020 \text{ J} - 0.454 \text{ J} = +0.57$ J. (Note that the kinetic energy is positive, as it must be.)

6-37 The rate at which work is being done against gravity is

$P = Fv = mgv = (700 \text{ kg})(9.80 \text{ m/s}^2)(2.5 \text{ m/s}) = 17.15$ kW.

This is the part of the engine power that is being used to make the airplane climb. The fraction this is of the total is

17.15 kw/75 kW = 0.23

6-39 Find the total mass that can be lifted:

$P_{\text{av}} = \dfrac{\Delta W}{\Delta t} = \dfrac{mgh}{t}$, so $m = \dfrac{P_{\text{av}}t}{gh}$

$P_{\text{av}} = (40 \text{ hp}) \left(\dfrac{746 \text{ W}}{1 \text{ hp}} \right) = 2.984 \times 10^4$ W

$m = \dfrac{P_{\text{av}}t}{gh} = \dfrac{(2.984 \times 10^4 \text{ W})(16.0 \text{ s})}{(9.80 \text{ m/s}^2)(20.0 \text{ m})} = 2.436 \times 10^3$ kg

This is the total mass of elevator plus passengers. The mass of the passengers is 2.436×10^3 kg $- 600$ kg $= 1.836 \times 10^3$ kg.

The number of passengers is $\dfrac{1.836 \times 10^3 \text{ kg}}{65.0 \text{ kg}} = 28.2.$

28 passengers can ride.

6-43 **a)** $F = ma$

$v = v_0 + at$ and $v_0 = 0$, so $v = at$.

The instantaneous power is $P = Fv = (ma)(at) = ma^2t$.

b) P is proportional to a^2.

Triple a says that increase P by a factor of 9.

c) $\dfrac{P}{t} = ma^2 = \text{constant}$, so $\dfrac{P_1}{t_1} = \dfrac{P_2}{t_2}$

$P_2 = P_1 \left(\dfrac{t_2}{t_1} \right) = 36 \text{ W} \left(\dfrac{15.0 \text{ s}}{5.0 \text{ s}} \right) = 108$ W.

6-47 **a)** $P = Fv$, so $F = P/v$.

$$P = (8.00 \text{ hp}) \left(\frac{746 \text{ W}}{1 \text{ hp}} \right) = 5968 \text{ W}$$

$$v = (60.0 \text{ km/h}) \left(\frac{1000 \text{ m}}{1 \text{ km}} \right) \left(\frac{1 \text{ h}}{3600 \text{ s}} \right) = 16.67 \text{ m/s}$$

$$F = \frac{P}{v} = \frac{5968 \text{ W}}{16.67 \text{ m/s}} = 358 \text{ N}.$$

b) The power required is the 8.00 hp of part (a) plus the power P_g required to lif the car against gravity.

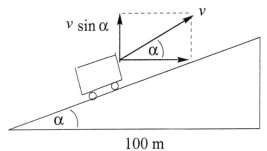

$$\tan \alpha = \frac{10 \text{ m}}{100 \text{ m}} = 0.10$$

$$\alpha = 5.71°$$

The vertical component of the velocity of the car is $v \sin \alpha = (16.67 \text{ m/s}) \sin 5.71° =$ 1.658 m/s.

Then $P_g = F(v \sin \alpha) = mgv \sin \alpha = (1800 \text{ kg})(9.80 \text{ m/s}^2)(1.658 \text{ m/s}) =$ 2.92×10^4 W

$$P_g = 2.92 \times 10^4 \text{ W} \left(\frac{1 \text{ hp}}{746 \text{ W}} \right) = 39.1 \text{ hp}$$

The total power required is 8.00 hp + 39.1 hp = 47.1 hp.

c) The power required from the engine is <u>reduced</u> by the rate at which gravity doe positive work.

The road incline angle α is given by $\tan \alpha = 0.0100$, so $\alpha = 0.5729°$.

$P_g = mg(v \sin \alpha) = (1800 \text{ kg})(9.80 \text{ m/s}^2)(16.67 \text{ m/s}) \sin 0.5729° = 2.94 \times 10^3$ W = 3.94 hp.

The power required from the engine is then 8.00 hp - 3.94 hp = 4.06 hp.

d) No power is needed from the engine if gravity does work at the rate of

$$P_g = 8.00 \text{ hp} = 5968 \text{ W}$$

$$P_g = mgv \sin \alpha, \text{ so } \sin \alpha = \frac{P_g}{mgv} = \frac{5968 \text{ W}}{(1800 \text{ kg})(9.80 \text{ m/s}^2)(16.67 \text{ m/s})} = 0.02030$$

$\alpha = 1.163°$ and $\tan \alpha = 0.0203$, a 2.03% grade.

Problems

6-49 Forces on the suitcase:

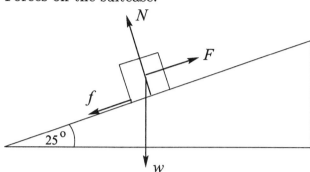

a) $W_F = (F \cos \phi)s$

Both $\vec{F}$ and $\vec{s}$ are parallel to the incline and in the same direction, so $\phi = 0°$ and
$W_F = Fs = (140 \text{ N})(3.80 \text{ m}) = 532 \text{ J}$

b)

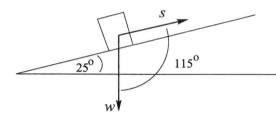

$W_w = (w \cos \phi)s$
$\phi = 115°$, so
$W_w = (196 \text{ N})(\cos 115°)(3.80 \text{ m})$
$W_w = -315 \text{ J}$

Alternatively, the component of w parallel to the incline is $w \sin 25°$. This component is down the incline so its angle with $\vec{s}$ is $\phi = 180°$.
$W_{w \sin 25°} = (196 \text{ N} \sin 25°)(\cos 180°)(3.80 \text{ m}) = -315 \text{ J}$.
The other component of w, $w \cos 25°$, is perpendicular to $\vec{s}$ and hence does no work.
Thus $W_w = W_{w \sin 25°} = -315 \text{ J}$, which agrees with the above.

c) The normal force is perpendicular to the displacement ($\phi = 90°$), so $W_N = 0$.

d) $N = w \cos 25°$ so $f_k = \mu_k N = \mu_k w \cos 25° = (0.30)(196 \text{ N}) \cos 25° = 53.3 \text{ N}$
$W_f = (f_k \cos \phi)s = (53.3 \text{ N})(\cos 180°)(3.80 \text{ m}) = -202 \text{ J}$

e) $W_{\text{tot}} = W_F + W_w + W_N + W_f = +532 \text{ J} - 315 \text{ J} + 0 - 202 \text{ J} = 15 \text{ J}$

f) $W_{\text{tot}} = K_2 - K_1$, $K_1 = 0$, so $K_2 = W_{\text{tot}}$
$\frac{1}{2}mv_2^2 = W_{\text{tot}}$ so $v_2 = \sqrt{\dfrac{2W_{\text{tot}}}{m}} = \sqrt{\dfrac{2(15 \text{ J})}{20.0 \text{ kg}}} = 1.2 \text{ m/s}$

6-55 Point 1 is at the top of the ramp and point 2 is where the package stops.

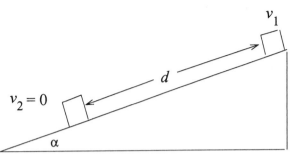

Apply $W_{\text{tot}} = K_2 - K_1$ to the motion of the package from point 1 to point 2.

$$K_2 = 0$$
$$K_1 = \tfrac{1}{2}mv_1^2$$

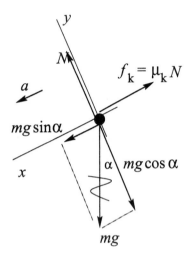

$$\sum F_y = ma_y$$
$$N - mg\cos\alpha = 0$$
$$N = mg\cos\alpha$$
$$f_k = \mu_k N = \mu_k mg\cos\alpha$$

$$W_{\text{tot}} = W_N + W_f + W_{mg}$$
$$W_N = 0, \quad W_f = -\mu_k mg\cos\alpha\, d, \quad W_{mg} = mg\sin\alpha\, d$$
Then $W_{\text{tot}} = K_2 - K_1$ gives $mg(\sin\alpha - \mu_k\cos\alpha)d = -\tfrac{1}{2}mv_1^2$.

$$d = \frac{-v_1^2}{2g(\sin\alpha - \mu_k\cos\alpha)} = \frac{-(2.20 \text{ m/s})^2}{2(9.80 \text{ m/s})(\sin 12.0° - (0.310)\cos 12.0°)} = 2.59 \text{ m}$$

6-57 $F = \alpha x^3$, with $\alpha = 4.00 \text{ N/m}^3$

a) For $x = 1.00$ m, $F = (4.00 \text{ N/m}^3)(1.00 \text{ m})^3 = 4.00$ N.

Force directed toward the origin means that $\vec{F}$ is in the $-x$-direction;
$\vec{F} = -(4.00 \text{ N})\hat{i}$.

b) For $x = 2.00$ m, $F = (4.00 \text{ N/m}^3)(2.00 \text{ m})^3 = 32.0$ N; $\quad \vec{F} = -(32.0 \text{ N})\hat{i}$.

c) $W = \int_{x_1}^{x_2} F_x\, dx$

$F_x = -\alpha x^3$, negative since $\vec{F}$ is toward the origin.

$$W = \int_{x_1}^{x_2}(-\alpha x^3)\, dx = -(\alpha/4)(x^4\big|_{x_1}^{x_2}) = -(\alpha/4)(x_2^4 - x_1^4) =$$
$$W = -\tfrac{1}{4}(4.00 \text{ N/m}^3)((2.00 \text{ m})^4 - (1.00 \text{ m})^4) = -15.0 \text{ J}$$

The work done is negative since the object moves away from the origin and the force is in the opposite direction, toward the origin.

6-59 a) Free-body diagram for the block:

$$\sum F_x = ma_x$$

$$T = m\frac{v^2}{R}$$

$$T = (0.120 \text{ kg})\frac{(0.70 \text{ m/s})^2}{0.40 \text{ m}} = 0.15 \text{ N}$$

b) $T = m\dfrac{v^2}{R} = (0.120 \text{ kg})\dfrac{(2.80 \text{ m/s})^2}{0.10 \text{ m}} = 9.4 \text{ N}$

c) The tension changes as the distance of the block from the hole changes. We could use $W = \int_{x_1}^{x_2} F_x\, dx$ to calculate the work. But a much simpler approach is to use $W_{\text{tot}} = K_2 - K_1$.

The only force doing work on the block is the tension in the cord, so $W_{\text{tot}} = W_T$.

$K_1 = \frac{1}{2}mv_1^2 = \frac{1}{2}(0.120 \text{ kg})(0.70 \text{ m/s})^2 = 0.0294 \text{ J}$
$K_2 = \frac{1}{2}mv_2^2 = \frac{1}{2}(0.120 \text{ kg})(2.80 \text{ m/s})^2 = 0.470 \text{ J}$

$W_{\text{tot}} = K_2 - K_1 = 0.470 \text{ J} - 0.029 \text{ J} = 0.44 \text{ J}$

This is the amount of work done by the person who pulled the cord.

6-61 a) $x(t) = \alpha t^2 + \beta t^3$

$v(t) = \dfrac{dx}{dt} = 2\alpha t + 3\beta t^2$

$t = 4.00 \text{ s}$: $v = 2(0.200 \text{ m/s}^2)(4.00 \text{ s}) + 3(0.0200 \text{ m/s}^3)(4.00 \text{ s})^2 = 2.56 \text{ m/s}.$

b) $a(t) = \dfrac{dv}{dt} = 2\alpha + 6\beta t$

$F = ma = m(2\alpha + 6\beta t)$

$t = 4.00 \text{ s}$: $F = 6.00 \text{ kg}(2(0.200 \text{ m/s}^2) + 6(0.0200 \text{ m/s}^3)(4.00 \text{ s})) = 5.28 \text{ N}$

c) $W_{\text{tot}} = K_2 - K_1$

At $t_1 = 0$, $v_1 = 0$ so $K_1 = 0$.

$W_{\text{tot}} = W_F$

$K_2 = \frac{1}{2}mv_2^2 = \frac{1}{2}(6.00 \text{ kg})(2.56 \text{ m/s})^2 = 19.7 \text{ J}$

Then $W_{\text{tot}} = K_2 - K_1$ gives that $W_F = 19.7$ J

6-63 a) $W_{\text{tot}} = K_2 - K_1$

$K_1 = \frac{1}{2}mv_1^2 = \frac{1}{2}(80.0 \text{ kg})(5.00 \text{ m/s})^2 = 1000$ J

$K_2 = \frac{1}{2}mv_2^2 = \frac{1}{2}(80.0 \text{ kg})(1.50 \text{ m/s})^2 = 90$ J

$W_{\text{tot}} = 90 \text{ J} - 1000 \text{ J} = -910$ J

b) Neglecting friction, work is done by you (with the force you apply to the pedals) and by gravity: $W_{\text{tot}} = W_{\text{you}} + W_{\text{gravity}}$.

The gravity force is $w = mg = (80.0 \text{ kg})(9.80 \text{ m/s}^2) = 784$ N, downward. The displacement is 5.20 m, upward. Thus $\phi = 180°$ and

$W_{\text{gravity}} = (F \cos \phi)s = (784 \text{ N})(5.20 \text{ m}) \cos 180° = -4077$ J

Then $W_{\text{tot}} = W_{\text{you}} + W_{\text{gravity}}$ gives

$W_{\text{you}} = W_{\text{tot}} - W_{\text{gravity}} = -910 \text{ J} - (-4077 \text{ J}) = +3170$ J

6-67

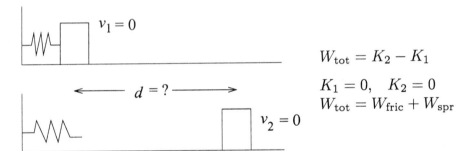

$W_{\text{tot}} = K_2 - K_1$

$K_1 = 0, \quad K_2 = 0$

$W_{\text{tot}} = W_{\text{fric}} + W_{\text{spr}}$

$W_{\text{spr}} = \frac{1}{2}kx^2$, where $x = 0.250$ m (Spring force is in direction of motion of block so it does positive work.)

$W_{\text{fric}} = -\mu_k mgd$

Then $W_{\text{tot}} = K_2 - K_1$ gives $\frac{1}{2}kx^2 - \mu_k mgd = 0$

$d = \dfrac{kx^2}{2\mu_k mg} = \dfrac{(250 \text{ N/m})(0.250 \text{ m})^2}{2(0.30)(2.50 \text{ kg})(9.80 \text{ m/s}^2)} = 1.1$ m, measured from the point where the block was released.

6-73 Apply $W_{\text{tot}} = K_2 - K_1$ to the system consisting of both blocks.

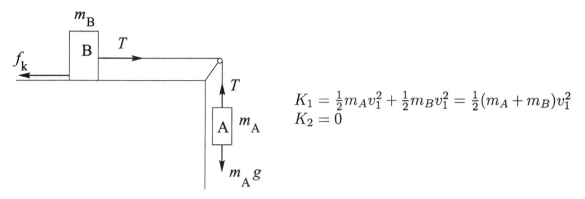

$$K_1 = \tfrac{1}{2}m_A v_1^2 + \tfrac{1}{2}m_B v_1^2 = \tfrac{1}{2}(m_A + m_B)v_1^2$$
$$K_2 = 0$$

The tension T in the rope does positive work on block B and the same magnitude of negative work on block A, so T does no net work on the system.

Gravity does work $W_{mg} = m_A gd$ on block A, where $d = 2.00$ m. (Block B moves horizontally, so no work is done on it by gravity.)

Friction does work $W_{\text{fric}} = -\mu_k m_B gd$ on block B.

Thus $W_{\text{tot}} = W_{mg} + W_{\text{fric}} = m_A gd - \mu_k m_B gd$.

Then $W_{\text{tot}} = K_2 - K_1$ gives $m_A gd - \mu_k m_B gd = -\tfrac{1}{2}(m_A + m_B)v_1^2$ and

$$\mu_k = \frac{m_A}{m_B} + \frac{\tfrac{1}{2}(m_A + m_B)v_1^2}{m_B gd} = \frac{6.00 \text{ kg}}{8.00 \text{ kg}} + \frac{(6.00 \text{ kg} + 8.00 \text{ kg})(0.900 \text{ m/s})^2}{2(8.00 \text{ kg})(9.80 \text{ m/s}^2)(2.00 \text{ m})} = 0.786$$

-77 1 day $= 8.64 \times 10^4$ s

Let t_{walk} be the time she spends walking and t_{other} be the time she spends in other activities; $t_{\text{other}} = 8.64 \times 10^4 \text{ s} - t_{\text{walk}}$.

The energy expended in each activity is the power output times the time, so

$E = Pt = (280 \text{ W})t_{\text{walk}} + (100 \text{ W})t_{\text{other}} = 1.1 \times 10^7 \text{ J}$

$(280 \text{ W})t_{\text{walk}} + (100 \text{ W})(8.64 \times 10^4 \text{ s} - t_{\text{walk}}) = 1.1 \times 10^7 \text{ J}$

$(180 \text{ W})t_{\text{walk}} = 2.36 \times 10^6 \text{ J}$

$t_{\text{walk}} = 1.31 \times 10^4 \text{ s} = 218 \text{ min} = 3.6 \text{ h}.$

-79 The power output is $P_{\text{av}} = 2000 \text{ MW} = 2.00 \times 10^9 \text{ W}$.

$P_{\text{av}} = \dfrac{\Delta W}{\Delta t}$ and 92% of the work done on the water by gravity is converted to electrical power output, so in 1.00 s the amount of work done on the water by gravity is

$$W = \frac{P_{\text{av}}\Delta t}{0.92} = \frac{(2.00 \times 10^9 \text{ W})(1.00 \text{ s})}{0.92} = 2.174 \times 10^9 \text{ J}$$

$W = mgh$, so the mass of water flowing over the dam in 1.00 s must be

$$m = \frac{W}{gh} = \frac{2.174 \times 10^9}{(9.80 \text{ m/s}^2)(170 \text{ m})} = 1.30 \times 10^6 \text{ kg}$$

$$\text{density} = \frac{m}{V} \text{ so } V = \frac{m}{\text{density}} = \frac{1.30 \times 10^6 \text{ kg}}{1.00 \times 10^3 \text{ kg/m}^3} = 1.30 \times 10^3 \text{ m}^3.$$

6-81 a) As in Example 6-11, $W = mgh$.

We need the mass of blood lifted; we are given the volume

$$V = (7500 \text{ L}) \left(\frac{1 \times 10^{-3} \text{ m}^3}{1 \text{ L}} \right) = 7.50 \text{ m}^3.$$

$m = \text{density X volume} = (1.05 \times 10^3 \text{ kg/m}^3)(7.50 \text{ m}^3) = 7.875 \times 10^3 \text{ kg}$

Then $W = mgh = (7.875 \times 10^3 \text{ kg})(9.80 \text{ m/s}^2)(1.63 \text{ m}) = 1.26 \times 10^5 \text{ J}.$

b) $P_{\text{av}} = \dfrac{\Delta W}{\Delta t} = \dfrac{1.26 \times 10^5 \text{ J}}{(24 \text{ h})(3600 \text{ s/h})} = 1.46 \text{ W}.$

6-85 a) $P = F_{\text{tot}}v$, with $F_{\text{tot}} = F_{\text{roll}} + F_{\text{air}}$

$F_{\text{air}} = \frac{1}{2}CA\rho v^2 = \frac{1}{2}(1.0)(0.463 \text{ m}^3)(1.2 \text{ kg/m}^3)(12.0 \text{ m/s})^2 = 40.0 \text{ N}$

$F_{\text{roll}} = \mu_r N = \mu_r w = (0.0045)(490 \text{ N} + 118 \text{ N}) = 2.74 \text{ N}$

$P = (F_{\text{roll}} + F_{\text{air}})v = (2.74 \text{ N} + 40.0 \text{ N})(12.0 \text{ s}) = 513 \text{ W}$

b) $F_{\text{air}} = \frac{1}{2}CA\rho v^2 = \frac{1}{2}(0.88)(0.366 \text{ m}^3)(1.2 \text{ kg/m}^3)(12.0 \text{ m/s})^2 = 27.8 \text{ N}$

$F_{\text{roll}} = \mu_r N = \mu_r w = (0.0030)(490 \text{ N} + 88 \text{ N}) = 1.73 \text{ N}$

$P = (F_{\text{roll}} + F_{\text{air}})v = (1.73 \text{ N} + 27.8 \text{ N})(12.0 \text{ s}) = 354 \text{ W}$

c) $F_{\text{air}} = \frac{1}{2}CA\rho v^2 = \frac{1}{2}(0.88)(0.366 \text{ m}^3)(1.2 \text{ kg/m}^3)(6.0 \text{ m/s})^2 = 6.96 \text{ N}$

$F_{\text{roll}} = \mu_r N = 1.73 \text{ N}$ (unchanged)

$P = (F_{\text{roll}} + F_{\text{air}})v = (1.73 \text{ N} + 6.96 \text{ N})(6.0 \text{ s}) = 52.1 \text{ W}$

Exercises

7-5 a) $K_1 + U_1 + W_{\text{other}} = K_2 + U_2$

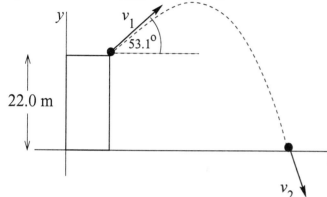

$W_{\text{other}} = 0$ (The only force on the ball while it is in the air is gravity.)

$K_1 = \frac{1}{2}mv_1^2$; $K_2 = \frac{1}{2}mv_2^2$

$U_1 = mgy_1$, $y = 22.0$ m

$U_2 = mgy_2 = 0$, since $y_2 = 0$ for our choice of coordinates.

$$\frac{1}{2}mv_1^2 + mgy_1 = \frac{1}{2}mv_2^2$$

$$v_2 = \sqrt{v_1^2 + 2gy_1} = \sqrt{(12.0 \text{ m/s})^2 + 2(9.80 \text{ m/s}^2)(22.0 \text{ m})} = 24.0 \text{ m/s}$$

Note that the projection angle of 53.1° doesn't enter into the calculation. The kinetic energy depends only on the magnitude of the velocity; it is independent of the direction of the velocity.

b) Nothing changes in the calculation. The expression derived in part (a) for v_2 is independent of the angle, so $v_2 = 24.0$ m/s, the same as in part (a).

c) The ball travels a shorter distance in part (b), so in that case air resistance will have less effect.

7-9 The forces on the object are gravity, the normal force N and friction.

The normal force is at all points in the motion perpendicular to the displacement, so it does no work. Hence $W_{\text{other}} = W_f$, the work done by friction.

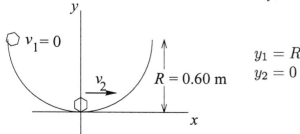

$y_1 = R$
$y_2 = 0$

$K_1 + U_1 + W_{other} = K_2 + U_2$

$K_1 = \frac{1}{2}mv_1^2 = 0.$ $K_2 = \frac{1}{2}mv_2^2$

$U_1 = mgy_1 = (0.20 \text{ kg})(9.80 \text{ m/s}^2)(0.50 \text{ m}) = 0.980 \text{ J},$ $U_2 = mgy_2 = 0$

$W_{other} = W_f = -0.22 \text{ J}$

Thus $0 + U_1 + W_f = K_2 + 0$

$0.980 \text{ J} - 0.22 \text{ J} = \frac{1}{2}(0.20 \text{ kg})v_2^2$

$$v_2 = \sqrt{\frac{2(0.760 \text{ J})}{0.20 \text{ kg}}} = 2.8 \text{ m/s}$$

7-11

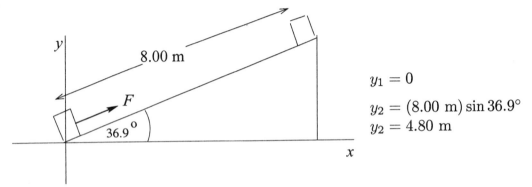

$y_1 = 0$

$y_2 = (8.00 \text{ m}) \sin 36.9°$

$y_2 = 4.80 \text{ m}$

a) $W_F = (F \cos \phi)s = (110 \text{ N})(\cos 0°)(8.00 \text{ m}) = 880 \text{ J}$

b) Use the free-body diagram for the oven to calculate the normal force N; then the friction force can be calculated from $f_k = \mu_k N$. For this calculation use coordinate parallel and perpendicular to the incline.

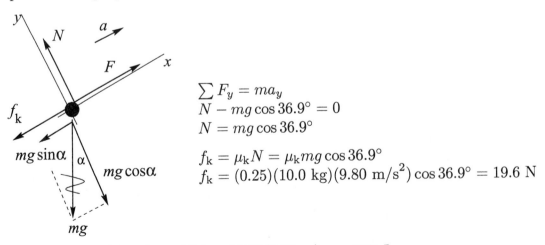

$\sum F_y = ma_y$

$N - mg \cos 36.9° = 0$

$N = mg \cos 36.9°$

$f_k = \mu_k N = \mu_k mg \cos 36.9°$

$f_k = (0.25)(10.0 \text{ kg})(9.80 \text{ m/s}^2) \cos 36.9° = 19.6 \text{ N}$

$W_f = (f_k \cos \phi)s = (19.6 \text{ N})(\cos 180°)(8.00 \text{ m}) = -157 \text{ J}$

c) $\Delta U = U_2 - U_1 = mg(y_2 - y_1) = (10.0 \text{ kg})(9.80 \text{ m/s}^2)(4.80 \text{ m} - 0) = 470 \text{ J}$

d) $K_1 + U_1 + W_{other} = K_2 + U_2$

$\Delta K = K_2 - K_1 = U_1 - U_2 + W_{other}$

$\Delta K = W_{other} - \Delta U$

$W_{other} = W_F + W_f = 880 \text{ J} - 157 \text{ J} = 723 \text{ J}$

$\Delta U = 470 \text{ J}$

Thus $\Delta K = 723 \text{ J} - 470 \text{ J} = 253 \text{ J}$.

e) We can use the free-body diagram that is in part (b):

$\sum F_x = ma_x$

$F - f_k - mg \sin 36.9° = ma$

$a = \dfrac{F - f_k - mg \sin 36.9°}{m} = \dfrac{110 \text{ N} - 19.6 \text{ N} - (10.0 \text{ kg})(9.80 \text{ m/s}^2) \sin 36.9°}{10.0 \text{ kg}} = 3.16 \text{ m/s}^2$

$v_1 = 0, \quad a = 3.16 \text{ m/s}^2, \quad x - x_0 = 8.00 \text{ m}, \quad v_2 = ?$

$v_2^2 = v_1^2 + 2a(x - x_0)$

$v_2 = \sqrt{2a(x - x_0)} = \sqrt{2(3.16 \text{ m/s}^2)(8.00 \text{ m})} = 7.11 \text{ m/s}^2$

Then $\Delta K = K_2 - K_1 = \frac{1}{2}mv_2^2 = \frac{1}{2}(10.0 \text{ kg})(7.11 \text{ m/s})^2 = 253 \text{ J}$; this agrees with the result calculated in part (d) using energy methods.

7-15 **a)** $U = \frac{1}{2}kx^2$

$x = \sqrt{\dfrac{2U}{k}} = \sqrt{\dfrac{2(3.20 \text{ J})}{1600 \text{ N/m}}} = 0.0632 \text{ m} = 6.32 \text{ cm}$

b)

$K_1 + U_1 + W_{other} = K_2 + U_2$

$W_{other} = 0$ (only work is that done by gravity and spring force)

$K_1 = 0, \ K_2 = 0$

$y = 0$ at final position of book
$U_1 = mg(h + d), \quad U_2 = \frac{1}{2}kd^2$

$0 + mg(h + d) + 0 = \frac{1}{2}kd^2$

The original gravitational potential energy of the system is converted into potential energy of the compressed spring.

$$\tfrac{1}{2}kd^2 - mgd - mgh = 0$$

$$d = \frac{1}{k}\left(mg \pm \sqrt{(mg)^2 + 4(\tfrac{1}{2}k)(mgh)}\right)$$

d must be positive, so $d = \dfrac{1}{k}\left(mg + \sqrt{(mg)^2 + 2kmgh}\right)$

$$d = \frac{1}{1600 \text{ N/m}}\big((1.20 \text{ kg})(9.80 \text{ m/s}^2) + \\ \sqrt{((1.20 \text{ kg})(9.80 \text{ m/s}^2)^2 + 2(1600 \text{ N/m})(1.20 \text{ kg})(9.80 \text{ m/s}^2)(0.80 \text{ m}))}$$

$d = 0.0074 \text{ m} + 0.1087 \text{ m} = 0.12 \text{ m} = 12 \text{ cm}$

7-17 $K_1 + U_1 + W_{\text{other}} = K_2 + U_2$

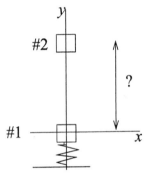

The spring force and gravity are the only forces doing work on the cheese, so $W_{\text{other}} = 0$ and $U = U_{\text{grav}} + U_{\text{el}}$.

Cheese released from rest implies $K_1 = 0$.

At the maximum height $v_2 = 0$ so $K_2 = 0$.

$U_1 = U_{1,\text{el}} + U_{1,\text{grav}}$

$y_1 = 0$ implies $U_{1,\text{grav}} = 0$

$U_{1,\text{el}} = \tfrac{1}{2}kx_1^2 = \tfrac{1}{2}(1800 \text{ N/m})(0.15 \text{ m})^2 = 20.25 \text{ J}$

(Here x_1 refers to the amount the spring is stretched or compressed when the chee is at poisition 1; it is not the x-coordinate of the cheese in the coordinate syste shown in the sketch.)

$U_2 = U_{2,\text{el}} + U_{2,\text{grav}}$

$U_{2,\text{grav}} = mgy_2$, where y_2 is the height we are solving for

$U_{2,\text{el}} = 0$ since now the spring is no longer compressed

Putting all this into $K_1 + U_1 + W_{\text{other}} = K_2 + U_2$ gives $U_{1,\text{el}} = U_{2,\text{grav}}$

The description in terms of energy is very simple; the elastic potential energy ori inally stored in the spring is converted into gravitational potential energy of t system.

$$y_2 = \frac{20.25 \text{ J}}{mg} = \frac{20.25 \text{ J}}{(1.20 \text{ kg})(9.80 \text{ m/s}^2)} = 1.72 \text{ m}$$

7-19 a) $K_1 + U_1 + W_{\text{other}} = K_2 + U_2$

point 1: the glider is at its initial position, where $x_1 = 0.100$ m and $v_1 = 0$
point 2: the glider is at $x = 0$

$K_1 = 0$ (released from rest), $\quad K_2 = \frac{1}{2}mv_2^2$
$U_1 = \frac{1}{2}kx_1^2, \quad U_2 = 0, \quad W_{\text{other}} = 0$ (only the spring force does work)

Thus $\frac{1}{2}kx_1^2 = \frac{1}{2}mv_2^2$. (The initial potential energy of the stretched spring is converted entirely into kinetic energy of the glider.)

$$v_2 = x_1\sqrt{\frac{k}{m}} = (0.100 \text{ m})\sqrt{\frac{5.00 \text{ N/m}}{0.200 \text{ kg}}} = 0.500 \text{ m/s}$$

b) The maximum speed occurs at $x = 0$, so the same equation applies.
$\frac{1}{2}kx_1^2 = \frac{1}{2}mv_2^2$

$$x_1 = v_2\sqrt{\frac{m}{k}} = 2.50 \text{ m/s}\sqrt{\frac{0.200 \text{ kg}}{5.00 \text{ N/m}}} = 0.500 \text{ m}$$

7-21 a) Choose point 1 as in Example 7-11 and let that be the origin,
so $y_1 = 0$. Let point 2 be 1.00 m below point 1, so $y_2 = -1.00$ m.

$K_1 + U_1 + W_{\text{other}} = K_2 + U_2$
$K_1 = \frac{1}{2}mv_1^2 = \frac{1}{2}(2000 \text{ kg})(25 \text{ m/s})^2 = 625,000 \text{ J}, \quad U_1 = 0$
$W_{\text{other}} = -f\,|\,y_2\,| = -(17.000 \text{ N})(1.00 \text{ m}) = -17,000 \text{ J}$
$K_2 = \frac{1}{2}mv_2^2$
$U_2 = U_{2,\text{grav}} + U_{2,\text{el}} = mgy_2 + \frac{1}{2}ky_2^2$
$U_2 = (2000 \text{ kg})(9.80 \text{ m/s}^2)(-1.00 \text{ m}) + \frac{1}{2}(1.41 \times 10^5 \text{ N/m})(1.00 \text{ m})^2$
$U_2 = -19,600 \text{ J} + 70,500 \text{ J} = +50,900 \text{ J}$

Thus $625,000 \text{ J} - 17,000 \text{ J} = \frac{1}{2}mv_2^2 + 50,900 \text{ J}$
$\frac{1}{2}mv_2^2 = 557,100 \text{ J}$

$$v_2 = \sqrt{\frac{2(557,100 \text{ J})}{2000 \text{ kg}}} = 23.6 \text{ m/s}$$

b) Free-body diagram for the elevator:

$F_{spr} = kd$, where d is the distance the spring is compressed

$$\sum F_y = ma_y$$
$$f_k + F_{spr} - mg = ma$$
$$f_k + kd - mg = ma$$

$$a = \frac{f_k + kd - mg}{m} = \frac{17,000\ \text{N} + (1.41 \times 10^5\ \text{N/m})(1.00\ \text{m}) - (2000\ \text{kg})(9.80\ \text{m/s}^2)}{2000\ \text{kg}}$$

$$= 69.2\ \text{m/s}^2$$

We calculate that a is positive, so the acceleration is upward.

7-25 $W = \int_1^2 \vec{F} \cdot d\vec{l}$, $\vec{F} = -\alpha x^2 \hat{i}$

a) $d\vec{l} = dy\hat{j}$ (x is constant; the displacement is in the $+y$-direction)

$\vec{F} \cdot d\vec{l} = 0$ (since $\hat{i} \cdot \hat{j} = 0$) and thus $W = 0$.

b) $d\vec{l} = dx\hat{i}$

$\vec{F} \cdot d\vec{l} = (-\alpha x^2) \cdot (dx\hat{i}) = -\alpha x^2\, dx$

$W = \int_{x_1}^{x_2}(-\alpha x^2)\, dx = -\frac{1}{3}\alpha x^3\big|_{x_1}^{x_2} = -\frac{1}{3}\alpha(x_2^3 - x_1^3) = -\frac{12\ \text{N/m}^2}{3}((0.30\ \text{m})^3$

$(0.10\ \text{m})^3) = -0.10\ \text{J}$

c) $d\vec{l} = dx\hat{i}$ as in part (b), but now $x_1 = 0.30$ m and $x_2 = 0.10$ m

$W = -\frac{1}{3}\alpha(x_2^3 - x_1^3) = +0.10\ \text{J}$

d) The total work for the displacement along the x-axis from 0.10 m to 0.30 m an
then back to 0.10 m is the sum of the results of parts (b) and (c), which is zero.

The total work is zero when the starting and ending points are the same, so t
force is conservative.

$W_{x_1 \to x_2} = -\frac{1}{3}\alpha(x_2^3 - x_1^3) = \frac{1}{3}\alpha x_1^3 - \frac{1}{3}\alpha x_2^3$

The definition of the potential energy function is $W_{x_1 \to x_2} = U_1 - U_2$. Comparisc
of the two expressions for W gives $U = \frac{1}{3}\alpha x^3$. This does correspond to $U = 0$ whe
$x = 0$.

7-29 $f = \mu_k mg = (0.25)(1.5\ \text{kg})(9.80\ \text{m/s}^2) = 3.675\ \text{N}$;

direction of $\vec{f}$ is opposite to the motion.

a)

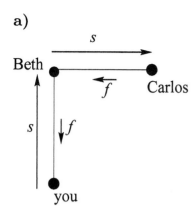

For the motion from you to Beth the friction force is directed opposite to the displacement $\vec{s}$ and $W_1 = -fs = -(3.675 \text{ N})(8.0 \text{ m}) = -29.4 \text{ J}$.

For the motion from Beth to Carlos the friction force is again directed opposite to the displacement and $W_2 = -29.4 \text{ J}$.

$$W_{\text{tot}} = W_1 + W_2 = -29.4 \text{ J} - 29.4 \text{ J} = -59 \text{ J}$$

b)

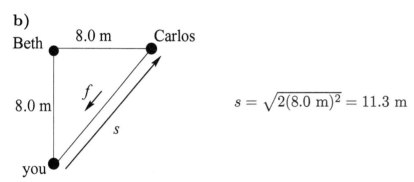

$$s = \sqrt{2(8.0 \text{ m})^2} = 11.3 \text{ m}$$

$\vec{f}$ is opposite to $\vec{s}$, so $W = -fs = -(3.675 \text{ N})(11.3 \text{ m}) = -42 \text{ J}$

c)

For the motion from you to Kim
$W = -fs$
$W = -(3.675 \text{ N})(8.0 \text{ m}) = -29.4 \text{ J}$

For the motion from Kim to you
$W = -fs = -29.4 \text{ J}$

The total work for the round trip is $-29.4 \text{ J} - 29.4 \text{ J} = -59 \text{ J}$.

c) Parts (a) and (b) show that for two different paths between you and Carlos, the work done by friction is different.

Part (c) shows that when the starting and ending points are the same, the tot.
work is not zero.

Both these results show that the friction force is nonconservative.

7-31 Use coordinates where the origin is at one atom.

The other atom then has coordinate x.

$$F_x = -\frac{dU}{dx} = -\frac{d}{dx}\left(-\frac{C_6}{x^6}\right) = +C_6\frac{d}{dx}\left(\frac{1}{x^6}\right) = -\frac{6C_6}{x^7}$$

The minus sign mean that F_x is directed in the $-x$-direction, toward the origin.

The force has magnitude $6C_6/x^7$ and is attractive.

7-33 $U(x, y) = k(x^2 + y^2) + k'xy$

$$F_x = -\frac{\partial U}{\partial x} = -2kx - k'y$$

$$F_y = -\frac{\partial U}{\partial y} = -2ky - k'x$$

$$\vec{F} = -(2kx + k'y)\hat{i} - (2ky + k'x)\hat{j}$$

7-35 a)

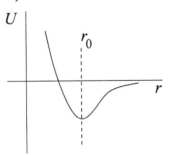
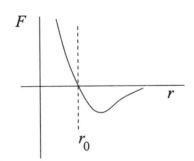

$$U = \frac{a}{r^{12}} - \frac{b}{r^6}$$

$$F = -\frac{dU}{dr} = +\frac{12a}{r^{13}} -$$

b) At equilibrium $F = 0$, so $\dfrac{dU}{dr} = 0$

$F = 0$ implies $\dfrac{+12a}{r^{13}} - \dfrac{6b}{r^7} = 0$

$6br^6 = 12a$; solution is the equilibrium distance $r_0 = (2a/b)^{1/6}$

U is a minimum at this r; the equilibrium is stable.

c) At $r = (2a/b)^{1/6}$, $U = a/r^{12} - b/r^6 = a(b/2a)^2 - b(b/2a) = -b^2/4a$.

At $r \to \infty$, $U = 0$. The energy that must be added is $-\Delta U = b^2/4a$.

d) $r_0 = (2a/b)^{1/6} = 1.13 \times 10^{-10}$ m gives that

$2a/b = 2.082 \times 10^{-60}$ m^6 and $b/4a = 2.402 \times 10^{59}$ m^{-6}

$b^2/4a = b(b/4a) = 1.54 \times 10^{-18}$ J

$b(2.402 \times 10^{59}$ m$^{-6}) = 1.54 \times 10^{-18}$ J and $b = 6.41 \times 10^{-78}$ J·m^6.

Then $2a/b = 2.082 \times 10^{-60}$ m^6 gives $a = (b/2)(2.082 \times 10^{-60}$ m$^6) =$
$\frac{1}{2}(6.41 \times 10^{-78}$ J·m$^6)(2.082 \times 10^{-60}$ m$^6) = 6.67 \times 10^{-138}$ J·m^{12}

Problems

7-39 $K_1 + U_1 + W_{\text{other}} = K_2 + U_2$

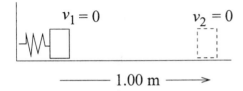

Work is done on the block by the spring and by friction, so $W_{\text{other}} = W_f$ and $U = U_{\text{el}}$.

$K_1 = K_2 = 0$

$U_1 = U_{1,\text{el}} = \frac{1}{2}kx_1^2 = \frac{1}{2}(100$ N/m$)(0.200)^2 = 2.00$ J

$U_2 = U_{2,\text{el}} = 0$, since after the block leaves the spring has given up all its stored energy

$W_{\text{other}} = W_f = (f_k \cos \phi)s = \mu_k mg(\cos \phi)s = -\mu_k mgs$, since $\phi = 180°$ (The friction force is directed opposite to the displacement and does negative work.)

Putting all this into $K_1 + U_1 + W_{\text{other}} = K_2 + U_2$ gives

$U_{1,\text{el}} + W_f = 0$

(The potential energy originally stored in the spring is taken out of the system by the negative work done by friction.)

$\mu_k mgs = U_{1,\text{el}}$

$\mu_k = \dfrac{U_{1,\text{el}}}{mgs} = \dfrac{2.00 \text{ J}}{(0.50 \text{ kg})(9.80 \text{ m/s}^2)(1.00 \text{ m})} = 0.41.$

7-41 $K_1 + U_1 + W_{\text{other}} = K_2 + U_2$

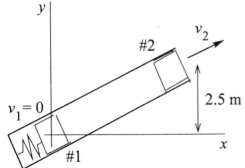

Take $y = 0$ at his initial position.

$K_1 = 0$, $K_2 = \frac{1}{2}mv_2^2$

$W_{\text{other}} = W_{\text{fric}} = -fs$

$W_{\text{other}} = -(40$ N$)(4.0$m$) = -160$ J

$U_{1,\text{grav}} = 0, \quad U_{1,\text{el}} = \frac{1}{2}kd^2$, where d is the distance the spring is initially compressed

$F = kd$ so $d = \dfrac{F}{k} = \dfrac{4400 \text{ N}}{1100 \text{ N/m}} = 4.00 \text{ m}$

and $U_{1,\text{el}} = \frac{1}{2}(1100 \text{ N/m})(4.00 \text{ m})^2 = 8800 \text{ J}$

$U_{2,\text{grav}} = mgy_2 = (60 \text{ kg})(9.80 \text{ m/s}^2)(2.5 \text{ m}) = 1470 \text{ J}, \; U_{2,\text{el}} = 0$

Then $K_1 + U_1 + W_{\text{other}} = K_2 + U_2$ gives
$8800 \text{ J} - 160 \text{ J} = \frac{1}{2}mv_2^2 + 1470 \text{ J}$

$\frac{1}{2}mv_2^2 = 7170 \text{ J}$ and $v_2 = \sqrt{\dfrac{2(7170 \text{ J})}{60 \text{ kg}}} = 15.5 \text{ m/s}$

7-43 $K_1 + U_1 + W_{\text{other}} = K_2 + U_2$

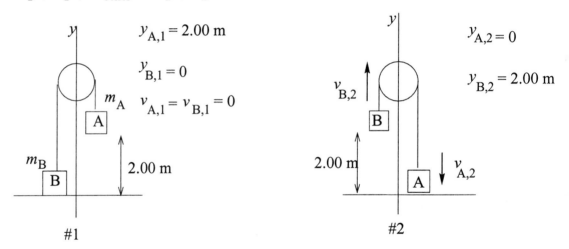

#1   #2

The tension force does positive work on the 4.0 kg bucket and an equal amount of negative work on the 12.0 kg bucket, so the net work done by the tension is zero.

Work is done on the system only by gravity, so $W_{\text{other}} = 0$ and $U = U_{\text{grav}}$

$K_1 = 0$

$K_2 = \frac{1}{2}m_A v_{A,2}^2 + \frac{1}{2}m_B v_{B,2}^2$ But since the two buckets are connected by a rope they move together and have the same speed: $v_{A,2} = v_{B,2} = v_2$.

Thus $K_2 = \frac{1}{2}(m_A + m_B)v_2^2 = (8.00 \text{ kg})v_2^2$.

$U_1 = m_A g y_{A,1} = (12.0 \text{ kg})(9.80 \text{ m/s}^2)(2.00 \text{ m}) = 235.2 \text{ J}$.

$U_2 = m_B g y_{B,2} = (4.0 \text{ kg})(9.80 \text{ m/s}^2)(2.00 \text{ m}) = 78.4 \text{ J}$.

Putting all this into $K_1 + U_1 + W_{\text{other}} = K_2 + U_2$ gives
$U_1 = K_2 + U_2$
$235.2 \text{ J} = (8.00 \text{ kg})v_2^2 + 78.4 \text{ J}$

$v_2 = \sqrt{\dfrac{235.2 \text{ J} - 78.4 \text{ J}}{8.00 \text{ kg}}} = 4.4 \text{ m/s}$

7-47 **a)** $K_1 + U_1 + W_{\text{other}} = K_2 + U_2$

#1

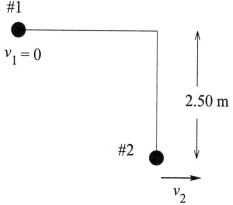

$v_1 = 0$

#2

v_2

2.50 m

$y_1 = 2.50$ m

$y_2 = 0$

The tension in the string is at all points in the motion perpendicular to the displacement, so $W_T = 0$

The only force that does work on the potato is gravity, so $W_{\text{other}} = 0$.

$K_1 = 0, \quad K_2 = \frac{1}{2}mv_2^2$

$U_1 = mgy_1, \quad U_2 = 0$

Thus $U_1 = K_2$.

$mgy_1 = \frac{1}{2}mv_2^2$

$v_2 = \sqrt{2gy_1} = \sqrt{2(9.80 \text{ m/s}^2)(2.50 \text{ m})} = 7.00$ m/s

b) Free-body diagram for the potato as it swings through its lowest point:

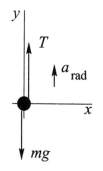

y

T

a_{rad}

x

mg

The acceleration $\vec{a}_{\text{rad}}$ is directed in toward the center of the circular path, so at this point it is upward

$\sum F_y = ma_y$

$T - mg = ma_{\text{rad}}$

$T = m(g + a_{\text{rad}}) = m(g + \frac{v_2^2}{R})$, where the radius R for the circular motion is the length L of the string.

It is instructive to use the algebraic expression for v_2 from part (a) rather than just putting in the numerical value:

$v_2 = \sqrt{2gy_1} = \sqrt{2gL}$, so $v_2^2 = 2gL$

Then $T = m(g + \frac{v_2^2}{L}) = m(g + \frac{2gL}{L}) = 3mg$; the tension at this point is three times the weight of the potato.

$T = 3mg = 3(0.100 \text{ kg})(9.80 \text{ m/s}^2) = 2.94$ N

7-49   a) Speed at ground if steps off platform at height h:

$$K_1 + U_1 + W_{\text{other}} = K_2 + U_2$$

$mgh = \frac{1}{2}mv_2^2$, so $v_2^2 = 2gh$

Motion from top to bottom of pole: (take $y = 0$ at bottom)

$$K_1 + U_1 + W_{\text{other}} = K_2 + U_2$$

$mgd - fd = \frac{1}{2}mv_2^2$

Use $v_2^2 = 2gh$ and get $mgd - fd = mgh$

$fd = mg(d - h)$

$f = mg(d - h)/d = mg(1 - h/d)$

For $h = d$ this gives $f = 0$ as it should (friction has no effect).

For $h = 0$, $v_2 = 0$ (no motion). The equation for f gives $f = mg$ in this specia
case. When $f = mg$ the forces on him cancel and he doesn't accelerate down th
pole, which agrees with $v_2 = 0$.

b) $f = mg(1 - h/d) = (75 \text{ kg})(9.80 \text{ m/s}^2)(1 - 1.0 \text{ m}/2.5 \text{ m}) = 441$ N.

c) Take $y = 0$ at bottom of pole, so $y_1 = d$ and $y_2 = y$.

$$K_1 + U_1 + W_{\text{other}} = K_2 + U_2$$

$0 + mgd - f(d - y) = \frac{1}{2}mv^2 + mgy$

$\frac{1}{2}mv^2 = mg(d - y) - f(d - y)$

Using $f = mg(1 - h/d)$ gives $\frac{1}{2}mv^2 = mg(d - y) - mg(1 - h/d)(d - y)$

$\frac{1}{2}mv^2 = mg(h/d)(d - y)$ and $v = \sqrt{2gh(1 - y/d)}$

(Note that this gives the correct results for $y = 0$ and for $y = d$.)

7-51

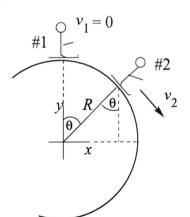

Let point 2 be where the skier loses contact
with the snowball.

Loses contact implies $N \to 0$.

$y_1 = R$, $y_2 = R\cos\theta$

First, analyze the forces on the skier when she is at point 2. For this use coordinate
that are in the tangential and radial directions. The skier moves in an arc of a circle
so her acceleration is $a_{\text{rad}} = v^2/R$, directed in towards the center of the snowball.

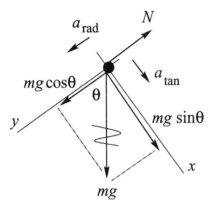

$$\sum F_y = ma_y$$
$$mg \cos \theta - N = mv_2^2/R$$
But $N = 0$ so $mg \cos \theta = mv_2^2/R$
$$v_2^2 = Rg \cos \theta$$

Now use conservation of energy to get another equation relating v_2 to θ:
$$K_1 + U_1 + W_{\text{other}} = K_2 + U_2$$
The only force that does work on the skier is gravity, so $W_{\text{other}} = 0$.
$$K_1 = 0, \quad K_2 = \tfrac{1}{2}mv_2^2$$
$$U_1 = mgy_1 = mgR \qquad U_2 = mgy_2 = mgR \cos \theta$$
Then $mgR = \tfrac{1}{2}mv_2^2 + mgR \cos \theta$
$$v_2^2 = 2gR(1 - \cos \theta)$$

Combine this with the $\sum F_y = ma_y$ equation:
$$Rg \cos \theta = 2gR(1 - \cos \theta)$$
$$\cos \theta = 2 - 2\cos \theta$$
$3 \cos \theta = 2$ so $\cos \theta = 2/3$ and $\theta = 48.2°$

7-53

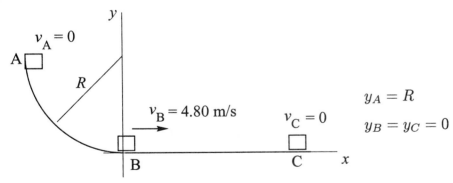

a) Apply conservation of energy to the motion from B to C:
$$K_B + U_B + W_{\text{other}} = K_C + U_C$$

The only force that does work on the package during this part of the motion is friction, so
$$W_{\text{other}} = W_f = f_k(\cos \phi)s = \mu_k mg(\cos 180°)s = -\mu_k mgs$$
$$K_B = \tfrac{1}{2}mv_B^2, \quad K_C = 0$$

$U_B = 0, \quad U_C = 0$

Thus $K_B + W_f = 0$ (The negative friction work takes away all the kinetic energy.

$\frac{1}{2}mv_B^2 - \mu_k mgs = 0$

$\mu_k = \dfrac{v_B^2}{2gs} = \dfrac{(4.80 \text{ m/s})^2}{2(9.80 \text{ m/s}^2)(3.00 \text{ m})} = 0.392$

b) Apply conservation of energy to the motion from A to B:

$K_A + U_A + W_{\text{other}} = K_B + U_B$

Work is done by gravity and by friction, so $W_{\text{other}} = W_f$.

$K_A = 0, \quad K_B = \frac{1}{2}mv_B^2 = \frac{1}{2}(0.200 \text{ kg})(4.80 \text{ m/s})^2 = 2.304 \text{ J}$

$U_A = mgy_A = mgR = (0.200 \text{ kg})(9.80 \text{ m/s}^2)(1.60 \text{ m}) = 3.136 \text{ J}, \quad U_B = 0$

Thus $U_A + W_f = K_B$

$W_f = K_B - U_A = 2.304 \text{ J} - 3.136 \text{ J} = -0.83 \text{ J}$

(W_f is negative as expected; the friction force does negative work since it is directe opposite to the displacement.)

7-55 $F_x = -\alpha x - \beta x^2$, $\alpha = 60.0 \text{ N/m}$ and $\beta = 18.0 \text{ N/m}^2$

a) $W_{F_x} = U_1 - U_2 = \int_{x_1}^{x_2} F_x(x)\, dx$

Let $x_1 = 0$ and $U_1 = 0$. Let x_2 be some arbitrary point x, so $U_2 = U(x)$. Then

$U(x) = -\int_0^x F_x(x)\, dx = -\int_0^x (-\alpha x - \beta x^2)\, dx = \int_0^x (\alpha x + \beta x^2)\, dx = \frac{1}{2}\alpha x^2 + \frac{1}{3}\beta x^3.$

b)

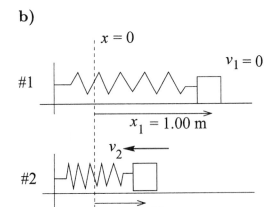

$K_1 + U_1 + W_{\text{other}} = K_2 + U_2$

The only force that does work on the object is the spring force, so $W_{\text{other}} = 0$.

$K_1 = 0, \quad K_2 = \frac{1}{2}mv_2^2$

$U_1 = U(x_1) = \frac{1}{2}\alpha x_1^2 + \frac{1}{3}\beta x_1^3 = \frac{1}{2}(60.0 \text{ N/m})(1.00 \text{ m})^2 + \frac{1}{3}(18.0 \text{ N/m}^2)(1.00 \text{ m})^3 =$ 36.0 J

$U_2 = U(x_2) = \frac{1}{2}\alpha x_2^2 + \frac{1}{3}\beta x_2^3 = \frac{1}{2}(60.0 \text{ N/m})(0.500 \text{ m})^2 + \frac{1}{3}(18.0 \text{ N/m}^2)(0.500 \text{ m})^3 =$ 8.25 J

Thus 36.0 J $= \frac{1}{2}mv_2^2 + 8.25$ J

$$v_2 = \sqrt{\frac{2(36.0\text{ J} - 8.25\text{ J})}{0.900\text{ kg}}} = 7.85 \text{ m/s}$$

7-61

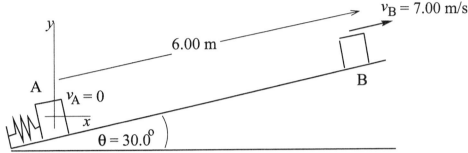

The normal force is $N = mg \cos\theta$, so $f_k = \mu_k N = \mu_k mg \cos\theta$.

$y_A = 0;\quad y_B = (6.00\text{ m})\sin 30.0° = 3.00$ m

Apply conservation of energy to the motion of the block from point A to point B:

$K_A + U_A + W_{\text{other}} = K_B + U_B$

Work is done by gravity, by the spring force, and by friction, so $W_{\text{other}} = W_f$ and $U = U_{\text{el}} + U_{\text{grav}}$

$K_A = 0,\quad K_B = \frac{1}{2}mv_B^2 = \frac{1}{2}(1.50\text{ kg})(7.00\text{ m/s})^2 = 36.75$ J

$U_A = U_{\text{el},A} + U_{\text{grav},A} = U_{\text{el},A}$, since $U_{\text{grav},A} = 0$

$U_B = U_{\text{el},B} + U_{\text{grav},B} = 0 + mgy_B = (1.50\text{ kg})(9.80\text{ m/s}^2)(3.00\text{ m}) = 44.1$ J

$W_{\text{other}} = W_f = (f_k \cos\phi)s = \mu_k mg \cos\theta(\cos 180°)s = -\mu_k mg \cos\theta\, s$

$W_{\text{other}} = -(0.50)(1.50\text{ kg})(9.80\text{ m/s}^2)(\cos 30.0°)(6.00\text{ m}) = -38.19$ J

Thus $U_{\text{el},A} - 38.19$ J $= 36.75$ J $+ 44.10$ J

$U_{\text{el},A} = 38.19$ J $+ 36.75$ J $+ 44.10$ J $= 119$ J

7-63 a) Apply $K_A + U_A + W_{\text{other}} = K_B + U_B$ to the motion from A to B.

$K_A = 0,\quad K_B = \frac{1}{2}mv_B^2$

$U_A = 0,\quad U_B = U_{\text{el},B} = \frac{1}{2}kx_B^2$, where $x_B = 0.25$ m

$W_{\text{other}} = W_F = Fx_B$

Thus $Fx_B = \frac{1}{2}mv_B^2 + \frac{1}{2}kx_B^2$. (The work done by F goes partly to the potential energy of the stretched spring and partly to the kinetic energy fo the block.)

$Fx_B = (20.0\text{ N})(0.25\text{ m}) = 5.0$ J and $\frac{1}{2}kx_B^2 = \frac{1}{2}(40.0\text{ N/m})(0.25\text{ m})^2 = 1.25$ J

Thus 5.0 J $= \frac{1}{2}mv_B^2 + 1.25$ J and $v_B = \sqrt{\dfrac{2(3.75\text{ J})}{0.500\text{ kg}}} = 3.87$ m/s

b) Let point C be where the block is closest to the wall. When the block is at poin C the spring is compressed an amount $| x_C |$, so the block is 0.60 m$- | x_C |$ from the wall, and the distance between B and C is $x_B + | x_C |$.

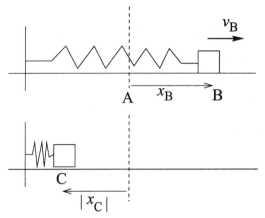

$$K_B + U_B + W_{other} = K_C + U_C$$

$W_{other} = 0$

$K_B = \frac{1}{2}mv_B^2 = 5.0 \text{ J} - 1.25 \text{ J} = 3.75 \text{ J}$
(from part (a))

$U_B = \frac{1}{2}kx_B^2 = 1.25 \text{ J}$

$K_C = 0$ (instantaneously at rest at point closest to wall)

$U_C = \frac{1}{2}k | x_C |^2$

Thus $3.75 \text{ J} + 1.25 \text{ J} = \frac{1}{2}k | x_C |^2$

$$| x_C | = \sqrt{\frac{2(5.0 \text{ J})}{40.0 \text{ N/m}}} = 0.50 \text{ m}$$

The distance of the block from the wall is $0.60 \text{ m} - 0.50 \text{ m} = 0.10 \text{ m}$.

7-67 $\vec{F} = -\alpha xy^2 \hat{j}, \quad \alpha = 2.50 \text{ N/m}^3$

a)

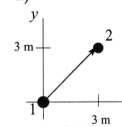

$d\vec{l} = dx\hat{i} + dy\hat{j}$
$\vec{F} \cdot d\vec{l} = -\alpha xy^2 \, dy$
On the path, $x = y$ so $\vec{F} \cdot d\vec{l} = -\alpha y^3 \, dy$

$W = \int_1^2 \vec{F} \cdot d\vec{l} = \int_{y_1}^{y_2}(-\alpha y^3)\, dy = -(\alpha/4)(y^4|_{y_1}^{y_2}) = -(\alpha/4)(y_2^4 - y_1^4)$

$y_1 = 0, y_2 = 3.00 \text{ m}$, so $W = -\frac{1}{4}(2.50 \text{ N/m}^3)(3.00 \text{ m})^4 = -50.6 \text{ J}$

b)

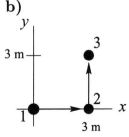

For the displacement from point 1 to point 2, $d\vec{l} = dx\hat{i}$, so $\vec{F} \cdot d\vec{l} = 0$ and $W = 0$
(The force is perpendicular to the displacement at each point along the path, s
$W = 0$.)

For the displacement from point 2 to point 3, $d\vec{l} = dy\hat{j}$, so $\vec{F} \cdot d\vec{l} = -\alpha xy^2 \, dy$. On this path, $x = 3.00$ m, so
$$\vec{F} \cdot d\vec{l} = -(2.50 \text{ N/m}^3)(3.00 \text{ m})y^2 \, dy = -(7.50 \text{ N/m}^2)y^2 \, dy.$$

$$W = \int_2^3 \vec{F} \cdot d\vec{l} = -(7.50 \text{ N/m}^2) \int_{y_2}^{y_3} y^2 \, dy = -(7.50 \text{ N/m}^2)\tfrac{1}{3}(y_3^3 - y_2^3)$$
$$W = -(7.50 \text{ N/m}^2)(\tfrac{1}{3})(3.00 \text{ m})^3 = -67.5 \text{ J}$$

c) For these two paths between the same starting and ending points the work done is different, so the force is nonconservative.

CHAPTER 8
MOMENTUM, IMPULSE, AND COLLISIONS

Exercises 5, 9, 11, 13, 17, 19, 23, 25, 29, 31, 33, 37, 39, 41, 43, 47, 51, 55
Problems 57, 63, 67, 69, 71, 75, 79, 83, 85, 87, 91

Exercises

8-5 $p_{1x} = mv_{1x} = (0.145 \text{ kg})(1.30 \text{ m/s}) = 0.1885 \text{ kg} \cdot \text{m/s}$

$p_{2x} = mv_{2x} = (0.057 \text{ kg})(-7.80 \text{ m/s}) = -0.4446 \text{ kg} \cdot \text{m/s}$

$P_x = p_{1x} + p_{2x} = 0.1885 \text{ kg} \cdot \text{m/s} - 0.4446 \text{ kg} \cdot \text{m/s} = -0.256 \text{ kg} \cdot \text{m/s}$

(minus sign means that is in the $-y$-direction)

8-9 Take the x-axis to be toward the right, so $v_{1x} = +3.00 \text{ m/s}$.

a) $J_x = p_{2x} - p_{1x}$

$J_x = F_x(t_2 - t_1) = (+25.0\text{N})(0.050 \text{ s}) = +1.25 \text{ kg} \cdot \text{m/s}$

Thus $p_{2x} = J_x + p_{1x} = +1.25 \text{ kg} \cdot \text{m/s} + (0.160 \text{ kg})(+3.00 \text{ m/s}) = +1.73 \text{ kg} \cdot \text{m/s}$

$$v_{2x} = \frac{p_{2x}}{m} = \frac{1.73 \text{ kg} \cdot \text{m/s}}{0.160 \text{ kg}} = +10.8 \text{ m/s (to the right)}$$

b) $J_x = F_x(t_2 - t_1) = (-12.0 \text{ N})(0.050 \text{ s}) = -0.600 \text{ kg} \cdot \text{m/s}$ (negative since force is to left)

$p_{2x} = J_x + p_{1x} = -0.600 \text{ kg} \cdot \text{m/s} + (0.160 \text{ kg})(+3.00 \text{ m/s}) = -0.120 \text{ kg} \cdot \text{m/s}$

$$v_{2x} = \frac{p_{2x}}{m} = \frac{-0.120 \text{ kg} \cdot \text{m/s}}{0.160 \text{ kg}} = -0.75 \text{ m/s (to the left)}$$

8-11 a) $\vec{J} = \int_{t_1}^{t_2} \vec{F} \, dt = \int_{t_1}^{t_2} [(1.60 \times 10^7 \text{ N/s})t - (6.00 \times 10^9 \text{ N/s}^2)t^2]\hat{i} \, dt$,

where $t_1 = 0$ and $t_2 = 2.50 \times 10^{-3}$ s.

$\vec{J} = [(1.60 \times 10^7 \text{ N/s})(\frac{1}{2}t_2^2) - (6.00 \times 10^9 \text{ N/s}^2)(\frac{1}{3}t_2^3)]\hat{i}$

$\vec{J} = [\frac{1}{2}(1.60 \times 10^7 \text{ N/s})(2.50 \times 10^{-3} \text{ s})^2 - \frac{1}{3}(6.00 \times 10^9 \text{ N/s}^2)(2.50 \times 10^{-3} \text{ s})^3]\hat{i} =$
$(18.8 \text{ kg} \cdot \text{m/s})\hat{i}$

b) $\vec{w} = -(mg)\hat{j}$ (constant)

$\vec{J} = \vec{w}(t_2 - t_1) = -mgt_2\hat{j} = -(0.145 \text{ kg})(9.80 \text{ m/s}^2)(2.50 \times 10^{-3} \text{ s})\hat{j}$

$\vec{J} = -(0.00355 \text{ kg} \cdot \text{m/s})\hat{j}$

c) $\vec{F}_{\text{av}} = \dfrac{\vec{J}}{(t_2 - t_1)} = \dfrac{(18.8 \text{ kg} \cdot \text{m/s})\vec{\imath}}{2.50 \times 10^{-3} \text{ s}} = (7520 \text{ N})\hat{\imath}$

d) $\vec{J} = \vec{p}_2 - \vec{p}_1$, so $\vec{p}_2 = \vec{p}_1 + \vec{J}$

$\vec{J} = (18.8 \text{ kg} \cdot \text{m/s})\hat{\imath} - (0.00335 \text{ kg} \cdot \text{m/s})\hat{\jmath}$

$\vec{p}_1 = m\vec{v}_1 = (0.145 \text{ kg})[(-40.0 \text{ m/s})\hat{\imath} - (5.0 \text{ m/s})\hat{\jmath}\,] = (-5.80 \text{ kg} \cdot \text{m/s})\hat{\imath} - (0.725 \text{ kg} \cdot \text{m/s})\hat{\jmath}$

Then $\vec{p}_2 = (-5.80 \text{ kg·m/s})\hat{\imath} - (0.725 \text{ kg·m/s})\hat{\jmath} + (18.8 \text{ kg·m/s})\hat{\imath} - (0.00335 \text{ kg·m/s})\hat{\jmath}$

$\vec{p}_2 = (13.0 \text{ kg} \cdot \text{m/s})\hat{\imath} - (0.73 \text{ kg} \cdot \text{m/s})\hat{\jmath}$

$\vec{v}_2 = \vec{p}_2/m = (89.7 \text{ m/s})\hat{\imath} - (5.0 \text{ m/s})\hat{\jmath}$

3-13   Take the x-axis to be toward the right, so $F_x = +(A + Bt^2)$

a) $J_x = \int_{t_1}^{t_2} F_x \, dt = \int_0^{t_2} (A + Bt^2) \, dt = (At + \frac{1}{3}Bt^3)\,\big|_0^{t_2} = At_2 + \frac{1}{3}Bt_2^3.$

The impulse has magnitude $J = At_2 + \frac{1}{3}Bt_2^3$ and is directed to the right.

b) $J_x = p_{2x} - p_{1x}$ so $p_{2x} = J_x + p_{1x}$

Initially at rest implies $p_{1x} = 0$.

$v_{2x} = p_{2x}/m = J_x/m = (A/m)t_2 + (B/3m)t_2^3$

Her speed is $v = (A/m)t_2 + (B/3m)t_2^3$ and she is moving to the right.

3-17   a)   Let Gretzky be object A and the defender be object B.

Let the x-direction be the direction in which Gretzky is moving initially. No horizontal external forces, so P_x is constant.

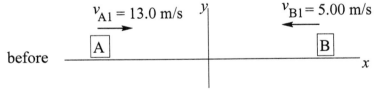

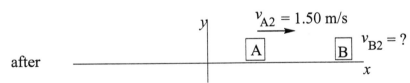

P_x constant implies $m_A v_{A1x} + m_B v_{B1x} = m_A v_{A2x} + m_B v_{B2x}$

If we multiply through by g we get an equation that uses the object's weight rather than its mass:

$w_A v_{A1x} + w_B v_{B1x} = w_A v_{A2x} + w_B v_{B2x}$

The components can be positive or negative depending on the directions of the velocities relative to our coordinate system.

Putting in the numbers:

$(756 \text{ N})(+13.0 \text{ m/s}) + (900 \text{ N})(-5.00 \text{ m/s}) = (756 \text{ N})(+1.50 \text{ m/s}) + (900 \text{ N})v_{B2x}$

$9828 \text{ N} \cdot \text{m/s} - 4500 \text{ N} \cdot \text{m/s} = 1134 \text{ N} \cdot \text{m/s} + (900 \text{ N})v_{B2x}$

$$v_{B2x} = \frac{9828 \text{ N} \cdot \text{m/s} - 4500 \text{ N} \cdot \text{m/s} - 1134 \text{ N} \cdot \text{m/s}}{900 \text{ N}} = 4.66 \text{ m/s}$$

After the collision the defender is moving at 4.66 m/s in the same direction a Gretzky.

b) $K_1 = \frac{1}{2}m_A v_{A1}^2 + \frac{1}{2}m_B v_{B1}^2$

$$K_1 = \frac{1}{2}\left(\frac{756 \text{ N}}{9.80 \text{ m/s}^2}\right)(13.0 \text{ m/s})^2 + \frac{1}{2}\left(\frac{900 \text{ N}}{9.80 \text{ m/s}^2}\right)(5.00 \text{ m/s})^2$$

$K_1 = 6519 \text{ J} + 1148 \text{ J} = 7667 \text{ J}$ (Note that the kinetic energy of an object is alway positive, and that it does not depend on the direction of the object's velocity.)

$K_2 = \frac{1}{2}m_A v_{A2}^2 + \frac{1}{2}m_b v_{B2}^2$

$$K_2 = \frac{1}{2}\left(\frac{756 \text{ N}}{9.80 \text{ m/s}^2}\right)(1.50 \text{ m/s})^2 + \frac{1}{2}\left(\frac{900 \text{ N}}{9.80 \text{ m/s}^2}\right)(4.66 \text{ m/s})^2$$

$K_2 = 87 \text{ J} + 997 \text{ J} = 1084 \text{ J}$

$\Delta K = K_2 - K_1 = 1084 \text{ J} - 7667 \text{ J} = -6580 \text{ J}$

The kinetic energy decreases by 6580 J.

8-19 Initially both blocks are at rest.

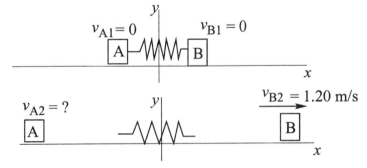

a) No horizontal force implies P_x is constant.

$m_A v_{A1x} + m_B v_{B1x} = m_A v_{A2x} + m_B v_{B2x}$

$0 = m_A v_{A2x} + m_B v_{B2x}$

$$v_{A2x} = -\left(\frac{m_B}{m_A}\right)v_{b2x} = -\left(\frac{3.00 \text{ kg}}{1.00 \text{ kg}}\right)(+1.20 \text{ m/s}) = -3.60 \text{ m/s}$$

Block A has a final speed of 3.60 m/s, and moves off in the opposite direction to E

b) Use energy conservation: $K_1 + U_1 + W_{\text{other}} = K_2 + U_2$

Only the spring force does work so $W_{\text{other}} = 0$ and $U = U_{\text{el}}$.

$K_1 = 0$ (the blocks initially are at rest)

$U_2 = 0$ (no potential energy is left in the spring)

$K_2 = \frac{1}{2}m_A v_{A2}^2 + \frac{1}{2}m_B v_{B2}^2 = \frac{1}{2}(1.00 \text{ kg})(3.60 \text{ m/s})^2 + \frac{1}{2}(3.00 \text{ kg})(1.20 \text{ m/s})^2 = 8.64 \text{ J}$

$U_1 = U_{1,\text{el}}$ the potential energy stored in the compresssed spring.

Thus $U_{1,\text{el}} = K_2 = 8.64 \text{ J}$

8-23 Let Ken be object A and Kim be object B.

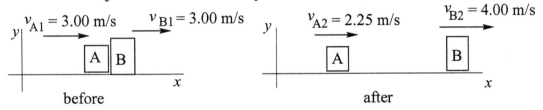

P_x constant implies $m_A v_{A1x} + m_B v_{B1x} = m_A v_{A2x} + m_B v_{B2x}$

$m_A(v_{A1x} - v_{A2x}) = m_B(v_{B2x} - v_{B1x})$

$m_B = m_A \left(\dfrac{v_{A1x} - v_{A2x}}{v_{B2x} - v_{B1x}} \right)$; multiply by g and get $w_B = w_A \left(\dfrac{v_{A1x} - v_{A2x}}{v_{B2x} - v_{B1x}} \right)$

$w_B = (700 \text{ N}) \left(\dfrac{3.00 \text{ m/s} - 2.25 \text{ m/s}}{4.00 \text{ m/s} - 3.00 \text{ m/s}} \right) = (700 \text{ N})(0.75) = 525 \text{ N}$

8-25 Take the x-axis to lie along the initial velocity of A.

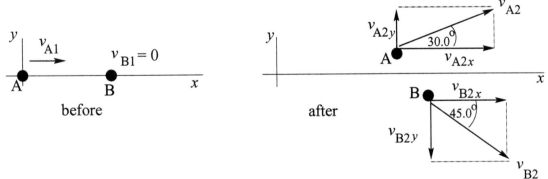

a) P_x is constant implies $m_A v_{A1x} + m_B v_{B1x} = m_A v_{A2x} + m_B v_{B2x}$

$m v_{A1} = m v_{A2} \cos 30.0° + m v_{B2} \cos 45.0°$

The mass m divides out and the equation becomes $40.0 \text{ m/s} = 0.8667 v_{A2} + 0.7071 v_{B2}$.

P_y is constant implies $m_A v_{A1y} + m_B v_{B1y} = m_A v_{A2y} + m_B v_{B2y}$

$0 = m v_{A2} \sin 30.0° - m v_{B2} \sin 45.0°$

The mass m divides out and the equation becomes $0 = 0.5000 v_{A2} - 0.7071 v_{B2}$.

Adding these two equations gives $1.366v_{A2} = 40.0$ m/s and $v_{A2} = 29.3$ m/s.

Then $v_{B2} = \left(\dfrac{0.5000}{0.7071}\right)v_{A2} = \left(\dfrac{0.5000}{0.7071}\right)(29.3$ m/s$) = 20.7$ m/s.

b) $K = K_{A1} = \frac{1}{2}mv_{A1}^2$

$K_2 = K_{A2} + K_{B2} = \frac{1}{2}mv_{A2}^2 + \frac{1}{2}mv_{B2}^2$

$\Delta K = K_2 - K_1$

The fraction of the original kinetic energy of puck A dissipated during the collision is

$$-\frac{\Delta K}{K_1} = -\frac{K_2 - K_1}{K_1} = -\left(\frac{\frac{1}{2}mv_{A2}^2 + \frac{1}{2}mv_{B2}^2 - \frac{1}{2}mv_{A1}^2}{\frac{1}{2}mv_{A1}^2}\right) = 1 - \left(\frac{v_{A2}}{v_{A1}}\right)^2 - \left(\frac{v_{B2}}{v_{A1}}\right)^2$$

$-\dfrac{\Delta K}{K_1} = 1 - ((29.3$ m/s$)/(40.0$ m/s$))^2 - ((20.7$ m/s$)/(40.0$ m/s$))^2 =$

$1 - 0.5366 - 0.2679 = 0.196;$ 19.6%.

8-29 Let the automobile be object A and the truck be object B.

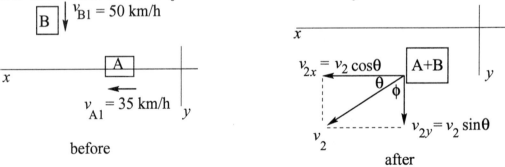

P_x is constant implies $m_A v_{A1x} + m_B v_{B1x} = (m_A + m_B)v_{2x}$

$m_A v_{A1x} = (m_A + m_B)v_{2x}$

$(1400$ kg$)(+35.0$ km/h$) = (1400$ kg $+ 2800$ kg$)v_{2x}$

$v_{2x} = 11.67$ km/h

P_y is constant implies $m_A v_{A1y} + m_B v_{B1y} = (m_A + m_B)v_{2y}$

$m_B v_{B1y} = (m_A + m_B)v_{2y}$

$(2800$ kg$)(+50.0$ km/h$) = (1400$ kg $+ 2800$ kg$)v_{2y}$

$v_{2y} = 33.33$ km/h

$v_2 = \sqrt{v_{2x}^2 + v_{2y}^2} = \sqrt{(11.67 \text{ km/h})^2 + (33.33 \text{ km/h})^2} = 35.3$ km/h

$\tan\theta = \dfrac{v_{2y}}{v_{2x}} = \dfrac{33.33 \text{ km/h}}{11.67 \text{ km/h}};$ $\theta = 70.7°$ and $\phi = 19.3°$

The wreckage is moving in the direction 19.3° west of south.

-31 Apply conservation of momentum to the collision between the bullet and the block. Let object A be the bullet and object B be the block. Let v_A be the speed of the bullet before the collision and let V be the speed of the block with the bullet inside just after the collision.

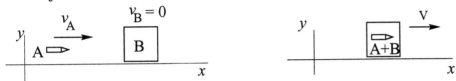

P_x is constant gives $m_A v_A = (m_A + m_B)V$

Apply conservation of energy to the motion of the block after the collision.

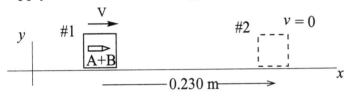

$K_1 + U_1 + W_{\text{other}} = K_2 + U_2$

Work is done by friction so $W_{\text{other}} = W_f = (f_k \cos \phi)s = -f_k s = -\mu_k mgs$

$U_1 = U_2 = 0$ (no work done by gravity)

$K_1 = \frac{1}{2}mV^2; \quad K_2 = 0$ (block has come to rest)

Thus $\frac{1}{2}mV^2 - \mu_k mgs = 0$

$V = \sqrt{2\mu_k gs} = \sqrt{2(0.20)(9.80 \text{ m/s}^2)(0.230 \text{ m})} = 0.9495 \text{ m/s}$

Use this in the conservation of momentum equation

$v_A = \left(\dfrac{m_A + m_B}{m_A}\right) V = \left(\dfrac{5.00 \times 10^{-3} \text{ kg} + 1.20 \text{ kg}}{5.00 \times 10^{-3} \text{ kg}}\right)(0.9495 \text{ m/s}) = 229 \text{ m/s}$

8-33 Elastic collision

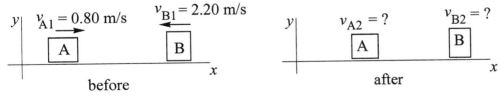

From conservation of x-component of momentum:

$m_A v_{A1x} + m_B v_{B1x} = m_A v_{A2x} + m_B v_{B2x}$

$m_A v_{A1} - m_B v_{B1} = m_A v_{A2x} + m_B v_{B2x}$

$(0.150 \text{ kg})(0.80 \text{ m/s}) - (0.300 \text{ kg})(2.20 \text{ m/s}) = (0.150 \text{ kg})v_{A2x} + (0.300 \text{ kg})v_{B2x}$

$$-3.60 \text{ m/s} = v_{A2x} + 2v_{B2x}$$

From the relative velocity equation for an elastic collision (Eq.8-27):

$$v_{B2x} - v_{A2x} = -(v_{B1x} - v_{A1x}) = -(-2.20 \text{ m/s} - 0.80 \text{ m/s}) = +3.00 \text{ m/s}$$
$$3.00 \text{ m/s} = -v_{A2x} + v_{B2x}$$

Adding the two equations gives $-0.60 \text{ m/s} = 3v_{B2x}$ and $v_{B2x} = -0.20 \text{ m/s}$.
Then $v_{A2x} = v_{B2x} - 3.00 \text{ m/s} = -3.20 \text{ m/s}$.

The 0.150 kg glider (A) is moving to the left at 3.20 m/s and the 0.300 kg glider (B) is moving to the left at 0.20 m/s,

8-37 a) Let A be the proton and B be the target nucleus.

The collision is elastic, all velocities lie along a line, and B is at rest before the collision. Hence the results of Eqs.(8-24) and (8-25) apply.

Eq.(8-24): $m_B(v + v_A) = m_A(v - v_A)$, where v is the velocity component of A before the collision and v_A is the velocity component of A after the colliison.

Here, $v = 1.50 \times 10^7 \text{ m/s}$ (take direction of incident beam to be positive) and $v_A = -1.20 \times 10^7 \text{ m/s}$ (negative since traveling in direction opposite to incident beam).

$$m_B = m_A \left(\frac{v - v_A}{v + v_A} \right) = m \left(\frac{1.50 \times 10^7 \text{ m/s} + 1.20 \times 10^7 \text{ m/s}}{1.50 \times 10^7 \text{ m/s} - 1.20 \times 10^7 \text{ m/s}} \right) =$$
$$m \left(\frac{2.70}{0.30} \right) = 9.00m$$

b) Eq.(8-25): $v_b = \left(\frac{2m_A}{m_A + m_B} \right) v = \left(\frac{2m}{m + 9.00m} \right) (1.50 \times 10^7 \text{ m/s}) =$
$3.00 \times 10^6 \text{ m/s}$

8-39 Apply Eq.(8-28) with the sun as mass 1 and Jupiter as mass 2.
Take the origin at the sun and let Jupiter lie on the positive x-axis.

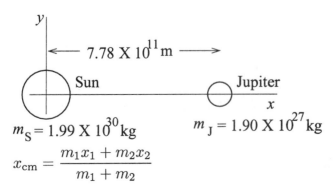

$$x_{cm} = \frac{m_1 x_1 + m_2 x_2}{m_1 + m_2}$$

$x_1 = 0$ and $x_2 = 7.78 \times 10^{11}$ m

$$x_{cm} = \frac{(1.90 \times 10^{27} \text{ kg})(7.78 \times 10^{11} \text{ m})}{1.99 \times 10^{30} \text{ kg} + 1.90 \times 10^{27} \text{ kg}} = 7.42 \times 10^8 \text{ m}$$

The center of mass is 7.42×10^8 m from the center of the sun and is on the line connecting the centers of the sun and Jupiter. The sun's radius is 6.96×10^8 m so the center of mass lies just outside the sun.

8-41 $x_{cm} = 2.0$ m, $y_{cm} = 0$; $\vec{v}_{cm} = (5.0 \text{ m/s})\hat{j}$

$$x_{cm} = \frac{m_1 x_1 + m_2 x_2}{m_1 + m_2} = \frac{m_1(0) + (0.10 \text{ kg})(8.0 \text{ m})}{m_1 + (0.10 \text{ kg})} = \frac{0.80 \text{ kg} \cdot \text{m}}{m_1 + 0.10 \text{ kg}}$$

$x_{cm} = 2.0$ m gives $2.0 \text{ m} = \dfrac{0.80 \text{ kg} \cdot \text{m}}{m_1 + 0.10 \text{ kg}}$

$m_1 + 0.10 \text{ kg} = \dfrac{0.80 \text{ kg} \cdot \text{m}}{2.0 \text{ m}} = 0.40$ kg

$m_1 = 0.30$ kg

b) $\vec{P} = M\vec{v}_{cm} = (0.10 \text{ kg} + 0.30 \text{ kg})(5.0 \text{ m/s})\hat{i} = (2.0 \text{ kg} \cdot \text{m/s})\hat{i}$

c) $v_{cm} = \dfrac{m_1 \vec{v}_1 + m_2 \vec{v}_2}{m_1 + m_2}$

particle 2 at rest says $v_2 = 0$

Then $\vec{v}_1 = \left(\dfrac{m_1 + m_2}{m_1}\right)\vec{v}_{cm} = \left(\dfrac{0.30 \text{ kg} + 0.10 \text{ kg}}{0.30 \text{ kg}}\right)(5.00 \text{ m/s})\hat{i} = (6.7 \text{ m/s})\hat{i}$.

8-43 **a)** $x_{cm} = \dfrac{m_1 x_1 + m_2 x_2}{m_1 + m_2}$

$m_1 + m_2 = \dfrac{m_1 x_1 + m_2 x_2}{x_{cm}} = \dfrac{m_1(0) + (0.50 \text{ kg})(6.0 \text{ m})}{2.4 \text{ m}} = 1.25$ kg

b) $\vec{a}_{cm} = \dfrac{d\vec{v}_{cm}}{dt} = (1.5 \text{ m/s}^3)t\hat{i}$

c) $\sum \vec{F}_{ext} = M\vec{a}_{cm} = (1.25 \text{ kg})(1.5 \text{ m/s}^3)t\hat{i}$

At $t = 3.0$ s, $\sum \vec{F}_{ext} = (1.25 \text{ kg})(1.5 \text{ m/s}^3)(3.0 \text{ s})\hat{i} = (5.6 \text{ N})\hat{i}$.

8-47 Apply Eq.(8-39): $a = -\dfrac{v_{ex}}{m}\dfrac{dm}{dt}$

$$\frac{dm}{dt} = -\frac{ma}{v_{ex}} = -\frac{(6000 \text{ kg})(25.0 \text{ m/s}^2)}{2000 \text{ m/s}} = -75.0 \text{ kg/s}$$

So in 1 s the rocket must eject 75.0 kg of gas.

8-51 Use Eq.(8-40): $v - v_0 = v_{ex} \ln(m_0/m)$

$v_0 = 0$ ("fired from rest"), so $v/v_{ex} = \ln(m_0/m)$

Thus $m_0/m = e^{v/v_{ex}}$, or $m/m_0 = e^{-v/v_{ex}}$

If v is the final speed then m is the mass left when all the fuel has been expended m/m_0 is the fraction of the initial mass that is not fuel.

a) $v = 1.00 \times 10^{-3}c = 3.00 \times 10^5$ m/s gives

$m/m_0 = e^{-(3.00 \times 10^5 \text{ m/s})/(2000 \text{ m/s})} = 7.2 \times 10^{-66}$

This is clearly not feasible, for so little of the initial mass to not be fuel.

b) $v = 3000$ m/s gives $m/m_0 = e^{-(3000 \text{ m/s})/(2000 \text{ m/s})} = 0.223$

22.3% of the total initial mass not fuel, so 77.7% is fuel; this is possible.

8-55 Let $\vec{p}_e$, $\vec{p}_{\bar{v}}$ and $\vec{p}_N$ be the final momenta of the electron, the antineutrino, and the recoiling ^{210}Po nucleus.

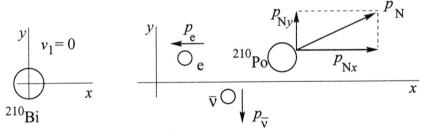

before   after

a) P_x is conserved and initially $P_x = 0$ implies $0 = p_{Nx} - p_e$

$p_{Nx} = p_e = 3.60 \times 10^{-22}$ kg · m/s

P_y is conserved and initially $P_y = 0$ implies $0 = p_{Ny} - p_{\bar{v}}$

$p_{Ny} = p_{\bar{v}} = 5.20 \times 10^{-22}$ kg · m/s

Then $p_N = \sqrt{p_{Nx}^2 + p_{Ny}^2} = \sqrt{(3.60 \times 10^{-22})^2 + (5.20 \times 10^{-22})^2}$ kg · m/s $=$

6.32×10^{-22} kg · m/s

b) $K_N = \frac{1}{2} m_N v_N^2$ and $p_N = m_N v_N$ implies $K_N = \dfrac{p_N^2}{2m_N}$

$$K_N = \frac{(6.32 \times 10^{-22} \text{ kg} \cdot \text{m/s})^2}{2(3.50 \times 10^{-25} \text{ kg})} = 5.71 \times 10^{-19} \text{ J}$$

Problems

8-57 $\vec{F} = (\alpha t^2)\hat{i} + (\beta + \gamma t)\hat{j}$; $\alpha = 25.0$ N/s², $\beta = 30.0$ N, $\gamma = 5.0$ N/s

$J_x = \int_{t_1}^{t_2} F_x(t)\, dt = \int_0^{t_2} (\alpha t^2)\, dt = \frac{1}{3}\alpha t_2^3 = \frac{1}{3}(25.0 \text{ N/s}^2)(0.500 \text{ s})^3 = 1.042 \text{ N}\cdot\text{s}$

$J_y = \int_{t_1}^{t_2} F_y(t)\, dt = \int_0^{t_2} (\beta + \gamma t)\, dt = (\beta t_2 + \frac{1}{2}\gamma t_2^2)$

$J_y = (30.0 \text{ N})(0.500 \text{ s}) + \frac{1}{2}(5.00 \text{ N/s})(0.500 \text{ s})^2 = 15.62 \text{ N}\cdot\text{s}$

$J_x = p_{2x} - p_{1x} = m(v_{2x} - v_{1x}) = mv_{2x}$, since $v_{1x} = 0$

$v_{2x} = \dfrac{J_x}{m} = \dfrac{1.042 \text{ N}\cdot\text{s}}{2.00 \text{ kg}} = 0.521 \text{ m/s}$

$J_y = p_{2y} - p_{1y} = m(v_{2y} - v_{1y}) = mv_{2y}$, since $v_{1y} = 0$

$v_{2y} = \dfrac{J_y}{m} = \dfrac{15.62 \text{ N}\cdot\text{s}}{2.00 \text{ kg}} = 7.81 \text{ m/s}$

Thus $\vec{v} = (0.521 \text{ m/s})\hat{i} + (7.81 \text{ m/s})\hat{j}$.

8-63 Use a coordinate system attached to the ground. Take the x-axis to be east (along the tracks) and the y-axis to be north (parallel to the ground and perpendicular to the tracks). Then P_x is conserved and P_y is <u>not</u> conserved, due to the sideways force exerted by the tracks, the force that keeps the handcar on the tracks.

a) Let A be the 25.0 kg mass and B be the car (mass 175 kg). After the mass is thrown sideways relative to the car it still has the same eastward component of velocity, 5.00 m/s, as it had before it was thrown.

P_x is conserved so $(m_A + m_B)v_1 = m_A v_{A2x} + m_B v_{B2x}$

$(200 \text{ kg})(5.00 \text{ m/s}) = (25.0 \text{ kg})(5.00 \text{ m/s}) + (175 \text{ kg})v_{B2x}$

$v_{B2x} = \dfrac{1000 \text{ kg}\cdot\text{m/s} - 125 \text{ kg}\cdot\text{m/s}}{175 \text{ kg}} = 5.00 \text{ m/s}.$

The final velocity of the car is 5.00 m/s, east (unchanged).

b) We need the final velocity of A relative to the ground.

$$\vec{v}_{A/E} = \vec{v}_{A/B} + \vec{v}_{B/E}$$

$v_{B/E} = +5.00$ m/s

$v_{A/B} = -5.00$ m/s (minus since the mass is moving west relative to the car). This gives $v_{A/E} = 0$; the mass is at rest relative to the earth after it is thrown backward from the car.

As in part (a), $(m_A + m_B)v_1 = m_A v_{A2x} + m_B v_{B2x}$.

Now $v_{A2x} = 0$, so $(m_A + m_B)v_1 = m_B v_{B2x}$.

$$v_{B2x} = \left(\frac{m_A + m_B}{m_B}\right) v_1 = \left(\frac{200 \text{ kg}}{175 \text{ kg}}\right)(5.00 \text{ m/s}) = 5.71 \text{ m/s}.$$

The final velocity of the car is 5.71 m/s, east.

c) Let A be the 25.0 kg mass and B be the car (mass $m_B = 200$ kg).

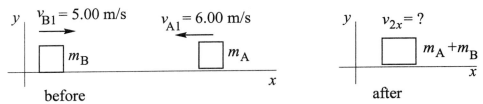

P_x is conserved so $m_A v_{A1x} + m_B v_{B1x} = (m_A + m_B)v_{2x}$

$-m_A v_{A1} + m_B v_{B1} = (m_A + m_B)v_{2x}$

$$v_{2x} = \frac{m_B v_{B1} - m_A v_{A1}}{m_A + m_B} = \frac{(200 \text{ kg})(5.00 \text{ m/s}) - (25.0 \text{ kg})(6.00 \text{ m/s})}{200 \text{ kg} + 25.0 \text{ kg}} = 3.78 \text{ m/s}.$$

The final velocity of the car is 3.78 m/s, east.

8-67 Use the free-body diagram for the frame when it hangs at rest on the end of the spring to find the force constant k of the spring. Let s be the amount the spring is stretched.

$a = 0$ ks (the spring force)

$\sum F_y = ma_y$

$-mg + ks = 0$

mg

$$k = \frac{mg}{s} = \frac{(0.150 \text{ kg})(9.80 \text{ m/s}^2)}{0.050 \text{ m}} = 29.4 \text{ N/m}$$

Next find the speed of the putty when it reaches the frame. The putty falls with acceleration $a = g$, downward.

$$v^2 = v_0^2 + 2a(y - y_0)$$

$$v = \sqrt{2a(y - y_0)} = \sqrt{2(9.80 \text{ m/s}^2)(0.300 \text{ m})} = 2.425 \text{ m/s}$$

Apply conservation of momentum to the collision between the putty (A) and the frame (B):

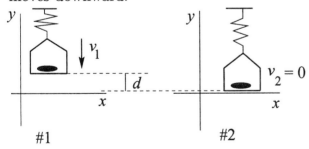

before   after

P_y is conserved, so $-m_A v_{A1} = -(m_A + m_B)v_2$

$$v_2 = \left(\frac{m_A}{m_A + m_B}\right) v_{A1} = \left(\frac{0.200 \text{ kg}}{0.350 \text{ kg}}\right)(2.425 \text{ m/s}) = 1.386 \text{ m/s}$$

Apply conservation of energy to the motion of the frame on the end of the spring after the collision. Let point 1 be just after the putty strikes and point 2 be when the frame has its maximum downward displcement. Let d be the amount the frame moves downward.

When the frame is at position 1 the spring is stretched a distance $x_1 = 0.050$ m. When the frame is at position 2 the spring is stretched a distance $x_2 = 0.050$ m $+ d$. Use coordinates with the y-direction upward and $y = 0$ at the lowest point reached by the frame, so that $y_1 = d$ and $y_2 = 0$. Work is done on the frame by gravity and by the spring force, so $W_{\text{other}} = 0$, and $U = U_{\text{el}} + U_{\text{gravity}}$.

$$K_1 + U_1 + W_{\text{other}} = K_2 + U_2$$

$$W_{\text{other}} = 0$$

$K_1 = \frac{1}{2}mv_1^2 = \frac{1}{2}(0.350 \text{ kg})(1.386 \text{ m/s})^2 = 0.3362 \text{ J}$

$U_1 = U_{1,\text{el}} + U_{1,\text{grav}} = \frac{1}{2}kx_1^2 + mgy_1 = \frac{1}{2}(29.4 \text{ N/m})(0.050 \text{ m})^2 + (0.350 \text{ kg})(9.80 \text{ m/s}^2)$

$U_1 = 0.03675 \text{ J} + (3.43 \text{ N})d$

$U_2 = U_{2,\text{el}} + U_{2,\text{grav}} = \frac{1}{2}kx_2^2 + mgy_2 = \frac{1}{2}(29.4 \text{ N/m})(0.050 \text{ m} + d)^2$

$U_2 = 0.03675 \text{ J} + (1.47 \text{ N})d + (14.7 \text{ N/m})d^2$

Thus $0.3362 \text{ J} + 0.03675 \text{ J} + (3.43 \text{ N})d = 0.03675 \text{ J} + (1.47 \text{ N})d + (14.7 \text{ N/m})d^2$

$(14.7 \text{ N/m})d^2 - (1.96 \text{ N})d - 0.3362 \text{ J} = 0$

$d = (1/29.4)[1.96 \pm \sqrt{(1.96)^2 - 4(14.7)(-0.3362)}] \text{ m} = 0.0667 \text{ m} \pm 0.1653 \text{ m}.$

The solution we want is a positive (downward) distance, so $d = 0.0667 \text{ m} + 0.1653 \text{ m} = 0.232 \text{ m}.$

8-69 Let A be the bullet and B be the stone.

a)

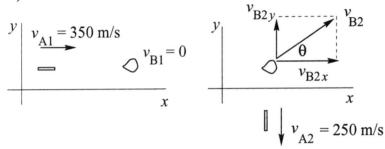

P_x is conserved so $m_A v_{A1x} + m_B v_{B1x} = m_A v_{A2x} + m_B v_{B2x}$

$m_A v_{A1} = m_B v_{B2x}$

$v_{B2x} = \left(\frac{m_A}{m_B}\right)v_{A1} = \left(\frac{6.00 \times 10^{-3} \text{ kg}}{0.100 \text{ kg}}\right)(350 \text{ m/s}) = 21.0 \text{ m/s}$

P_y is conserved so $m_A v_{A1y} + m_B v_{B1y} = m_A v_{A2y} + m_B v_{B2y}$

$0 = -m_A v_{A2} + m_B v_{B2y}$

$v_{B2y} = \left(\frac{m_A}{m_B}\right)v_{A2} = \left(\frac{6.00 \times 10^{-3} \text{ kg}}{0.100 \text{ kg}}\right)(250 \text{ m/s}) = 15.0 \text{ m/s}$

$v_{B2} = \sqrt{v_{B2x}^2 + v_{B2y}^2} = \sqrt{(21.0 \text{ m/s})^2 + (15.0 \text{ m/s})^2} = 25.8 \text{ m/s}$

$\tan\theta = \frac{v_{B2y}}{v_{B2x}} = \frac{15.0 \text{ m/s}}{21.0 \text{ m/s}} = 0.7143; \quad \theta = 35.5° \text{ (defined in the sketch)}$

b) To answer this question compare K_1 and K_2 for the system:

$K_1 = \frac{1}{2}m_A v_{A1}^2 + \frac{1}{2}m_B v_{B1}^2 = \frac{1}{2}(6.00 \times 10^{-3} \text{ kg})(350 \text{ m/s})^2 = 368 \text{ J}$

$K_2 = \frac{1}{2}m_A v_{A2}^2 + \frac{1}{2}m_B v_{B2}^2 = \frac{1}{2}(6.00 \times 10^{-3} \text{ kg})(250 \text{ m/s})^2 + \frac{1}{2}(0.100 \text{ kg})(25.8 \text{ m/s})^2 =$

221 J

$\Delta K = K_2 - K_1 = 221\ \text{J} - 368\ \text{J} = -147\ \text{J}$

The kinetic energy of the system decreases by 147 J as a result of the collision; the collision is not elastic.

8-71 Apply conservation of momentum to the collision between the bullet and the block and apply conservation of energy to the motion of the block after the collision.

Collision between the bullet and the block: Let object A be the bullet and object B be the block. Apply momentum conservation to find the speed v_{B2} of the block just after the collision.

before   after

P_x is conserved so $m_A v_{A1x} + m_B v_{B1x} = m_A v_{A2x} + m_B v_{B2x}$

$m_A v_{A1} = m_A v_{A2} + m_B v_{B2x}$

$$v_{B2x} = \frac{m_A(v_{A1} - v_{A2})}{m_B} = \frac{4.00 \times 10^{-3}\ \text{kg}(400\ \text{m/s} - 120\ \text{m/s})}{0.800\ \text{kg}} = 1.40\ \text{m/s}$$

Motion of the block after the collision:

Let point 1 in the motion be just after the collision, where the block has the speed 1.40 m/s calculated above, and let point 2 be where the block has come to rest.

$K_1 + U_1 + W_{\text{other}} = K_2 + U_2$

Work is done on the block by friction, so $W_{\text{other}} = W_f$.

$W_{\text{other}} = W_f = (f_k \cos\phi)s = -f_k s = -\mu_k mgs$, where $s = 0.450\ \text{m}$

$U_1 = 0,\quad U_2 = 0$

$K_1 = \frac{1}{2}mv_1^2,\quad K_2 = 0$ (block has come to rest)

Thus $\frac{1}{2}mv_1^2 - \mu_k mgs = 0$.

$$\mu_k = \frac{v_1^2}{2gs} = \frac{(1.40\ \text{s})^2}{2(9.80\ \text{m/s}^2)(0.450\ \text{m})} = 0.222$$

b) For the bullet,

$K_1 = \frac{1}{2}mv_1^2 = \frac{1}{2}(4.00 \times 10^{-3}\ \text{kg})(400\ \text{m/s})^2 = 320\ \text{J}$

$K_2 = \frac{1}{2}mv_2^2 = \frac{1}{2}(4.00 \times 10^{-3}\text{ kg})(120\text{ m/s})^2 = 28.8\text{ J}$

$\Delta K = K_2 - K_1 = 28.8\text{ J} - 320\text{ J} = -291\text{ J}$

The kinetic energy of the bullet decreases by 291 J.

c) Immediately after the collision the speed of the block is 1.40 m/s so its kinetic energy is $K = \frac{1}{2}mv^2 = \frac{1}{2}(0.800\text{ kg})(1.40\text{ m/s})^2 = 0.784\text{ J}$.

Note that the collision is highly inelastic. The bullet loses 291 J of kinetic energy but only 0.784 J is gained by the block. But momentum is conserved in the collision. All the momentum lost by the bullet is gained by the block.

8-75 a) $K = \frac{1}{2}m_A v_A^2 + \frac{1}{2}m_B v_B^2$

Note $\vec{v}'_A$ and $\vec{v}'_B$ as defined in the problem are the velocities of A and B in coordinates moving with the center of mass. Note also that $m_A \vec{v}'_A + m_B \vec{v}'_B = M\vec{v}'_{cm}$ where $\vec{v}'_{cm}$ is the velocity of the car in these coordinates. But that's zero, so $m_A \vec{v}'_A + m_B \vec{v}'_B = 0$; we can use this in the proof.

$\vec{v}_A = \vec{v}'_A + \vec{v}_{cm}$, so $v_A^2 = v_A'^2 + v_{cm}^2 + 2\vec{v}'_A \cdot \vec{v}_{cm}$

$\vec{v}_B = \vec{v}'_B + \vec{v}_{cm}$, so $v_B^2 = v_B'^2 + v_{cm}^2 + 2\vec{v}'_B \cdot \vec{v}_{cm}$

(We have used that for a vector $\vec{A}$, $A^2 = \vec{A} \cdot \vec{A}$.)

Thus $K = \frac{1}{2}m_A v_A'^2 + \frac{1}{2}m_A v_{cm}^2 + m_A \vec{v}'_A \cdot \vec{v}_{cm} + \frac{1}{2}m_B v_B'^2 + \frac{1}{2}m_B v_{cm}^2 + m_B \vec{v}'_B \cdot \vec{v}_{cm}$

$K = \frac{1}{2}(m_A + m_B)v_{cm}^2 + \frac{1}{2}(m_A v_A'^2 + m_B v_B'^2) + (m_A \vec{v}'_A + m_B \vec{v}'_B) \cdot \vec{v}_{cm}$

But $m_A + m_B = M$ and as noted earlier $m_A \vec{v}'_A + m_B \vec{v}'_B = 0$, so

$K = \frac{1}{2}Mv_{cm}^2 + \frac{1}{2}(m_A v_A'^2 + m_B v_B'^2)$. This is the result the problem asked us to derive.

b) In the collision $\vec{P} = M\vec{v}_{cm}$ is constant, so $\frac{1}{2}Mv_{cm}^2$ stays constant. The asteroids can lose all their relative kinetic energy but the $\frac{1}{2}Mv_{cm}^2$ must remain.

8-79

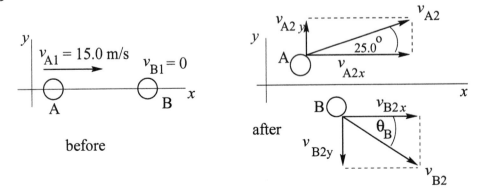

The result of Problem 8-78 part (d) applies here and says that $25.0° + \theta_B = 90.0°$ so that $\theta_B = 65.0°$. (A and B move off in perpendicular directions.)

P_x is conserved so $m_A v_{A1x} + m_B v_{B1x} = m_A v_{A2x} + m_B v_{B2x}$

But $m_A = m_B$ so $v_{A1} = v_{A2} \cos 25.0° + v_{B2} \cos 65.0°$

P_y is conserved so $m_A v_{A1y} + m_B v_{B1y} = m_A v_{A2y} + m_B v_{B2y}$

$0 = v_{A2y} + v_{B2y}$

$0 = v_{A2} \sin 25.0° + v_{B2} \sin 65.0°$

$v_{B2} = (\sin 25.0° / \sin 65.0°) v_{A2}$

This result in the first equation gives $v_{A1} = v_{A2} \cos 25.0° + \left(\dfrac{\sin 25.0° \cos 65.0°}{\sin 65.0°} \right) v_{A2}$

$v_{A1} = 1.103 v_{A2}$

$v_{A2} = v_{A1} / 1.103 = 15.0 \text{ m/s} / 1.103 = 13.6 \text{ m/s}$

And then $v_{B2} = (\sin 25.0° / \sin 65.0°)(13.6 \text{ m/s}) = 6.34 \text{ m/s}$.

8-83 Use coordinates fixed to the ice, with the direction you walk as the x-direction. $\vec{v}_{cm}$ is constant and initially $\vec{v}_{cm} = 0$.

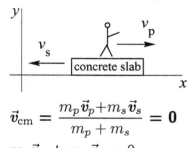

$\vec{v}_{cm} = \dfrac{m_p \vec{v}_p + m_s \vec{v}_s}{m_p + m_s} = \mathbf{0}$

$m_p \vec{v}_p + m_s \vec{v}_s = 0$

$m_p v_{px} + m_s v_{sx} = 0$

$v_{sx} = -(m_p / m_s) v_{px} = -(m_p / 5 m_p) 2.00 \text{ m/s} = -0.400 \text{ m/s}$

The slab moves at 0.400 m/s, in the direction opposite to the direction you are walking.

8-85 Apply momentum conservation in the x and y directions:

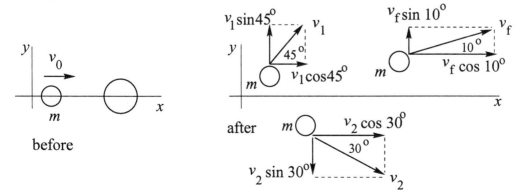

P_x is conserved so $m v_0 = m(v_1 \cos 45° + v_f \cos 10° + v_2 \cos 30°)$

$v_0 - v_f \cos 10° = v_1 \cos 45° + v_2 \cos 30°$

$1030.4 \text{ m/s} = v_1 \cos 45° + v_2 \cos 30°$

P_x is conserved so $0 = m(v_1 \sin 45° - v_2 \sin 30° + v_f \sin 10°)$
$v_1 \sin 45° = v_2 \sin 30° - 347.3 \text{ m/s}$

$\sin 45° = \cos 45°$ so
$1030.4 \text{ m/s} = v_2 \sin 30° - 347.3 \text{ m/s} + v_2 \cos 30°$

$$v_2 = \frac{1030.4 \text{ m/s} + 347.3 \text{ m/s}}{\sin 30° + \cos 30.0°} = 1010 \text{ m/s}$$

And then $v_1 = \dfrac{v_2 \sin 30° - 347.3 \text{ m/s}}{\sin 45°} = 223 \text{ m/s}.$

The two emitted neutrons have speeds of 223 m/s and 1010 m/s.

The speeds of the Ba and Kr nuclei are related by P_z conservation. P_z is constant implies that $0 = m_{\text{Ba}} v_{\text{Ba}} - m_{\text{Kr}} v_{\text{Kr}}$

$$v_{\text{Kr}} = \left(\frac{m_{\text{Ba}}}{m_{\text{Kr}}} \right) v_{\text{Ba}} = \left(\frac{2.3 \times 10^{-25} \text{ kg}}{1.5 \times 10^{-25} \text{ kg}} \right) v_{\text{Ba}} = 1.5 v_{\text{Ba}}.$$

We can't say what these speeds are but they must satisfy this relation. The value of v_{Ba} depends on energy considerations.

8-87 a) Objects stick together says that the relative speed after the collision is zero, so $\epsilon = 0$.

b) In an elastic collision the relative velociy of the two bodies has the same magnitude before and after the collision, so $\epsilon = 1$.

c) Speed of ball just before collsion: $mgh = \frac{1}{2} m v_1^2$
$v_1 = \sqrt{2gh}$

Speed of ball just after collision: $mgH_1 = \frac{1}{2} m v_2^2$
$v_2 = \sqrt{2gH_1}$

The second object (the surface) is stationary, so $\epsilon = v_2/v_1 = \sqrt{H_1/h}.$

d) $\epsilon = \sqrt{H_1/h}$ implies $H_1 = h\epsilon^2 = (1.2 \text{ m})(0.85)^2 = 0.87 \text{ m}$

e) $H_1 = h\epsilon^2$
$H_2 = h_1 \epsilon^2 = h\epsilon^4$
$H_3 = H_2 \epsilon^2 = (h\epsilon^4)\epsilon^2 = h\epsilon^6$
Generalize to $H_n = H_{n-1}\epsilon^2 = h\epsilon^{2(n-1)}\epsilon^2 = h\epsilon^{2n}$

f) 8th bounce implies $n = 8$

$H_8 = h\epsilon^{16} = 1.2 \text{ m}(0.85)^{16} = 0.089 \text{ m}$

8-91 **a)** Eq.(8-40): $v - v_0 = v_{ex} \ln(m_0/m)$

$v_0 = 0$ so $v = v_{ex} \ln(m_0/m)$

The total initial mass of the rocket is $m_0 = 12{,}000 \text{ kg} + 1000 \text{ kg} = 13{,}000 \text{ kg}$. Of this, $9000 \text{ kg} + 700 \text{ kg} = 9700 \text{ kg}$ is fuel, so the mass m left after all the fuel is burned is $13{,}000 \text{ kg} - 9700 \text{ kg} = 3300 \text{ kg}$.

$v = v_{ex} \ln(13{,}000 \text{ kg}/3300 \text{ kg}) = 1.37 v_{ex}$

b) First stage: $v = v_{ex} \ln(m_0/m)$

$m_0 = 13{,}000 \text{ kg}$

The first stage has 9000 kg of fuel, so the mass left after the first stage fuel has burned is $13{,}000 \text{ kg} - 9000 \text{ kg} = 4000 \text{ kg}$.

$v = v_{ex} \ln(13{,}000 \text{ kg}/4000 \text{ kg}) = 1.18 v_{ex}$

c) Second stage:

$m_0 = 1000 \text{ kg}, \quad m = 1000 \text{ kg} - 700 \text{ kg} = 300 \text{ kg}$

$v = v_0 + v_{ex} \ln(m_0/m) = 1.18 v_{ex} + v_{ex} \ln(1000 \text{ kg}/300 \text{ kg}) = 2.38 v_{ex}$

d) $v = 7.00 \text{ km/s}$

$v_{ex} = v/2.38 = (7.00 \text{ km/s})/2.38 = 2.94 \text{ km/s}$

CHAPTER 9
ROTATION OF RIGID BODIES

Exercises 5, 7, 9, 11, 17, 19, 23, 25, 29, 31, 33, 35, 41, 43, 45, 47, 49, 53
Problems 57, 59, 61, 63, 65, 69, 71, 73, 81

Exercises

9-5 $\theta = \gamma t + \beta t^3$; $\gamma = 0.400$ rad/s, $\beta = 0.0120$ rad/s^3

a) $\omega = \dfrac{d\theta}{dt} = \gamma + 3\beta t^2$

b) At $t = 0$, $\omega = \gamma = 0.400$ rad/s

c) At $t = 5.00$ s, $\omega = 0.400$ rad/s $+ 3(0.0120$ rad/s$^2)(5.00$ s$)^2 = 1.30$ rad/s

$$\omega_{av} = \frac{\Delta\theta}{\Delta t} = \frac{\theta_2 - \theta_1}{t_2 - t_1}$$

For $t_1 = 0$, $\theta_1 = 0$.

For $\theta_2 = 5.00$ s, $\theta_2 = (0.400$ rad/s$)(5.00$ s$) + (0.012$ rad/s$^3)(5.00$ s$)^3 = 3.50$ rad

So $\omega_{av} = \dfrac{3.50 \text{ rad} - 0}{5.00 \text{ s} - 0} = 0.700$ rad/s.

ω at 5.00 s is larger than ω_{av} for the time interval 0 to 5.00 s. The angular velocit
is larger than its average value during the interval.

9-7 **a)** $\theta = a + bt^2 - ct^3$

$$\omega = \frac{d\theta}{dt} = 2bt - 3ct^2$$

$$\alpha = \frac{d\omega}{dt} = 2b - 6ct$$

b) The angular velocity is instantaneously not changing when $\alpha = 0$.
$\alpha = 0$ implies $2b - 6ct = 0$ and $t = 2b/6c = b/3c$.

9-9 **a)** $\omega_0 = (500$ rev/min$)(1$ min/60 s$) = 8.333$ rev/s
$\omega_0 = (200$ rev/min$)(1$ min/60 s$) = 3.333$ rev/s, $t = 4.00$ s, $\alpha = ?$
$\omega = \omega_0 + \alpha t$
$$\alpha = \frac{\omega - \omega_0}{t} = \frac{3.333 \text{ rev/s} - 8.333 \text{ rev/s}}{4.00 \text{ s}} = -1.25 \text{ rev/s}^2$$

$\theta - \theta_0 = ?$

$\theta - \theta_0 = \omega_0 t + \frac{1}{2}\alpha t^2 = (8.333 \text{ rev/s})(4.00 \text{ s}) + \frac{1}{2}(-1.25 \text{ rev/s}^2)(4.00 \text{ s})^2 = 23.3 \text{ rev}$

b) $\omega = 0$ (comes to rest); $\quad \omega_0 = 3.333 \text{ rev/s}; \quad \alpha = -1.25 \text{ rev/s}^2; \quad t = ?$

$\omega = \omega_0 + \alpha t$

$t = \dfrac{\omega - \omega_0}{\alpha} = \dfrac{0 - 3.333 \text{ rev/s}}{-1.25 \text{ rev/s}^2} = 2.67 \text{ s}$

9-11 a) $\alpha = 1.50 \text{ rad/s}^2; \quad \omega_0 = 0$ (starts from rest); $\quad \omega = 36.0 \text{ rad/s}; \quad t = ?$

$\omega = \omega_0 + \alpha t$

$t = \dfrac{\omega - \omega_0}{\alpha} = \dfrac{36.0 \text{ rad/s} - 0}{1.50 \text{ rad/s}^2} = 24.0 \text{ s}$

b) $\theta - \theta_0 = ?$

$\theta - \theta_0 = \omega_0 t + \frac{1}{2}\alpha t^2 = 0 + \frac{1}{2}(1.50 \text{ rad/s}^2)(24.0 \text{ s})^2 = 432 \text{ rad}$

$\theta - \theta_0 = 432 \text{ rad}(1 \text{ rev}/2\pi \text{ rad}) = 68.8 \text{ rev}$

9-17 a) Consider the motion from $t = 0$ to $t = 2.00$ s:

$\theta - \theta_0 = ?; \quad \omega_0 = 24.0 \text{ rad/s}; \quad \alpha = 30.0 \text{ rad/s}^2; \quad t = 2.00 \text{ s}$

$\theta - \theta_0 = \omega_0 t + \frac{1}{2}\alpha t^2 = (24.0 \text{ rad/s})(2.00 \text{ s}) + \frac{1}{2}(30.0 \text{ rad/s}^2)(2.00 \text{ s})^2$

$\theta - \theta_0 = 48.0 \text{ rad} + 60.0 \text{ rad} + 108 \text{ rad}$

Total angular displacement from $t = 0$ until stops: $108 \text{ rad} + 432 \text{ rad} = 540 \text{ rad}$

Note: At $t = 2.00$ s, $\omega = \omega_0 + \alpha t = 24.0 \text{ rad/s} + (30.0 \text{ rad/s}^2)(2.00 \text{ s}) = 84.0 \text{ rad/s}$; angular speed when breaker trips.

b) Consider the motion from when the circuit trips until wheel stops. For this calculation let $t = 0$ when the breaker trips.

$t = ?; \quad \theta - \theta_0 = 432 \text{ rad}; \quad \omega = 0; \quad \omega_0 = 84.0 \text{ rad/s}$ (from part (a)

$\theta - \theta_0 = \left(\dfrac{\omega_0 + \omega}{2}\right) t$

$t = \dfrac{2(\theta - \theta_0)}{\omega_0 + \omega} = \dfrac{2(432 \text{ rad})}{84.0 \text{ rad/s} + 0} = 10.3 \text{ s}$

The wheel stops 10.3 s after the breaker trips so $2.00 \text{ s} + 10.3 \text{ s} = 12.3 \text{ s}$ from the beginning.

c) $\alpha = ?$; consider the same motion as in part (b):

$\omega = \omega_0 + \alpha t$

$$\alpha = \frac{\omega - \omega_0}{t} = \frac{0 - 84.0 \text{ rad/s}}{10.3 \text{ s}} = -8.16 \text{ rad/s}^2$$

9-19 The tangential speed of a blade tip is $v = r\omega$.

$$\omega = (90.0 \text{ rev/min}) \left(\frac{2\pi \text{ rad}}{1 \text{ rev}}\right) \left(\frac{1 \text{ min}}{60 \text{ s}}\right) = 9.425 \text{ rad/s}$$

$v = r\omega = (5.0 \text{ m})(9.425 \text{ rad/s}) = 47.1 \text{ m/s}$

The upward velocity of the entire blade has magnitude 4.00 m/s. The tangentia‍l velocity and the upward velocity of a blade tip are perpendicular, so their resultan‍t has magnitude

$$v_{\text{res}} = \sqrt{(47.1 \text{ m/s})^2 + (4.00 \text{ m/s})^2} = 47.3 \text{ m/s}$$

9-23 **a)** <u>at the start</u>

$t = 0$

flywheel starts from rest so $\omega = \omega_0 = 0$

$a_{\text{tan}} = r\alpha = (0.300 \text{ m})(0.600 \text{ rad/s}^2) = 0.180 \text{ m/s}^2$

$a_{\text{rad}} = r\omega^2 = 0$

$a = \sqrt{a_{\text{rad}}^2 + a_{\text{tan}}^2} = 0.180 \text{ m/s}^2$

b) $\underline{\theta - \theta_0 = 60°}$

$a_{\text{tan}} = r\alpha = 0.180 \text{ m/s}^2$

Calculate ω:

$\theta - \theta_0 = 60°(\pi \text{ rad}/180°) = 1.047 \text{ rad}; \quad \omega_0 = 0; \quad \alpha = 0.600 \text{ rad/s}^2; \quad \omega = ?$

$\omega^2 = \omega_0^2 + 2\alpha(\theta - \theta_0)$

$\omega = \sqrt{2\alpha(\theta - \theta_0)} = \sqrt{2(0.600 \text{ rad/s}^2)(1.047 \text{ rad})} = 1.121 \text{ rad/s}$

Then $a_{\text{rad}} = r\omega^2 = (0.300 \text{ m})(1.121 \text{ rad/s})^2 = 0.377 \text{ m/s}^2$.

$a = \sqrt{a_{\text{rad}}^2 + a_{\text{tan}}^2} = \sqrt{(0.377 \text{ m/s}^2)^2 + (0.180 \text{ m/s}^2)^2} = 0.418 \text{ m/s}^2$

c) $\underline{\theta - \theta_0 = 120°}$

$a_{\text{tan}} = r\alpha = 0.180 \text{ m/s}^2$

Calculate ω:

$\theta - \theta_0 = 120°\,(\pi \text{ rad}/180°) = 2.094 \text{ rad}; \quad \omega_0 = 0; \quad \alpha = 0.600 \text{ rad/s}^2; \quad \omega = ?$

$\omega^2 = \omega_0^2 + 2\alpha(\theta - \theta_0)$

$\omega = \sqrt{2\alpha(\theta - \theta_0)} = \sqrt{2(0.600 \text{ rad/s}^2)(2.094 \text{ rad})} = 1.585 \text{ rad/s}$

Then $a_{\text{rad}} = r\omega^2 = (0.300 \text{ m})(1.585 \text{ rad/s})^2 = 0.754 \text{ m/s}^2$.

$a = \sqrt{a_{\text{rad}}^2 + a_{\text{tan}}^2} = \sqrt{(0.754 \text{ m/s}^2)^2 + (0.180 \text{ m/s}^2)^2} = 0.775 \text{ m/s}^2$

9-25 $a_{\text{rad}} = r\omega^2$ so $r = a_{\text{rad}}/\omega^2$, where ω must be in rad/s

$a_{\text{rad}} = 3000g = 3000(9.80 \text{ m/s}^2) = 29,400 \text{ m/s}^2$

$\omega = (5000 \text{ rev/min})\left(\dfrac{1 \text{ min}}{60 \text{ s}}\right)\left(\dfrac{2\pi \text{ rad}}{1 \text{ rev}}\right) = 523.6 \text{ rad/s}$

Then $r = \dfrac{a_{\text{rad}}}{\omega^2} = \dfrac{29,400 \text{ m/s}^2}{(523.6 \text{ rad/s})^2} = 0.107 \text{ m}.$

The diameter is then 0.214 m, which is larger than 0.127 m, so the claim is <u>not</u> realistic.

9-29 a) $a_{\text{rad}} = r\omega^2$

$F_{\text{rad}} = ma_{\text{rad}} = mr\omega^2$

$\dfrac{F_{\text{rad},2}}{F_{\text{rad},1}} = \left(\dfrac{\omega_2}{\omega_1}\right)^2 = \left(\dfrac{640 \text{ rev/min}}{423 \text{ rev/min}}\right)^2 = 2.29$

(Note: since a ratio is used the untis cancel and there is no need to convert ω to rad/s.)

b) $v = r\omega$

$\dfrac{v_2}{v_1} = \dfrac{\omega_2}{\omega_1} = \dfrac{640 \text{ rev/min}}{423 \text{ rev/min}} = 1.51$

c) $v = r\omega$

$\omega = (640 \text{ rev/min})\left(\dfrac{1 \text{ min}}{60 \text{ s}}\right)\left(\dfrac{2\pi \text{ rad}}{1 \text{ rev}}\right) = 67.0 \text{ rad/s}$

Then $v = r\omega = (0.235 \text{ m})(67.0 \text{ rad/s}) = 15.7 \text{ m/s}.$

$a = r\omega^2 = (0.235 \text{ m})(67.0 \text{ rad/s})^2 = 1060 \text{ m/s}^2$

$\dfrac{a}{g} = \dfrac{1060 \text{ m/s}^2}{9.80 \text{ m/s}^2} = 108; \quad a = 108g$

9-31

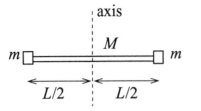

$$I = I_{\text{rod}} + 2I_{\text{cap}}$$

$$I = \tfrac{1}{12}ML^2 + 2(m)(L/2)^2 = (\tfrac{1}{12}M + \tfrac{1}{2}m)L^2$$

9-33 a)

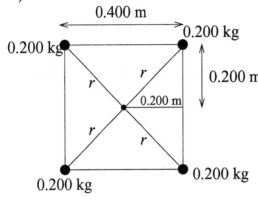

$$r = \sqrt{(0.200 \text{ m})^2 + (0.200 \text{ m})^2} = 0.2828 \text{ m}$$

$$I = \sum m_i r_i^2 = 4(0.200 \text{ kg})(0.2828 \text{ m})^2$$

$$I = 0.0640 \text{ kg} \cdot \text{m}^2$$

b)

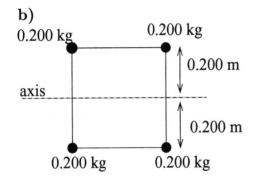

$$r = 0.200 \text{ m}$$

$$I = \sum m_i r_i^2 = 4(0.200 \text{ kg})(0.200 \text{ m})^2$$

$$I = 0.0320 \text{ kg} \cdot \text{m}^2$$

c)

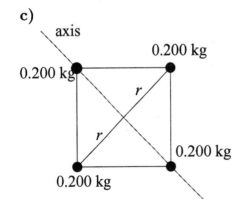

$$r = 0.2828 \text{ m}$$

$$I = \sum m_i r_i^2 = 2(0.200 \text{ kg})(0.2828 \text{ m})^2$$

$$I = 0.0320 \text{ kg} \cdot \text{m}^2$$

9-35 $I = \sum m_i r_i^2$ implies $I = I_{\text{rim}} + I_{\text{spokes}}$

$I_{\text{rim}} = MR^2 = (1.40 \text{ kg})(0.300 \text{ m})^2 = 0.126 \text{ kg} \cdot \text{m}^2$

Each spoke can be treated as a slender rod with the axis through one end, so
$I_{\text{spokes}} = 8(\frac{1}{3}ML^2) = \frac{8}{3}(0.280 \text{ kg})(0.300 \text{ m})^2 = 0.0672 \text{ kg} \cdot \text{m}^2$

$I = I_{\text{rim}} + I_{\text{spokes}} = 0.126 \text{ kg} \cdot \text{m}^2 + 0.0672 \text{ kg} \cdot \text{m}^2 = 0.193 \text{ kg} \cdot \text{m}^2$

9-41 $K = \frac{1}{2}I\omega^2$

$a_{\text{rad}} = R\omega^2$, so $\omega = \sqrt{a_{\text{rad}}/R} = \sqrt{(3500 \text{ m/s}^2)/1.20 \text{ m}} = 54.0 \text{ rad/s}$

For a disk, $I = \frac{1}{2}MR^2 = \frac{1}{2}(70.0 \text{ kg})(1.20 \text{ m})^2 = 50.4 \text{ kg} \cdot \text{m}^2$

Thus $K = \frac{1}{2}I\omega^2 = \frac{1}{2}(50.4 \text{ kg} \cdot \text{m}^2)(54.0 \text{ rad/s})^2 = 7.35 \times 10^4 \text{ J}$

9-43 a) $K = \frac{1}{2}I\omega^2, \quad \omega = 2\pi/T$

$$K = \frac{1}{2}I\left(\frac{2\pi}{T}\right)^2 = \frac{2\pi^2 I}{T^2}$$

b) $\dfrac{dK}{dt} = 2I\pi^2 \left(\dfrac{dT^{-2}}{dt}\right) = 2I\pi^2 \left(-\dfrac{2}{T^3}\right)\dfrac{dT}{dt} = -\dfrac{4\pi^2 I}{T^3}\dfrac{dT}{dt}$

c) $K = \dfrac{2\pi^2 I}{T^2} = \dfrac{2\pi^2 (8.0 \text{ kg} \cdot \text{m}^2)}{(1.5 \text{ s})^2} = 70 \text{ J}$

d) $\dfrac{dK}{dt} = -\dfrac{4\pi^2 I}{T^3}\dfrac{dT}{dt} = -\dfrac{4\pi^2 (8.0 \text{ kg} \cdot \text{m}^2)}{(1.5 \text{ s})^3}(0.0060) = -0.56 \text{ J/s}$

9-45 Eq.(9-18) says $U = Mgy_{\text{cm}}$

The positive work done by the wrestler must equal in magnitude the negative work done by gravity.

$W = -W_{\text{grav}} = U_2 - U_1 = Mg(\Delta y_{\text{cm}})$

$W = (120 \text{ kg})(9.80 \text{ m/s}^2)(0.700 \text{ m}) = 823 \text{ J}$

9-47 For a thin-walled hollow sphere, axis along a diameter, $I = \frac{2}{3}MR^2$.

For a solid sphere with mass M and radius R, $I_{\text{cm}} = \frac{2}{5}MR^2$, for an axis along a diameter.

Find d such that $I_P = I_{\text{cm}} + Md^2$ with $I_P = \frac{2}{3}MR^2$:

$\frac{2}{3}MR^2 = \frac{2}{5}MR^2 + Md^2$

The factors of M divide out and the equation bercomes $(\frac{2}{3} - \frac{2}{5})R^2 = d^2$

$d = \sqrt{(10 - 6)/15}R = 2R/\sqrt{15} = 0.516R.$

The axis is parallel to a diameter and is $0.516R$ from the center.

9-49 $I_P = I_{cm} + Md^2$

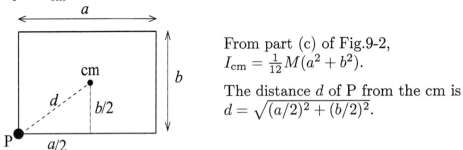

From part (c) of Fig.9-2,
$I_{cm} = \frac{1}{12}M(a^2 + b^2)$.

The distance d of P from the cm is
$d = \sqrt{(a/2)^2 + (b/2)^2}$.

Thus $I_P = I_{cm} + Md^2 = \frac{1}{12}M(a^2 + b^2) + M(\frac{1}{4}a^2 + \frac{1}{4}b^2) = (\frac{1}{12} + \frac{1}{4})M(a^2 + b^2) =$
$\frac{1}{3}M(a^2 + b^2)$

9-53 Eq.(9-20): $I = \int r^2\, dm$

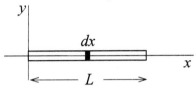

Take the x-axis to lie along the rod, with the origin at the left end. Consider a thin
slice at coordinate x and width dx. The mass per unit length for this rod is M/L
so the mass of this slice is $dm = (M/L)dx$.

$I = \int_0^L x^2(M/L)\, dx = (M/L)\int_0^L x^2\, dx = (M/L)(L^3/3) = \frac{1}{3}ML^2$
This result agrees with Table 9-2.

Problems

9-57 $\theta(t) = \gamma t^2 - \beta t^3$; $\gamma = 3.20$ rad/s^2, $\beta = 0.500$ rad/s^3

a) $\omega(t) = \dfrac{d\theta}{dt} = \dfrac{d(\gamma t^2 - \beta t^3)}{dt} = 2\gamma t - 3\beta t^2$

b) $\alpha(t) = \dfrac{d\omega}{dt} = \dfrac{d(2\gamma t - 3\beta t^2)}{dt} = 2\gamma - 6\beta t$

c) The maximum angular velocity occurs when $\alpha = 0$.

$2\gamma - 6\beta t = 0$ implies $t = \dfrac{2\gamma}{6\beta} = \dfrac{\gamma}{3\beta} = \dfrac{3.20 \text{ rad/s}^2}{3(0.500 \text{ rad/s}^3)} = 2.133$ s

At this t, $\omega = 2\gamma t - 3\beta t^2 = 2(3.20 \text{ rad/s}^2)(2.133 \text{ s}) - 3(0.500 \text{ rad/s}^3)(2.133 \text{ s})^2 =$
6.83 rad/s
The maximum positive angular velocity is 6.83 rad/s and it occurs at 2.13 s.

9-59 **a)** $\dfrac{v_{toy}}{v_{scale}} = \dfrac{L_{toy}}{L_{real}}$

$v_{toy} = v_{scale}\left(\dfrac{L_{toy}}{L_{real}}\right) = (700 \text{ km/h})\left(\dfrac{0.150 \text{ m}}{3.0 \text{ m}}\right) = 35.0 \text{ km/h}$

$v_{toy} = (35.0 \text{ km/h})(1000 \text{ m/1 km})(1 \text{ h/3600 s}) = 9.72 \text{ m/s}$

b) $K = \frac{1}{2}mv^2 = \frac{1}{2}(0.180 \text{ kg})(9.72 \text{ m/s})^2 = 8.50 \text{ J}$

c) $K = \frac{1}{2}I\omega^2$ gives that $\omega = \sqrt{\dfrac{2K}{I}} = \sqrt{\dfrac{2(8.50 \text{ J})}{4.00 \times 10^{-5} \text{ kg} \cdot \text{m}^2}} = 652 \text{ rad/s}$

9-61 **a)** $W_{tot} = K_2 - K_1$ so $K_2 = K_1 + W_{tot}$

$W_{tot} = -4000 \text{ J}$ (the amount of energy given up by the flywheel)

$K_1 = \frac{1}{2}I\omega_1^2$, but ω_1 must be in rad/s.

$\omega_1 = (300 \text{ rev/min})(2\pi \text{ rad/1 rev})(1 \text{ min/60 s}) = 31.42 \text{ rad/s}$

$K_1 = \frac{1}{2}(16.0 \text{ kg} \cdot \text{m}^2)(31.42 \text{ rad/s})^2 = 7898 \text{ J}$

Then $K_2 = K_1 + W_{tot} = 7898 \text{ J} - 4000 \text{ J} = 3898 \text{ J}$

and $K_2 = \frac{1}{2}I\omega_2^2$ gives $\omega_2 = \sqrt{\dfrac{2K_2}{I}} = \sqrt{\dfrac{2(3898 \text{ J})}{16.0 \text{ kg} \cdot \text{m}^2}} = 22.1 \text{ rad/s}$

$\omega_2 = 22.1 \text{ rad/s } (1 \text{ rev/}2\pi \text{ rad})(60 \text{ s/1 min}) = 211 \text{ rev/min.}$

b) The 4000 J of energy must be restored to the flywheel,

$P_{av} = \dfrac{\Delta W}{\Delta t} = \dfrac{4000 \text{ J}}{5.00 \text{ s}} = 800 \text{ W}$

9-63

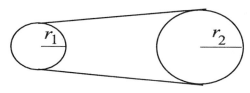

a) $v_1 = r_1\omega_1$

$\omega_1 = (60.0 \text{ rev/s})(2\pi \text{ rad/1 rev}) = 377 \text{ rad/s}$

$v_1 = r_1\omega_1 = (0.45 \times 10^{-2} \text{ m})(377 \text{ rad/s}) = 1.70 \text{ m/s}$

b) $v_1 = v_2$

$r_1\omega_1 = r_2\omega_2$

$\omega_2 = (r_1/r_2)\omega_1 = (0.45 \text{ cm/2.00 cm})(377 \text{ rad/s}) = 84.8 \text{ rad/s}$

9-65 $a_{\text{rad}} = r\omega^2, \quad a_{\text{rad},1} = r\omega_1^2, \quad a_{\text{rad},2} = r\omega_2^2$

$\Delta a_{\text{rad}} = a_{\text{rad},2} - a_{\text{rad},1} = r(\omega_2^2 - \omega_1^2)$

One of the constant acceleration equations can be written

$\omega_2^2 = \omega_1^2 + 2\alpha(\theta_2 - \theta_1)$, or $\omega_2^2 - \omega_1^2 = 2\alpha(\theta_2 - \theta_1)$

Thus $\Delta a_{\text{rad}} = r 2\alpha(\theta_2 - \theta_1) = 2r\alpha(\theta_2 - \theta_1)$, as was to be shown.

b) $\alpha = \dfrac{\Delta a_{\text{rad}}}{2r(\theta_2 - \theta_1)} = \dfrac{85.0 \text{ m/s}^2 - 25.0 \text{ m/s}^2}{2(0.250 \text{ m})(15.0 \text{ rad})} = 8.00 \text{ rad/s}^2$

Then $a_{\text{tan}} = r\alpha = (0.250 \text{ m})(8.00 \text{ rad/s}^2) = 2.00 \text{ m/s}^2$

c) $K = \frac{1}{2}I\omega^2$

$K_2 = \frac{1}{2}I\omega_2^2, \ K_1 = \frac{1}{2}I\omega_1^2$

$\Delta K = K_2 - K_1 = \frac{1}{2}I(\omega_2^2 - \omega_1^2) = \frac{1}{2}I(2\alpha(\theta_2 - \theta_1)) = I\alpha(\theta_2 - \theta_1)$, as was to be shown

d) $I = \dfrac{\Delta K}{\alpha(\theta_2 - \theta_1)} = \dfrac{45.0 \text{ J} - 20.0 \text{ J}}{(8.00 \text{ rad/s}^2)(15.0 \text{ rad})} = 0.208 \text{ kg} \cdot \text{m}^2$

9-69

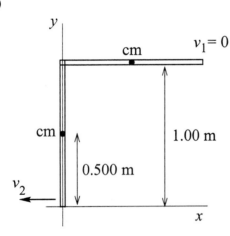

Take the origin of coordinates at the lowest point reached by the stick and take the positiv y-direction to be upward.

a) Use Eq.(9-18): $U = Mgy_{\text{cm}}$

$\Delta U = U_2 - U_1 = Mg(y_{\text{cm2}} - y_{\text{cm1}})$

The center of mass of the meter stick is at its geometrical center, so

$y_{\text{cm1}} = 1.00 \text{ m}$ and $y_{\text{cm2}} = 0.50 \text{ m}$

Then $\Delta U = (0.160 \text{ kg})(9.80 \text{ m/s}^2)(0.50 \text{ m} - 1.00 \text{ m}) = -0.784 \text{ J}$

b) Use conservation of energy: $K_1 + U_1 + W_{\text{other}} = K_2 + U_2$

Gravity is the only force that does work on the meter stick, so $W_{\text{other}} = 0$.

$K_1 = 0$.

Thus $K_2 = U_1 - U_2 = -\Delta U$, where ΔU was calculated in part (a).

$K_2 = \frac{1}{2}I\omega_2^2$ so $\frac{1}{2}I\omega_2^2 = -\Delta U$ and $\omega_2 = \sqrt{2(-\Delta U)/I}$

For stick pivoted about one end, $I = \frac{1}{3}ML^2$ where $L = 1.00$ m, so

$$\omega_2 = \sqrt{\frac{6(-\Delta U)}{ML^2}} = \sqrt{\frac{6(0.784 \text{ J})}{(0.160 \text{ kg})(1.00 \text{ m})^2}} = 5.42 \text{ rad/s}$$

c) $v = r\omega = (1.00 \text{ m})(5.42 \text{ rad/s}) = 5.42 \text{ m/s}$

d) $v_{0y} = 0; \quad y - y_0 = -1.00$ m; $\quad a_y = -9.80$ m/s²; $\quad v = ?$

$v^2 = v_{0y}^2 + 2a_y(y - y_0)$

$v = -\sqrt{2a_y(y - y_0)} = -\sqrt{2(-9.80 \text{ m/s}^2)(-1.00 \text{ m})} = -4.43 \text{ m/s}$

The magnitude of the answer in part (c) is larger.

9-71

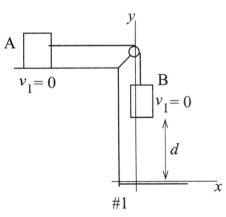 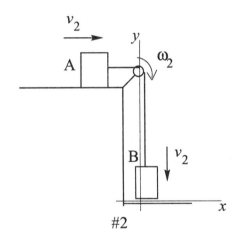

#1   #2

Use the work-energy relation $K_1 + U_1 + W_{\text{other}} = K_2 + U_2$. Use coordinates where $+y$ is upward and where the origin is at the position of block B after it has descended.

The tension in the rope does positive work on block A and negative work of the same magnitude on block B, so the net work done by the tension in the rope is zero.

Gravity does work on block B and kinetic friction does work on block A. Therefore $W_{\text{other}} = W_f = -\mu_k m_A g d$.

$K_1 = 0$ (system is released from rest)

$U_1 = m_B g y_{B1} = m_B g d; \quad U_2 = m_B g y_{B2} = 0$

$K_2 = \frac{1}{2}m_A v_2^2 + \frac{1}{2}m_B v_2^2 + \frac{1}{2}I\omega_2^2$.

But $v(\text{blocks}) = R\omega(\text{pulley})$, so $\omega_2 = v_2/R$ and

$K_2 = \frac{1}{2}(m_A + m_B)v_2^2 + \frac{1}{2}I(v_2/R)^2 = \frac{1}{2}(m_A + m_B + I/R^2)v_2^2$

Putting all this into the work-energy relation gives

$$m_B g d - \mu_k m_A g d = \tfrac{1}{2}(m_A + m_B + I/R^2)v_2^2$$
$$(m_A + m_B + I/R^2)v_2^2 = 2gd(m_B - \mu_k m_A)$$

$$v_2 = \sqrt{\frac{2gd(m_B - \mu_k m_A)}{m_A + m_B + I/R^2}}$$

9-73 The center of mass of the hoop is at its geometrical center.

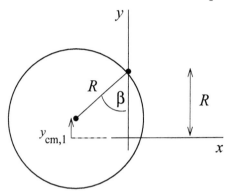

Take the origin to be at the original location of the center of the hoop, before it is rotated to one side.

$$y_{cm1} = R - R\cos\beta = R(1 - \cos\beta)$$

$y_{cm2} = 0$ (at equilibrium position hoop is at original position)

$$K_1 + U_1 + W_{other} = K_2 + U_2$$

$W_{other} = 0$ (only gravity does work)

$K_1 = 0$ (released from rest), $K_2 = \tfrac{1}{2}I\omega_2^2$

For a hoop, $I_{cm} = MR^2$, so $I = Md^2 + MR^2$ with $d = R$ and $I = 2MR^2$, for an axis at the edge. Thus $K_2 = \tfrac{1}{2}(2MR^2)\omega_2^2 = MR^2\omega_2^2$.

$$U_1 = Mgy_{cm1} = MgR(1 - \cos\beta), \qquad U_2 = mgy_{cm2} = 0$$

Thus $K_1 + U_1 + W_{other} = K_2 + U_2$ gives

$$MgR(1 - \cos\beta) = MR^2\omega_2^2 \text{ and } \omega_2 = \sqrt{g(1 - \cos\beta)/R}$$

9-81 **a)** Let L be the length of the cylinder. Divide the cylinder into thin cylindrical shells of inner radius r and outer radius $r + dr$. An end view is

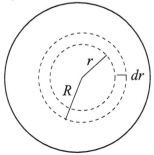

$$\rho = \alpha r$$

The mass of the thin cylindrical shell is
$$dm = \rho\, dV = \rho(2\pi r\, dr)L = 2\pi\alpha L r^2\, dr$$

$I = \int r^2 \, dm = 2\pi\alpha L \int_0^R r^4 \, dr = 2\pi\alpha L(\frac{1}{5}R^5) = \frac{2}{5}\pi\alpha L R^5$

Relate M to α:

$M = \int dm = 2\pi\alpha L \int_0^R r^2 \, dr = 2\pi\alpha L(\frac{1}{3}R^3) = \frac{2}{3}\pi\alpha L R^3$, so $\pi\alpha L R^3 = 3M/2$.

Using this in the above result for I gives
$I = \frac{2}{5}(3M/2)R^2 = \frac{3}{5}MR^2$.

b) For a cylinder of uniform density $I = \frac{1}{2}MR^2$. The answer in (a) is larger than this. Since the density increases with distance from the axis the cylinder in (a) has more mass farther from the axis than for a cylinder of uniform density.

CHAPTER 10
DYNAMICS OF ROTATIONAL MOTION

Exercises 1, 3, 5, 9, 11, 15, 19, 21, 27, 29, 31, 33, 37, 41
Problems 47, 49, 53, 55, 57, 59, 61, 65, 71, 73, 75, 77, 81

Exercises

10-1   a)

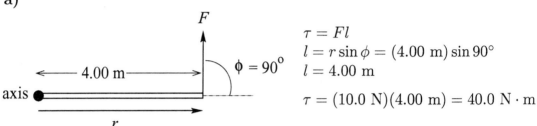

$\tau = Fl$
$l = r\sin\phi = (4.00 \text{ m})\sin 90°$
$l = 4.00 \text{ m}$

$\tau = (10.0 \text{ N})(4.00 \text{ m}) = 40.0 \text{ N} \cdot \text{m}$

This force tends to produce a counterclockwise rotation about the axis; by the right-hand rule the vector $\vec{\tau}$ is directed out of the plane of the figure.

b)

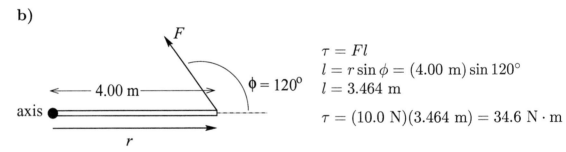

$\tau = Fl$
$l = r\sin\phi = (4.00 \text{ m})\sin 120°$
$l = 3.464 \text{ m}$

$\tau = (10.0 \text{ N})(3.464 \text{ m}) = 34.6 \text{ N} \cdot \text{m}$

This force tends to produce a counterclockwise rotation about the axis; by the right-hand rule the vector $\vec{\tau}$ is directed out of the plane of the figure.

c)

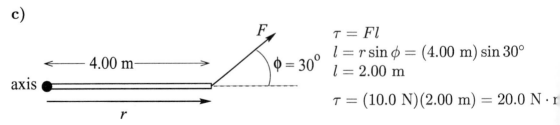

$\tau = Fl$
$l = r\sin\phi = (4.00 \text{ m})\sin 30°$
$l = 2.00 \text{ m}$

$\tau = (10.0 \text{ N})(2.00 \text{ m}) = 20.0 \text{ N} \cdot \text{m}$

This force tends to produce a counterclockwise rotation about the axis; by the right-hand rule the vector $\vec{\tau}$ is directed out of the plane of the figure.

d)

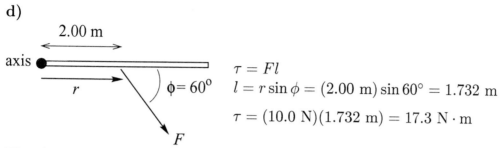

$$\tau = Fl$$
$$l = r \sin\phi = (2.00 \text{ m}) \sin 60° = 1.732 \text{ m}$$
$$\tau = (10.0 \text{ N})(1.732 \text{ m}) = 17.3 \text{ N} \cdot \text{m}$$

This force tends to produce a clockwise rotation about the axis; by the right-hand rule the vector $\vec{\tau}$ is directed into the plane of the figure.

e)

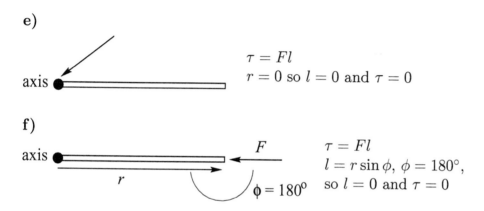

$$\tau = Fl$$
$$r = 0 \text{ so } l = 0 \text{ and } \tau = 0$$

f)

axis

$$\tau = Fl$$
$$l = r \sin\phi, \ \phi = 180°,$$
$$\text{so } l = 0 \text{ and } \tau = 0$$

10-3

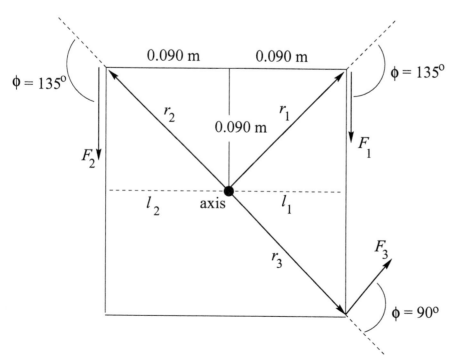

Let counterclockwise be the positive sense of rotation.

$$r_1 = r_2 = r_3 = \sqrt{(0.090 \text{ m})^2 + (0.090 \text{ m})^2} = 0.1273 \text{ m}$$

$$\tau_1 = -F_1 l_1$$

$$l_1 = r_1 \sin_1 = (0.1273 \text{ m}) \sin 135° = 0.0900 \text{ m}$$

$$\tau_1 = -(18.0 \text{ N})(0.0900 \text{ m}) = -1.62 \text{ N} \cdot \text{m}$$

$\vec{\tau}_1$ is directed into paper

$$\tau_2 = +F_2 l_2$$

$$l_2 = r_2 \sin_2 = (0.1273 \text{ m}) \sin 135° = 0.0900 \text{ m}$$

$$\tau_2 = +(26.0 \text{ N})(0.0900 \text{ m}) = +2.34 \text{ N} \cdot \text{m}$$

$\vec{\tau}_2$ is directed out of paper

$$\tau_3 = +F_3 l_3$$

$$l_3 = r_3 \sin_3 = (0.1273 \text{ m}) \sin 90° = 0.1273 \text{ m}$$

$$\tau_3 = +(14.0 \text{ N})(0.1273 \text{ m}) = +1.78 \text{ N} \cdot \text{m}$$

$\vec{\tau}_3$ is directed out of paper

$$\sum \tau = \tau_1 + \tau_2 + \tau_3 = -1.62 \text{ N} \cdot \text{m} + 2.34 \text{ N} \cdot \text{m} + 1.78 \text{ N} \cdot \text{m} = 2.50 \text{ N} \cdot \text{m}$$

Net torque is positive, which means it tends to produce a counterclockwise rotation the vector torque is directed out of the plane of the paper.

10-5 $\vec{r} = (-0.450 \text{ m})\hat{\imath} + (0.150 \text{ m})\hat{\jmath};$ $\vec{F} = (-5.00 \text{ N})\hat{\imath} + (4.00 \text{ N})\hat{\jmath}$

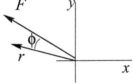

b) When the fingers of your right hand curl from the direction of $\vec{r}$ into the direction of $\vec{F}$ (through the smaller of the two angles, angle ϕ) your thumb points into the page (the direction of $\vec{\tau}$, the $-z$-direction).

c) $\vec{\tau} = \vec{r} \times \vec{F} = [(-0.450 \text{ m})\hat{\imath} + (0.150 \text{ m})\hat{\jmath}] \times [(-5.00 \text{ N})\hat{\imath} + (4.00 \text{ N})\hat{\jmath}]$

$\vec{\tau} = +(2.25 \text{ N} \cdot \text{m})\hat{\imath} \times \hat{\imath} - (1.80 \text{ N} \cdot \text{m})\hat{\imath} \times \hat{\jmath} - (0.750 \text{ N} \cdot \text{m})\hat{\jmath} \times \hat{\imath} + (0.600 \text{ N} \cdot \text{m})\hat{\jmath} \times \hat{\jmath}$

$\hat{\imath} \times \hat{\imath} = \hat{\jmath} \times \hat{\jmath} = 0$

$\hat{\imath} \times \hat{\jmath} = \hat{k}, \quad \hat{\jmath} \times \hat{\imath} = -\hat{k}$

Thus $\vec{\tau} = -(1.80 \text{ N} \cdot \text{m})\hat{k} - (0.750 \text{ N} \cdot \text{m})(-\hat{k}) = (-1.05 \text{ N} \cdot \text{m})\hat{k}$.

Note: The calculation gives that $\vec{\tau}$ is in the $-z$-direction. This agrees with what we got from the right-hand rule.

10-9 **a)**

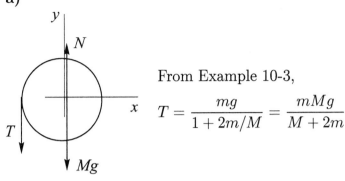

From Example 10-3,

$$T = \frac{mg}{1 + 2m/M} = \frac{mMg}{M + 2m}$$

$$\sum F_y = ma_y$$

$$N - T - Mg = 0$$

$$N = T + Mg = \frac{mMg}{M + 2m} + Mg = \frac{mMg + M(M + 2m)g}{M + 2m} = \frac{M(M + 3m)g}{M + 2m}$$

$$N = \left(\frac{M + 3m}{M + 2m}\right)Mg = \left(\frac{M + 3m}{1 + 2m/M}\right)g$$

b) Re-write the expression for N:

$$N = \frac{mMg}{M + 2m} + Mg = Mg + mg\left(\frac{M}{M + 2m}\right) = Mg + mg + mg\left(\frac{M}{M + 2m} - 1\right)$$

$$N = (M + m)g - mg\left(\frac{M + 2m - M}{M + 2m}\right) = (M + m)g - \left(\frac{2m}{2m + M}\right)mg; \ N \text{ is less}$$

than $(M + m)g$.

The block is accelerating downward, so the tension is less than its weight. Or, you could say that part of the system is accelerating downward and the rest has no vertical acceleration, so the total upward force N on the system must be less than the total downward force $(M + m)g$ on the system.

c) The force diagrams in Example 10-3 are unchanged, so this has no effect on T and N.

10-11 Use the kinematic information to solve for the angular acceleration of the grindstone. Assume that the grindstone is rotating counterclockwise and let that be the positive sense of rotation.

$\omega_0 = 850 \text{ rev/min}(2\pi \text{ rad/1 rev})(1 \text{ min/60 s}) = 89.0 \text{ rad/s}$

$t = 7.50 \text{ s}; \quad \omega = 0 \text{ (comes to rest)}; \quad \alpha = ?$

$\omega = \omega_0 + \alpha t$

$$\alpha = \frac{0 - 89.0 \text{ rad/s}}{7.50 \text{ s}} = -11.9 \text{ rad/s}^2$$

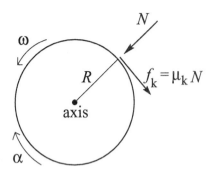

The normal force has zero moment arm for rotation about an axis at the center of the grindstone, and therefore zero torque. The only torque on the grindstone is that due to the friction froce f_k exerted by the ax; for this force the moment arm is $l = R$ and the torque is negative.

$\sum \tau = -f_k R = -\mu_k N R$

$I = \frac{1}{2} M R^2$ (solid disk, axis through center)

Thus $\sum \tau = I\alpha$ gives $-\mu_k N R = (\frac{1}{2} M R^2)\alpha$

$$\mu_k = -\frac{MR\alpha}{2N} = -\frac{(50.0 \text{ kg})(0.260 \text{ m})(-11.9 \text{ rad/s}^2)}{2(160 \text{ N})} = 0.483$$

10-15

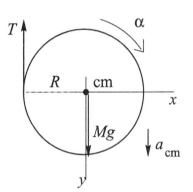

For translational motion of the center of mass of the hoop take the origin of coordinates at the cm and take $+y$ downward. Then $\sum F_y = m a_y$ gives $Mg - T = M a_{cm}$.

Apply $\sum \tau = I\alpha$ for rotation about the center of mass with the clockwise sense of rotation positive. The weight Mg has zero moment arm and therefore zero torque.

Thus $TR = I_{cm}\alpha$

For a hoop $I_{cm} = MR^2$ so $TR = MR^2\alpha$

$T = MR\alpha$

$a_{cm} = R\alpha$ so the equation becomes $T = M a_{cm}$

Combine these two equations to eliminate T: $Mg - M a_{cm} = M a_{cm}$

The mass M divides out and we get $2a_{cm} = g$, so $a_{cm} = g/2$.

Then $T = M(g/2) = \frac{1}{2} Mg = \frac{1}{2}(0.180 \text{ kg})(9.80 \text{ m/s}^2) = 0.882 \text{ N}$

b) Apply the constant acceleration kinematic equations to the motion of the center of mass:

$v_{0y} = 0;$ $y - y_0 = 0.750$ m; $a_y = g/2 = 4.90$ m/s^2; $t = ?$

$y - y_0 = v_{0y}t + \frac{1}{2}a_yt^2$

$$t = \sqrt{\frac{2(y - y_0)}{a_y}} = \sqrt{\frac{2(0.750 \text{ m})}{4.90 \text{ m/s}^2}} = 0.553 \text{ s}$$

c) We can use the constant angular acceleration equations for the rotational motion:

$t = 0.553$ s; $\omega_0 = 0;$ $\alpha = a_{cm}/R = \frac{1}{2}(9.80 \text{ m/s}^2)/0.0800 \text{ m} = 61.25$ rad/s^2;

$\omega = ?$

$\omega = \omega_0 + \alpha t = 0 + (61.25 \text{ m/s}^2)(0.553 \text{ s}) = 33.9$ rad/s

Alternatively, we can find v_{cm} at this point in the motion and then use $\omega = v_{cm}/R$:

$v_{0y} = 0;$ $v_y = ?;$ $a_y = 4.90$ m/s^2; $y - y_0 = 0.750$ m

$v_y^2 = v_{0y}^2 + 2a_y(y - y_0)$

$v_y = \sqrt{2a_y(y - y_0)} = \sqrt{2(4.90 \text{ m/s}^2)(0.750 \text{ m})} = 2.71$ m/s

Then $\omega = v_{cm}/R = (2.71 \text{ m/s})/0.0800$ m $= 33.9$ rad/s, the same as before.

10-19

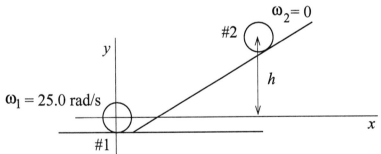

$\omega_2 = 0$

#2

y

h

$\omega_1 = 25.0$ rad/s

#1

x

Take $y = 0$ at the center of the wheel when it is at the bottom of the hill.

$K_1 + U_1 + W_{other} = K_2 + U_2$

$W_{other} = W_{fric} = -3500$ J (the friction work is negative)

$K_1 = \frac{1}{2}I\omega_1^2 + \frac{1}{2}Mv_1^2;$ $v = R\omega$ and $I = 0.800MR^2$ so

$K_1 = \frac{1}{2}(0.800)MR^2\omega_1^2 + \frac{1}{2}MR^2\omega_1^2 = 0.900MR^2\omega_1^2$

$K_2 = 0,$ $U_1 = 0,$ $U_2 = Mgh$

Thus $0.900MR^2\omega_1^2 + W_{fric} = Mgh$

$M = w/g = 392$ N/$(9.80 \text{ m/s}^2) = 40.0$ kg

$$h = \frac{0.900MR^2\omega_1^2 + W_{fric}}{Mg}$$

$$h = \frac{(0.900)(40.0 \text{ kg})(0.600 \text{ m})^2(25.0 \text{ rad/s})^2 - 3500 \text{ J}}{(40.0 \text{ kg})(9.80 \text{ m/s}^2)} = 11.7 \text{ m}$$

10-21 a)

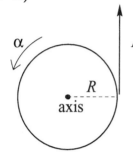

Apply $\sum \tau = I\alpha$ to find the angular acceleration:
$$FR = I\alpha$$
$$\alpha = \frac{FR}{I} = \frac{(18.0 \text{ N})(2.40 \text{ m})}{2100 \text{ kg} \cdot \text{m}^2} = 0.02057 \text{ rad/s}^2$$

Use the constant α kinematic equations to find ω:

$\omega = ?$; ω_0 (initially at rest); $\alpha = 0.02057 \text{ rad/s}^2$; $t = 15.0 \text{ s}$

$\omega = \omega_0 + \alpha t = 0 + (0.02057 \text{ rad/s}^2)(15.0 \text{ s}) = 0.309 \text{ rad/s}$

b) This question can be answered in either of two ways:

(1) $W = \tau \Delta\theta$ (Eq.(10-24))

$\Delta\theta = \theta - \theta_0 = \omega_0 t + \frac{1}{2}\alpha t^2 = 0 + \frac{1}{2}(0.02057 \text{ rad/s}^2)(15.0 \text{ s})^2 = 2.314 \text{ rad}$

$\tau = FR = (18.0 \text{ N})(2.40 \text{ m}) = 43.2 \text{ N} \cdot \text{m}$

Then $W = \tau \Delta\theta = (43.2 \text{ N} \cdot \text{m})(2.314 \text{ rad}) = 100 \text{ J}$.

<u>or</u>

(2) $W_{\text{tot}} = K_2 - K_1$ (the work-energy relation from chapter 6)

$W_{\text{tot}} = W$, the work done by the child

$K_1 = 0$; $K_2 = \frac{1}{2}I\omega^2 = \frac{1}{2}(2100 \text{ kg} \cdot \text{m}^2)(0.309 \text{ rad/s})^2 = 100 \text{ J}$

Thus $W = 100 \text{ J}$, the same as before.

c) $P_{\text{av}} = \dfrac{\Delta W}{\Delta t} = \dfrac{100 \text{ J}}{15.0 \text{ s}} = 6.67 \text{ W}$

10-27 Use Eq.(10-26): $P = \tau\omega$, where ω must be in rad/s.

$\omega = (4000 \text{ rev/min})(2\pi \text{ rad/1 rev})(1 \text{ min/ } 60 \text{ s}) = 418.9 \text{ rad/s}$

$$\tau = \frac{P}{\omega} = \frac{1.50 \times 10^5 \text{ W}}{418.9 \text{ rad/s}} = 358 \text{ N} \cdot \text{m}$$

b)

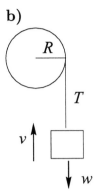

v constant implies $a = 0$ and $T = w$

$\tau = TR$ implies $T = \tau/R = 358$ N $\cdot$ m/0.200 m $= 1790$ N

Thus a weight $w = 1790$ N can be lifted.

c) $v = R\omega$ and the drum has $\omega = 418.9$ rad/s, so

$v = (0.200$ m$)(418.9$ rad/s$) = 83.8$ m/s

10-29 a) Use $L = mvr \sin \phi$ (Eq.(10-28)):

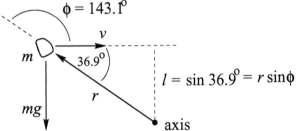

$L = mvr \sin \phi =$
$(2.00$ kg$)(12.0$ m/s$)(8.00$ m$)\sin 143.1°$
$L = 115$ kg$\cdot$m^2/s

To find the direction of $\vec{L}$ apply the right-hand rule by turning $\vec{r}$ into the direction of $\vec{v}$ by pushing on it with the fingers of your right hand. Your thumb points into the page, in the direction of $\vec{L}$.

b) By Eq.(10-29) the rate of change of the angular momentum of the rock equals the torque of the net force acting on it.

$\tau = mg(8.00$ m$)\cos 36.9° = 125$ kg$\cdot$m^2/s^2

To find the direction of $\vec{\tau}$ and hence of $d\vec{L}/dt$, apply the right-hand rule by turning $\vec{r}$ into the direction of the gravity force by pushing on it with the fingers of your right hand. Your thumb points out of the page, in the direction of $d\vec{L}/dt$.

10-31 Use $L = I\omega$

The second hand makes 1 revolution in 1 minute, so

$\omega = (1.00$ rev/min$)(2\pi$ rad/1 rev$)(1$ min/60 s$) = 0.1047$ rad/s

For a slender rod, with the axis about one end,

$I = \frac{1}{3}ML^2 = \frac{1}{3}(6.00 \times 10^{-3}$ kg$)(0.150$ m$)^2 = 4.50 \times 10^{-5}$ kg$\cdot$m^2

Then $L = I\omega = (4.50 \times 10^{-5}$ kg $\cdot$ m$^2)(0.1047$ rad/s$) = 4.71 \times 10^{-6}$ kg$\cdot$m^2/s.

10-33 a) Yes, angular momentum is conserved. The moment arm for the tension in the cord is zero so this force exerts no torque and there is no net torque on the block

b) $L_1 = L_2$ so $I_1\omega_1 = I_2\omega_2$ Block treated as a point mass, so $I = mr^2$, where r i
the distance of the block from the hole.

$mr_1^2\omega_1 = mr_2^2\omega_2$

$$\omega_2 = \left(\frac{r_1}{r_2}\right)^2 \omega_1 = \left(\frac{0.300 \text{ m}}{0.150 \text{ m}}\right)^2 (1.75 \text{ rad/s}) = 7.00 \text{ rad/s}$$

c) $K_1 = \frac{1}{2}I_1\omega_1^2 = \frac{1}{2}mr_1^2\omega_1^2 = \frac{1}{2}mv_1^2$
$v_1 = r_1\omega_1 = (0.300 \text{ m})(1.75 \text{ rad/s}) = 0.525 \text{ m/s}$
$K_1 = \frac{1}{2}mv_1^2 = \frac{1}{2}(0.0250 \text{ kg})(0.525 \text{ m/s})^2 = 0.00345 \text{ J}$
$K_2 = \frac{1}{2}mv_2^2$
$v_2 = r_2\omega_2 = (0.150 \text{ m})(7.00 \text{ rad/s}) = 1.05 \text{ m/s}$
$K_2 = \frac{1}{2}mv_2^2 = \frac{1}{2}(0.0250 \text{ kg})(1.05 \text{ m/s})^2 = 0.01378 \text{ J}$

$\Delta K = K_2 - K_1 = 0.01378 \text{ J} - 0.00345 \text{ J} = 0.0103 \text{ J}$

d) $W_{\text{tot}} = \Delta K$
But $W_{\text{tot}} = W$, the work done by the tension in the cord, so $W = 0.0103$ J

10-37 a) $L_1 = L_2$
$I_1\omega_1 = I_2\omega_2$, so $\omega_2 = (I_1/I_2)\omega_1$

$I_1 = I_{tt} = \frac{1}{2}MR^2 = \frac{1}{2}(120 \text{ kg})(2.00 \text{ m})^2 = 240 \text{ kg}\cdot\text{m}^2$
$I_2 = I_{tt} + I_{bag} = 240 \text{ kg}\cdot\text{m}^2 + mR^2 = 240 \text{ kg}\cdot\text{m}^2 + (70 \text{ kg})(2.00 \text{ m})^2 = 520 \text{ kg}\cdot\text{m}$
$\omega_2 = (I_1/I_2)\omega_1 = (240 \text{ kg}\cdot\text{m}^2/520 \text{ kg}\cdot\text{m}^2)(3.00 \text{ rad/s}) = 1.38 \text{ rad/s}$

b) $K_1 = \frac{1}{2}I_1\omega_1^2 = \frac{1}{2}(240 \text{ kg}\cdot\text{m}^2)(3.00 \text{ rad/s})^2 = 1080 \text{ J}$
$K_2 = \frac{1}{2}I_2\omega_2^2 = \frac{1}{2}(520 \text{ kg}\cdot\text{m}^2)(1.38 \text{ rad/s})^2 = 495 \text{ J}$
The kinetic energy decreases because of the negative work done on the turntabl
and the parachutist by the friction force between these two objects.

10-41 a) By the work-energy relation $W = \Delta K$ the work done on the gyroscope
equals its increase in kinetic energy.
Thus $P_{\text{av}} = W/t = \Delta K/t$ and $t = \Delta K/P$

Since $K_1 = 0$, $\Delta K = K_2$, the amount of energy stored in the gyroscope when it i
up to speed.
Thus $\Delta K = K_2 = \frac{1}{2}I\omega^2$.
For a solid disk $I = \frac{1}{2}MR^2 = \frac{1}{2}(60,000 \text{ kg})(2.00 \text{ m})^2 = 1.20 \times 10^5 \text{ kg}\cdot\text{m}^2$
$\omega = 500 \text{ rev/min}(2\pi \text{ rad/1 rev})(1 \text{ min/60 s}) = 52.36 \text{ rad/s}$

Thus $\Delta K = \frac{1}{2}I\omega^2 = \frac{1}{2}(1.20 \times 10^5 \text{ kg} \cdot \text{m}^2)(52.36 \text{ rad/s})^2 = 1.645 \times 10^8 \text{ J}$

and $t = \dfrac{\Delta K}{P} = \dfrac{1.645 \times 10^8 \text{ J}}{7.46 \times 10^4 \text{ W}} = 2205 \text{ s} = 36.8 \text{ min.}$

b) Eq.(10-36) says $\Omega = \tau/L = \tau/I\omega$, so $\tau = I\omega\Omega$.

$\Omega = 1.00°/\text{s}(\pi \text{ rad}/180°) = 0.01745 \text{ rad/s}$

$\tau = I\omega\Omega = (1.20 \times 10^5 \text{ kg} \cdot \text{m}^2)(52.36 \text{ rad/s})(0.01745 \text{ rad/s}) = 1.10 \times 10^5 \text{ N} \cdot \text{m}$

Problems

10-47 a) $P = \tau\omega$

Since α constant, $\omega = \omega_0 + \alpha t$. $\omega_0 = 0$.

And $\tau = I\alpha$ says $\alpha = \tau/I$ and then $\omega = \tau t/I$.

Thus $P = \tau(\tau t/I) = \tau^2 t/I$

b) $\dfrac{P}{\tau^2} = \dfrac{t}{I} = \text{constant}$, so $\dfrac{P_1}{\tau_1^2} = \dfrac{P_2}{\tau_2^2}$

$P_2 = P_1 \left(\dfrac{\tau_2}{\tau_1}\right)^2 = 500 \text{ W} \left(\dfrac{60.0 \text{ N} \cdot \text{m}}{20.0 \text{ N} \cdot \text{m}}\right)^2 = 4500 \text{ W}$

c) $P = \tau\omega$

$\omega^2 = \omega_0^2 + 2\alpha(\theta - \theta_0) = 2(\tau/I)(\theta - \theta_0)$

$\omega = \sqrt{(2\tau/I)(\theta - \theta_0)}$

Thus $P = \tau\omega = \tau\sqrt{(2\tau/I)(\theta - \theta_0)} = (\sqrt{2(\theta - \theta_0)/I})\tau^{3/2}$.

d) $\dfrac{P}{\tau^{3/2}} = \sqrt{2(\theta - \theta_0)/I} = \text{constant}$, so $\dfrac{P_1}{\tau_1^{3/2}} = \dfrac{P_2}{\tau_2^{3/2}}$

$P_2 = P_1 \left(\dfrac{\tau_2}{\tau_1}\right)^{3/2} = 500 \text{ W} \left(\dfrac{60.0 \text{ N} \cdot \text{m}}{20.0 \text{ N} \cdot \text{m}}\right)^{3/2} = 2600 \text{ W}$

e) There is no contradiction. When the net torque is varied, keeping t constant is different from keeping $\theta - \theta_0$ constant.

10-49 Use $\sum \tau = I\alpha$ to find α, and then use the constant α kinematic equations to solve for t.

$F = 220$ N

$l = 1.25$ m

axis at hinge

$\sum \tau = Fl = (220 \text{ N})(1.25 \text{ m}) = 275 \text{ N} \cdot \text{m}$

From Table 9-2(d), $I = \frac{1}{3}Ml^2$

$I = \frac{1}{3}(750 \text{ N}/9.80 \text{ m/s}^2)(1.25 \text{ m})^2 = 39.9 \text{ kg} \cdot \text{m}^2$

$\sum \tau = I\alpha$ so $\alpha = \dfrac{\sum \tau}{I} = \dfrac{275 \text{ N} \cdot \text{m}}{39.9 \text{ kg} \cdot \text{m}^2} = 6.89 \text{ rad/s}^2$

$\alpha = 6.89 \text{ rad/s}^2$; $\theta - \theta_0 = 90°(\pi \text{ rad}/180°) = \pi/2 \text{ rad}$; $\omega_0 = 0$ (door initially a rest); $t = ?$

$\theta - \theta_0 = \omega_0 t + \frac{1}{2}\alpha t^2$

$t = \sqrt{\dfrac{2(\theta - \theta_0)}{\alpha}} = \sqrt{\dfrac{2(\pi/2 \text{ rad})}{6.89 \text{ rad/s}^2}} = 0.675$ s

10-53 Force diagram for the crate

$\sum F_y = ma_y$

$T - mg = ma$

$T = m(g + a) = 50 \text{ kg}(9.80 \text{ m/s}^2 + 0.80 \text{ m/s}^2) = 530 \text{ N}$

Force diagram for the cylinder:

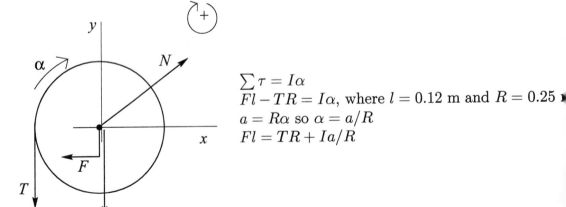

$\sum \tau = I\alpha$

$Fl - TR = I\alpha$, where $l = 0.12$ m and $R = 0.25$

$a = R\alpha$ so $\alpha = a/R$

$Fl = TR + Ia/R$

$F = T\left(\dfrac{R}{l}\right) + \dfrac{Ia}{Rl} = 530 \text{ N}\left(\dfrac{0.25 \text{ m}}{0.12 \text{ m}}\right) + \dfrac{(2.9 \text{ kg} \cdot \text{m}^2)(0.80 \text{ m/s}^2)}{(0.25 \text{ m})(0.12 \text{ m})} = 1200 \text{ N}$

10-55 Apply $\sum \tau = I\alpha$ to the rotation of the flywheel about the axis:

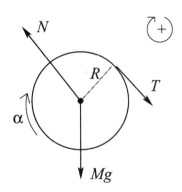

The forces N and Mg act at the axis so have zero torque.

$$\sum \tau = TR$$
$$TR = I\alpha$$

Apply $\sum \vec{F} = m\vec{a}$ to the translational motion of the block:

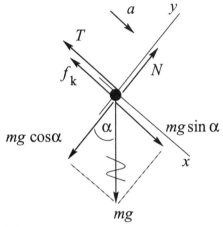

$$\sum F_y = ma_y$$
$$N - mg \cos 36.9° = 0$$
$$N = mg \cos 36.9°$$

$$f_k = \mu_k N = \mu_k mg \cos 36.9°$$

$$\sum F_x = ma_x$$
$$mg \sin 36.0° - T - \mu_k mg \cos 36.9° = ma$$
$$mg(\sin 36.9° - \mu_k \cos 36.9°) - T = ma$$

But we also know that $a_{\text{block}} = R\alpha_{\text{wheel}}$, so $\alpha = a/R$.
Using this in the $\sum \tau = I\alpha$ equation gives $TR = Ia/R$ and $T = (I/R^2)a$.
Use this to replace T in the $\sum F_y = ma_y$ equation:
$$mg(\sin 36.9° - \mu_k \cos 36.9°) - (I/R^2)a = ma$$

$$a = \frac{mg(\sin 36.9° - \mu_k \cos 36.9°)}{m + I/R^2}$$

$$a = \frac{(5.00 \text{ kg})(9.80 \text{ m/s}^2)(\sin 36.9° - (0.25) \cos 36.9°)}{5.00 \text{ kg} + 0.500 \text{ kg} \cdot \text{m}^2/(0.200 \text{ m})^2} = 1.12 \text{ m/s}^2$$

b) $T = \dfrac{0.500 \text{ kg} \cdot \text{m}^2}{(0.200 \text{ m})^2}(1.12 \text{ m/s}^2) = 14.0 \text{ N}$

10-57

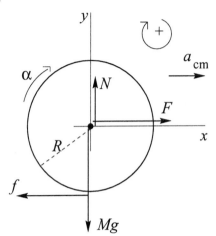

Apply $\sum \vec{F} = m\vec{a}$ to the translational motion of the center of mass:
$$\sum F_x = ma_x$$
$$F - f = Ma_{cm}$$

Apply $\sum \tau = I\alpha$ to the rotation about the center of mass:
$$\sum \tau = fR$$
thin-walled hollow cylinder: $I = MR^2$
Then $\sum \tau = I\alpha$ implies $fR = MR^2\alpha$.
But $a_{cm} = R\alpha$, so $f = Ma_{cm}$.
Using this in the $\sum F_x = ma_x$ equation gives $F - Ma_{cm} = Ma_{cm}$
$$a_{cm} = F/2M$$
And then $f = Ma_{cm} = M(F/2M) = F/2$.

10-59 **a)** This problem can be done either of two ways. We will do it both ways.
(1) <u>Conservation of energy:</u> $K_1 + U_1 + W_{other} = K_2 + U_2$

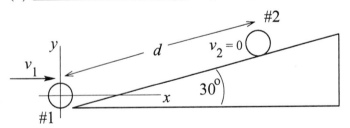

Take position 1 to be the location of the disk at the base of the ramp and 2 to be where the disk momentaril stops before rolling back dow

Take the origin of coordinates at the center of the disk at position 1 and take $+$ to be upward. Then $y_1 = 0$ and $y_2 = d\sin 30°$, where d is the distance that th disk rolls up the ramp.

"rolls without slipping" and neglect rolling friction says $W_f = 0$; only gravity doe work on the disk, so $W_{other} = 0$
$$U_1 = Mgy_1 = 0$$
$$K_1 = \tfrac{1}{2}Mv_1^2 + \tfrac{1}{2}I_{cm}\omega_1^2 \text{ (Eq.10-11). But } \omega_1 = v_1/R \text{ and } I_{cm} = \tfrac{1}{2}MR^2, \text{ so}$$
$$\tfrac{1}{2}I_{cm}\omega_1^2 = \tfrac{1}{2}(\tfrac{1}{2}MR^2)(v_1/R)^2 = \tfrac{1}{4}Mv_1^2.$$
Thus $K_1 = \tfrac{1}{2}Mv_1^2 + \tfrac{1}{4}Mv_1^2 = \tfrac{3}{4}Mv_1^2.$

$U_2 = Mgy_2 = Mgd\sin 30°$

$K_2 = 0$ (disk is at rest at point 2).

Thus $\frac{3}{4}Mv_1^2 = Mgd\sin 30°$

$$d = \frac{3v_1^2}{4g\sin 30°} = \frac{3(2.50 \text{ m/s})^2}{4(9.80 \text{ m/s}^2)\sin 30°} = 0.957 \text{ m}$$

(2) <u>force and acceleration</u>

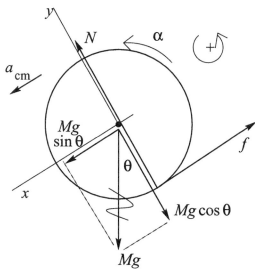

Apply $\sum F_x = ma_x$ to the translational motion of the center of mass:
$Mg\sin\theta - f = Ma_{cm}$

Apply $\sum \tau = I\alpha$ to the rotation about the center of mass:
$fR = (\frac{1}{2}MR^2)\alpha$
$f = \frac{1}{2}MR\alpha$

But $a_{cm} = R\alpha$ in this equation gives $f = \frac{1}{2}Ma_{cm}$.

Use this in the $\sum F_x = ma_x$ equation to eliminate f.

$Mg\sin\theta - \frac{1}{2}Ma_{cm} = Ma_{cm}$

M divides out and $\frac{3}{2}a_{cm} = g\sin\theta$. $a_{cm} = \frac{2}{3}g\sin\theta = \frac{2}{3}(9.80 \text{ m/s}^2)\sin 30° = 3.267 \text{ m/s}^2$

Apply the constant acceleration equations to the motion of the center of mass. Note that in our coordinates the positive x-direction is down the incline.

$v_{0x} = -2.50$ m/s (directed up the incline); $a_x = +3.267 \text{ m/s}^2$;

$v_x = 0$ (momentarily comes to rest); $x - x_0 = ?$

$v_x^2 = v_{0x}^2 + 2a_x(x - x_0)$

$$x - x_0 = -\frac{v_{0x}^2}{2a_x} = -\frac{(-2.50 \text{ m/s})^2}{2(3.267 \text{ m/s}^2)} = -0.957 \text{ m}$$

b) The mass M enters both in the linear inertia and in the gravity force so divides out. The mass M and radius R enter in both the rotational inertia and the gravitational torque so divide out.

10-61

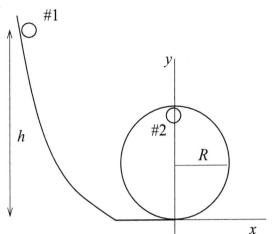

Take the origin at the lowest point in the track.

$y_{cm,1} = h$
$y_{cm,2} = 2R - r$
$I_{cm} = \frac{2}{5}mr^2$

$K_1 + U_1 + W_{\text{other}} = K_2 + U_2$

$W_{\text{other}} = 0; \quad K_1 = 0$

$K_2 = \frac{1}{2}mv^2 + \frac{1}{2}I_{cm}\omega^2 = \frac{1}{2}mv^2 + \frac{1}{2}(\frac{2}{5}mr^2)\omega^2$

Rolls without slipping implies $v = r\omega$ or $\omega = v/r$.

Thus $K_2 = \frac{1}{2}mv^2 + \frac{1}{2}(\frac{2}{5}mr^2)(v^2/r^2) = mv^2(\frac{1}{2} + \frac{1}{5}) = \frac{7}{10}mv^2$.

$U_1 = mgy_{cm,1} = mgh$

$U_2 = mgy_{cm,2} = mg(2R - r)$

Thus $K_1 + U_1 + W_{\text{other}} = K_2 + U_2$ gives

$mgh = \frac{7}{10}mv^2 + mg(2R - r)$

$\frac{7}{10}v^2 = g(h - 2R + r)$

Force diagram for the marble at point 2:

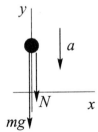

The radius of the circle in which the cm travels is $R - r$. When the marble just starts to leave the track (minimum h) $N \to 0$

$\sum F_y = ma_y$
$mg = mv^2/(R - r)$
$v^2 = g(R - r)$

Using this in the conservation of energy equation gives $\frac{7}{10}g(R - r) = g(h - 2R + r)$

and $h = \frac{7}{10}R + 2R - \frac{7}{10}r - r = (27R - 17r)/10$.

b) No friction implies that marble slides without rolling, so $K_1 = \frac{1}{2}mv^2$, no $\frac{1}{2}I_{cm}\omega^2$ term.

In the conservation of energy equation this gives $\frac{1}{2}v^2 = g(h - 2R + r)$.

Combining this with $v^2 = g(R - r)$ gives $\frac{1}{2}g(R - r) = g(h - 2R + r)$.

$h = \frac{5}{2}R - \frac{3}{2}r = (5R - 3r)/2$.

Note that when $r = 0$ this result is the same as for Problem 7-40 part (a).

10-65 $U_{el} = \frac{1}{2}kx^2 = \frac{1}{2}(400 \text{ N} \cdot \text{m})(0.15 \text{ m})^2 = 4.50 \text{ J}$

$K_1 = 0.800 U_{el} = 3.60 \text{ J}$

$U_2 = 0.900 K_1 = 3.24 \text{ J}$

a) $K_1 = \frac{1}{2}mv^2 + \frac{1}{2}I_{cm}\omega^2$

rolling without slipping says $\omega = v/R$

$I_{cm} = \frac{2}{5}mR^2$

Thus $K_1 = \frac{1}{2}mv^2 + (\frac{2}{5}mR^2)(v/R)^2 = mv^2(\frac{1}{2} + \frac{1}{5}) = \frac{7}{10}mv^2$

and $v = \sqrt{\dfrac{10K_1}{7m}} = \sqrt{\dfrac{10(3.60 \text{ J})}{7(0.0590 \text{ kg})}} = 9.34 \text{ m/s}.$

b)

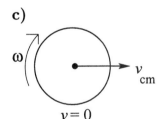

$v = 2v_{cm}$

From Fig.(10-13), at the top of the ball

$v = 2v_{cm} = 18.7 \text{ m/s}$

c)

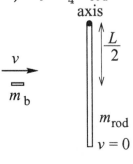

From Fig.(10-13), $v = 0$ at the bottom of the ball.

d) $U_2 = mgh = 3.24 \text{ J}$

$h = \dfrac{3.24 \text{ J}}{mg} = \dfrac{3.24 \text{ J}}{(0.0590 \text{ kg})(9.80 \text{ m/s}^2)} = 5.60 \text{ m}$

10-71 **a)** $m_b = \frac{1}{4}m_{rod}$

axis

before   after

$L_1 = m_b vr = \frac{1}{4}m_{rod}v(L/2)$

$L_1 = \frac{1}{8}m_{rod}vL$

$L_2 = (I_{rod} + I_b)\omega$

$I_{rod} = \frac{1}{3}m_{rod}L^2$

$I_b = m_b r^2 = \frac{1}{4}m_{rod}(L/2)^2$

$I_b = \frac{1}{16}m_{rod}L^2$

Thus $L_1 = L_2$ gives $\frac{1}{8}m_{\text{rod}}vL = (\frac{1}{3}m_{\text{rod}}L^2 + \frac{1}{16}m_{\text{rod}}L^2)\omega$

$\frac{1}{8}v = \frac{19}{48}L\omega$

$\omega = \frac{6}{19}v/L$

b) $K_1 = \frac{1}{2}mv^2 = \frac{1}{8}m_{\text{rod}}v^2$

$K_2 = \frac{1}{2}I\omega^2 = \frac{1}{2}(I_{\text{rod}} + I_b)\omega^2 = \frac{1}{2}(\frac{1}{3}m_{\text{rod}}L^2 + \frac{1}{16}m_{\text{rod}}L^2)(6v/19L)^2$

$K_2 = \frac{1}{2}(\frac{19}{48})(\frac{6}{19})^2 m_{\text{rod}}v^2 = \frac{3}{152}m_{\text{rod}}v^2$

Then $\dfrac{K_2}{K_1} = \dfrac{\frac{3}{152}m_{\text{rod}}v^2}{\frac{1}{8}m_{\text{rod}}v^2} = 3/19.$

10-73   **a)** Apply conservation of angular momentum to the collision between the bullet and the board:

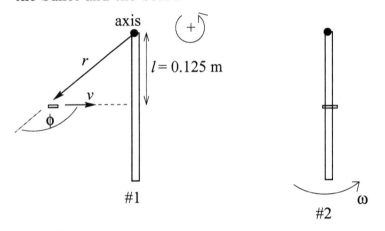

$L_1 = L_2$

$L_1 = mvr\sin\phi = mvl = (1.90 \times 10^{-3}\text{ kg})(360\text{ m/s})(0.125\text{ m}) = 0.0855\text{ kg·m}^2/\text{s}$

$L_2 = I_2\omega_2$

$I_2 = I_{\text{board}} + I_{\text{bullet}} = \frac{1}{3}ML^2 + mr^2$

$I_2 = \frac{1}{3}(0.750\text{ kg})(0.250\text{ m})^2 + (1.90 \times 10^{-3}\text{ kg})(0.125\text{ m})^2 = 0.01565\text{ kg}\cdot\text{m}^2$

Then $L_1 = L_2$ gives that $\omega_2 = \dfrac{L_1}{I_2} = \dfrac{0.0855\text{ kg}\cdot\text{m}^2/\text{s}}{0.01565\text{ kg}\cdot\text{m}^2} = 5.46\text{ rad/s}$

b) Apply conservation of energy to the motion of the board after the collision. Take the origin of coordinates at the center of the board and $+y$ to be upward, s $y_{\text{cm},1} = 0$ and $y_{\text{cm},2} = h$, the height being asked for.

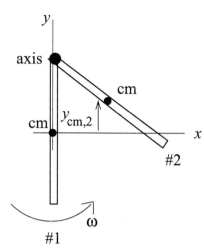

$$K_1 + U_1 + W_{\text{other}} = K_2 + U_2$$

Only gravity does work, so $W_{\text{other}} = 0$.

$$K_1 = \tfrac{1}{2}I\omega^2$$
$$U_1 = mgy_{\text{cm},1} = 0$$
$$K_2 = 0$$
$$U_2 = mgy_{\text{cm},2} = mgh$$

Thus $\tfrac{1}{2}I\omega^2 = mgh$.

$$h = \frac{I\omega^2}{2mg} = \frac{(0.01565 \ \text{kg} \cdot \text{m}^2)(5.46 \ \text{rad/s})^2}{2(0.750 \ \text{kg} + 1.90 \times 10^{-3} \ \text{kg})(9.80 \ \text{m/s}^2)} = 0.0317 \ \text{m} = 3.17 \ \text{cm}$$

c)

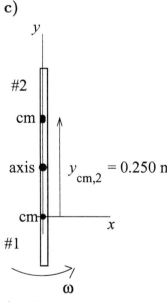

Apply conservation of energy as in part (b), except now we want $y_{\text{cm},2} = h = 0.250 \ \text{m}$. Solve for the ω after the collision that is required for this to happen.

$$\tfrac{1}{2}I\omega^2 = mgh$$

$$\omega = \sqrt{\frac{2mgh}{I}} = \sqrt{\frac{2(0.750 \ \text{kg} + 1.90 \times 10^{-3} \ \text{kg})(9.80 \ \text{m/s}^2)(0.250 \ \text{m})}{0.01565 \ \text{kg} \cdot \text{m}^2}}$$

$$\omega = 15.34 \ \text{rad/s}$$

Now go back to the equation that results from applying conservation of angular momentum to the collision and solve for the initial speed of the bullet.

$$L_1 = L_2 \text{ implies } m_{\text{bullet}} v l = I_2 \omega_2$$

$$v = \frac{I_2\omega_2}{m_{\text{bullet}}l} = \frac{(0.01565 \text{ kg} \cdot \text{m}^2)(15.34 \text{ rad/s})}{(1.90 \times 10^{-3} \text{ kg})(0.125 \text{ m})} = 1010 \text{ m/s}$$

10-75 $I_A = \frac{1}{3}I_B$, so $I_B = 3I_A$

There is no external torque on the system consisting of the two disks, so we can apply conservation of angular momentum, $L_1 = L_2$.

L_1 is the initial angular momentum of disk A, $L_1 = I_A\omega_0$.

L_2 is the final angular momentum of the two disks after they are connected and reach a common angular velocity ω, $L_2 = (I_A + I_B)\omega = 4I_A\omega$.

Then $L_1 = L_2$ implies $I_A\omega_0 = 4I_A\omega$.

$\omega = \omega_0/4$

Energy considerations: $W_{\text{tot}} = K_2 - K_1$

$W_{\text{tot}} = W_f = -2400 \text{ J}$

$K_1 = \frac{1}{2}I_A\omega_0^2$

$K_2 = \frac{1}{2}(I_A + I_B)\omega^2 = \frac{1}{2}(4I_A)(\omega_0/4)^2 = \frac{1}{4}(\frac{1}{2}I_A\omega_0^2) = \frac{1}{4}K_1$

Thus $-2400 \text{ J} = \frac{1}{4}K_1 - K_1$

$\frac{3}{4}K_1 = 2400 \text{ J}$ and $K_1 = \frac{4}{3}(2400 \text{ J}) = 3200 \text{ J}$

10-77 $L_1 = L_2$, counterclockwise positive

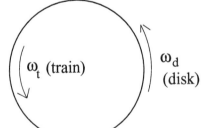

$L_1 = 0$ (before you switch on the train's engine; both the train and the platform are at rest)

$L_2 = L_{\text{train}} + L_{\text{disk}}$

The train is $\frac{1}{2}(0.95 \text{ m}) = 0.475 \text{ m}$ from the axis of rotation, so for it

$I_t = m_t R_t^2 = (1.20 \text{ kg})(0.475 \text{ m})^2 = 0.2708 \text{ kg} \cdot \text{m}^2$

$\omega_{\text{rel}} = v_{\text{rel}}/R_t = (0.600 \text{ m/s})/0.475 \text{ s} = 1.263 \text{ rad/s}$

This is the angular velocity of the train relative to the disk. Relative to the earth,

$\omega_t = \omega_{\text{rel}} + \omega_d$.

Thus $L_{\text{train}} = I_t\omega_t = I_t(\omega_{\text{rel}} + \omega_d)$.

$L_2 = L_1$ says $L_{\text{disk}} = -L_{\text{train}}$

$L_{\text{disk}} = I_d\omega_d$, where $I_d = \frac{1}{2}m_d R_d^2$

$\frac{1}{2}m_d R_d^2\omega_d = -I_t(\omega_{\text{rel}} + \omega_d)$

$$\omega_d = -\frac{I_t \omega_{\text{rel}}}{\frac{1}{2} m_d R_d^2 + I_t} = -\frac{(0.2708 \text{ kg} \cdot \text{m}^2)(1.263 \text{ rad/s})}{\frac{1}{2}(7.00 \text{ kg})(0.500 \text{ m})^2 + 0.2708 \text{ kg} \cdot \text{m}^2} = -0.30 \text{ rad/s}.$$

The minus sign tells us that the disk is rotating clockwise relative to the earth.

10-81 Eq.(10-36): $\Omega = \tau/I\omega$

The problem is asking for Ω.

The capsizing torque is $\tau = wr = (50.0 \text{ kg})(9.80 \text{ m/s}^2)(0.040 \text{ m}) = 19.6 \text{ N} \cdot \text{m}$.

$I = 0.085 \text{ kg} \cdot \text{m}^2$

$v_{\text{cm}} = 6.00 \text{ m/s}$, so the angular velocity of the wheel about its axle is

$\omega = v_{\text{cm}}/r = (6.00 \text{ m/s})/0.33 \text{ m} = 18.18 \text{ rad/s}$

Thus $\omega = \dfrac{\tau}{I\omega} = \dfrac{19.6 \text{ N} \cdot \text{m}}{(0.085 \text{ kg} \cdot \text{m}^2)(18.18 \text{ rad/s})} = 12.7 \text{ rad/s}.$

CHAPTER 11
EQUILIBRIUM AND ELASTICITY

Exercises 1, 3, 7, 11, 13, 15, 17, 19, 21, 25, 31, 33, 35
Problems 39, 43, 47, 51, 61, 63, 65, 67, 69, 71, 73, 75, 77, 79

Exercises

11-1

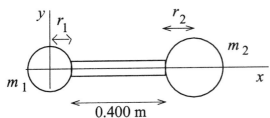

Use coordinates with the origin at the center of the 1.00 kg ball and the x-axi along the rod.

The x-coordinate of the center of gravity is given by $x_{cm} = \dfrac{m_1 x_1 + m_2 x_2}{m_1 + m_2}$.

The center of gravity of ball 1 is at $x_1 = 0$ and the center of gravity of ball 2 is a
$x_2 = 0.400 + r_1 + r_2 = 0.400 \text{ m} + 0.080 \text{ m} + 0.100 \text{ m} = 0.580 \text{ m}$.

Then $x_{cm} = \dfrac{0 + (2.00 \text{ kg})(0.580 \text{ m})}{1.00 \text{ kg} + 2.00 \text{ kg}} = 0.387 \text{ m}$.

The center of gravity is 0.387 m to the right of the center of the 1.00 kg ball.

11-3 Let cg be the center of gravity of the plank, at the geometrical center of the plan

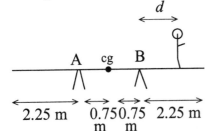

Let the child be a distance d to the right of the right-hand sawhorse.

Let M be the mass of the plank and m the mass of the child. Let the sawhorse touch the plank at points A and B.

Free-body diagram for the plank:

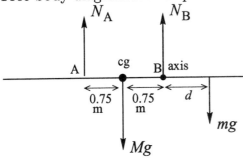

When the child is at the point where the plank just starts to tip, the plank starts to lose contact with the sawhorse at A, so $N_A \to 0$

Use $\tau = 0$, with axis at B and counterclockwise torques positive:

$+Mg(0.75 \text{ m}) - mg(d) = 0$

$d = (0.75 \text{ m})M/m = (0.75 \text{ m})(90 \text{ kg}/60 \text{ kg}) = 1.125 \text{ m}$; this distance from B is halfway to the end of the plank.

1-7   Free-body diagram for the board:

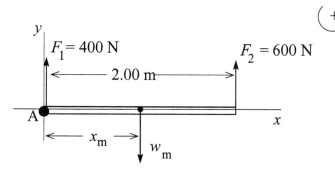

Let w_m be the weight of the. motor. Take the origin of coordinates at the end where the 400 N force is applied (point A).

$\sum F_y = ma_y$

$F_1 + F_2 - w_m = 0$

$w_m = F_1 + F_2 = 400 \text{ N} + 600 \text{ N} = 1000 \text{ N}$

$\sum \tau_A = 0$

$+F_2(2.00 \text{ m}) - w_m x_m = 0$

$x_m = 2.00 \text{ m}(F_2/w_m) = 2.00 \text{ m}(600 \text{ N}/1000 \text{ N}) = 1.20 \text{ m}$

The center of gravity of the motor is 1.20 m from the end where the 400 N force is applied.

1-11 a) Free-body diagram for the diving board:

Take the origin of coordinates at the left-hand end of the board (point A).

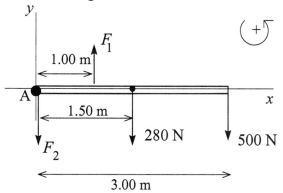

$\vec{F}_1$ is the force applied at the support point and $\vec{F}_2$ is the force at the end that is held down.

$\sum \tau_A = 0$ gives $+F_1(1.0 \text{ m}) - (500 \text{ N})(3.00 \text{ m}) - (280 \text{ N})(1.50 \text{ m}) = 0$

$F_1 = \dfrac{(500 \text{ N})(3.00 \text{ m}) + (280 \text{ N})(1.50 \text{ m})}{1.00 \text{ m}} = 1920 \text{ N}$

b) $\sum F_y = ma_y$

$F_1 - F_2 - 280 \text{ N} - 500 \text{ N} = 0$

$F_2 = F_1 - 280 \text{ N} - 500 \text{ N} = 1920 \text{ N} - 280 \text{ N} - 500 \text{ N} = 1140 \text{ N}$

11-13 a) Free-body diagram for the strut.

Take the origin of coordinates at the hinge (point A) and $+y$ upward. Let F_h an F_v be the horizontal and vertical components of the force $\vec{F}$ exerted on the stru by the pivot. The tension in the vertical cable is the weight w of the suspende object. The weight w of the strut can be taken to act at the center of the stru Let L be the length of the strut.

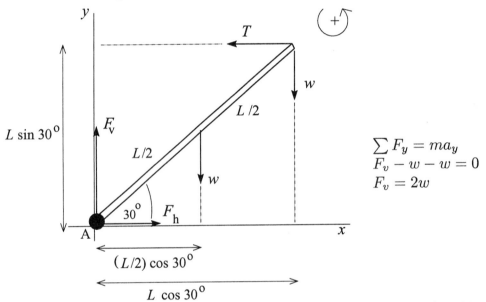

$\sum F_y = ma_y$
$F_v - w - w = 0$
$F_v = 2w$

Sum torques about point A. The pivot force has zero moment for this axis and doesn't enter into the torque equation.

$\tau_A = 0$

$TL \sin 30.0° - w((L/2) \cos 30.0°) - w(L \cos 30.0°) = 0$

$T \sin 30.0° - (3w/2) \cos 30.0° = 0$

$T = \dfrac{3w \cos 30.0°}{2 \sin 30.0°} = 2.60w$

Then $\sum F_x = ma_x$ implies $T - F_h = 0$ and $F_h = 2.60w$.

We now have the components of $\vec{F}$ so can find its magnitude and direction:

$F = \sqrt{F_h^2 + F_v^2}$

$F = \sqrt{(2.60w)^2 + (2.00w)^2}$

$F = 3.28w$

$\tan \theta = \dfrac{F_v}{F_h} = \dfrac{2.00w}{2.60w}$

$\theta = 37.6°$

b) Free-body diagram for the strut:

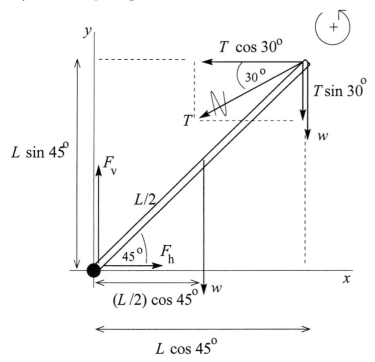

The tension T has been replaced by its x and y components. The torque due to T equals the sum of the torques of its components, and the latter are easier to calculate.

$$\sum \tau_A = 0 + (T \cos 30.0°)(L \sin 45.0°) - (T \sin 30.0°)(L \cos 45.0°) - w((L/2) \cos 45.0°) - w(L \cos 45.0°) = 0$$

The length L divides out of the equation. The equation can also be simplified by noting that $\sin 45.0° = \cos 45.0°$.

Then $T(\cos 30.0° - \sin 30.0°) = 3w/2$.

$$T = \frac{3w}{2(\cos 30.0° - \sin 30.0°)} = 4.10w$$

$\sum F_x = ma_x$

$F_h - T \cos 30.0° = 0$

$F_h = T \cos 30.0° = (4.10w)(\cos 30.0°) = 3.55w$

$\sum F_y = ma_y$

$F_v - w - w - T \sin 30.0° = 0$

$F_v = 2w + (4.10w) \sin 30.0° = 4.05w$

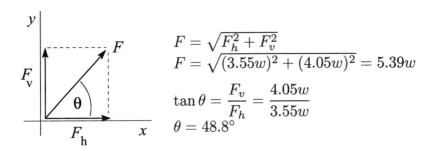

$$F = \sqrt{F_h^2 + F_v^2}$$
$$F = \sqrt{(3.55w)^2 + (4.05w)^2} = 5.39w$$

$$\tan\theta = \frac{F_v}{F_h} = \frac{4.05w}{3.55w}$$
$$\theta = 48.8°$$

11-15 Free-body diagram for the door:

Let $\vec{H}_1$ and $\vec{H}_2$ be the forces exerted by the upper and lower hinges. Take the origin of coordinates at the bottom hinge (point A) and $+y$ upward.

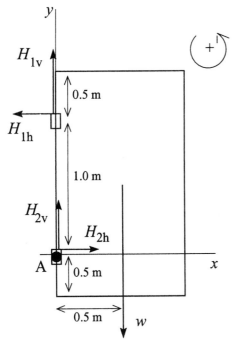

We are given that
$H_{1v} = H_{2v} = w/2 = 140$ N.

$$\sum F_x = ma_x$$
$$H_{2h} - H_{1h} = 0$$
$$H_{1h} = H_{2h}$$

The horizontal components of the hinge forces are equal in magnitude and opposite in direction.

Sum torques about point A. H_{1v}, H_{2v}, and H_{2h} all have zero moment arm and hence zero torque about an axis at this point.

Thus $\tau_A = 0$ gives $H_{1h}(1.00 \text{ m}) - w(0.50 \text{ m}) = 0$

$$H_{1h} = w\left(\frac{0.50 \text{ m}}{1.00 \text{ m}}\right) - \tfrac{1}{2}(280 \text{ N}) = 140 \text{ N}.$$

The horizontal component of each hinge force is 140 N.

11-17 Force diagram for the rod:

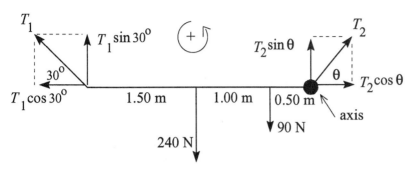

$\sum \tau = 0$, axis at right end of rod, counterclockwise torques positive

$(240 \text{ N})(1.50 \text{ m}) + (90 \text{ N})(0.50 \text{ m}) - (T_1 \sin 30.0°)(3.00 \text{ m}) = 0$

$$T_1 = \frac{360 \text{ N} \cdot \text{m} + 45 \text{ N} \cdot \text{m}}{1.50 \text{ m}} = 270 \text{ N}$$

$\sum F_x = ma_x$

$T_2 \cos \theta - T_1 \cos 30° = 0$

$T_2 \cos \theta = 234 \text{ N}$

$\sum F_y = ma_y$

$T_1 \sin 30° + T_2 \sin \theta - 240 \text{ N} - 90 \text{ N} = 0$

$T_2 \sin \theta = 330 \text{ N} - (270 \text{ N}) \sin 30° = 195 \text{ N}$

Then $\dfrac{T_2 \sin \theta}{T_2 \cos \theta} = \dfrac{195 \text{ N}}{234 \text{ N}}$

$\tan \theta = 0.8333$ and $\theta = 40°$

And $T_2 = \dfrac{195 \text{ N}}{\sin 40°} = 303 \text{ N}.$

11-19 a)

$\tau_1 = F_1 l_1 = +(8.00 \text{ N})(3.00 \text{ m})$

$\tau_1 = +24.0 \text{ N} \cdot \text{m}$

$\tau_2 = -F_2 l_2 = -(8.00 \text{ N})(l + 3.00 \text{ m})$

$\tau_2 = -24.0 \text{ N} \cdot \text{m} - (8.00 \text{ N})l$

$\sum \tau = \tau_1 + \tau_2 = +24.0 \text{ N} \cdot \text{m} - 24.0 \text{ N} \cdot \text{m} - (8.00 \text{ N})l = -(8.00 \text{ N})l$

Want l that makes $\sum \tau = -6.40 \text{ N} \cdot \text{m}$ (net torque must be clockwise)

$-(8.00 \text{ N})l = -6.40 \text{ N} \cdot \text{m}$

$l = (6.40 \text{ N} \cdot \text{m})/8.00 \text{ N} = 0.800$

b) $|\tau_2| > |\tau_1|$ since F_2 has a larger moment arm; the net torque is clockwise.

c)

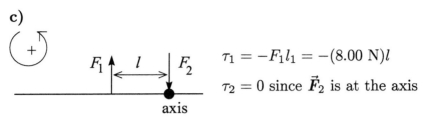

$T_1 = -F_1 l_1 = -(8.00 \text{ N})l$

$T_2 = 0$ since $\vec{F}_2$ is at the axis

$\sum \tau = -6.40 \text{ N} \cdot \text{m}$ gives $-(8.00 \text{ N})l = -6.40 \text{ N} \cdot \text{m}$

$l = 0.800$ m, same as in part (a).

11-21 $Y = \dfrac{l_0 F_\perp}{A \, \Delta l}$

$A = \dfrac{l_0 F_\perp}{Y \, \Delta l}$ (A is the cross-section area of the wire)

For steel, $Y = 2.0 \times 10^{11}$ Pa (Table 11-1)

Thus $A = \dfrac{(2.00 \text{ m})(400 \text{ N})}{(2.0 \times 10^{11} \text{ Pa})(0.25 \times 10^{-2} \text{ m})} = 1.6 \times 10^{-6} \text{ m}^2$.

$A = \pi r^2$, so $r = \sqrt{A/\pi} = \sqrt{1.6 \times 10^{-6} \text{ m}^2/\pi} = 7.1 \times 10^{-4}$ m

$d = 2r = 1.4 \times 10^{-3}$ m $= 1.4$ mm

11-25 Calculate the tensions T_1 and T_2.

0.50 m ↓ T_1

$m_1 = 6.0$ kg

0.50 m ↓ T_2

$m_2 = 10.0$ kg

Free-body-diagram for m_2:

y

T_2

$m_2 g$

x

$\sum F_y = ma_y$

$T_2 - m_2 g = 0$

$T_2 = 98.0$ N

Free-body-diagram for m_1:

$$\sum F_y = ma_y$$
$$T_1 - T_2 - m_1 g = 0$$
$$T_1 = T_2 + m_1 g$$
$$T_1 = 98.0 \text{ N} + 58.8 \text{ N} = 157 \text{ N}$$

a) $Y = \dfrac{\text{stress}}{\text{stain}}$ so strain $= \dfrac{\text{stress}}{Y} = \dfrac{F_\perp}{AY}$

upper wire: strain $= \dfrac{T_1}{AY} = \dfrac{157 \text{ N}}{(2.5 \times 10^{-7} \text{ m}^2)(2.0 \times 10^{11} \text{ Pa})} = 3.1 \times 10^{-3}$

lower wire: strain $= \dfrac{T_2}{AY} = \dfrac{98 \text{ N}}{(2.5 \times 10^{-7} \text{ m}^2)(2.0 \times 10^{11} \text{ Pa})} = 2.0 \times 10^{-3}$

b) strain $= \Delta l / l_0$ so $\Delta l = l_0(\text{strain})$
upper wire: $\Delta l = (0.50 \text{ m})(3.1 \times 10^{-3}) = 1.6 \times 10^{-4} \text{ m} = 1.6 \text{ mm}$
lower wire: $\Delta l = (0.50 \text{ m})(2.0 \times 10^{-3}) = 1.0 \times 10^{-3} \text{ m} = 1.0 \text{ mm}$

11-31 $B = -\dfrac{\Delta p}{\Delta V / V_0} = -\dfrac{(3.6 \times 10^6 \text{ Pa})(600 \text{ cm}^3)}{(-0.45 \text{ cm}^3)} = +4.8 \times 10^9 \text{ Pa}$

$k = 1/B = 1/4.8 \times 10^9 \text{ Pa} = 2.1 \times 10^{-10} \text{ Pa}^{-1}$

11-33 Same material implies same S

$S = \dfrac{\text{stress}}{\text{strain}}$ so strain $= \dfrac{\text{stress}}{S} = \dfrac{F_\parallel / A}{S}$

Same forces implies same $F_\parallel$
For the smaller object, $(\text{strain})_1 = F_\parallel / A_1 S$
For the larger object, $(\text{strain})_2 = F_\parallel / A_2 S$

$\dfrac{(\text{strain})_2}{(\text{strain})_1} = \left(\dfrac{F_\parallel}{A_2 S} \right) \left(\dfrac{A_1 S}{F_\parallel} \right) = \dfrac{A_1}{A_2}$

Larger solid has triple each edge length, so $A_2 = 9A_1$, and $\dfrac{(\text{strain})_2}{(\text{strain})_1} = \dfrac{1}{9}$

11-35 Tensile stress $= \dfrac{F_\perp}{A} = \dfrac{F_\perp}{\pi r^2} = \dfrac{90.8 \text{ N}}{\pi (0.92 \times 10^{-3} \text{ m})^2} = 3.41 \times 10^7 \text{ Pa}$

Problems

11-39 a) Free-body diagram for the car:

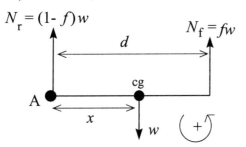

Let the cg be a distance x from the rear axle. Sum torques about the rear axle (point A).

$\sum \tau_A = 0$ so $N_f d - xw = 0$

$fwd - wx = 0$ and $x = fd$, as was to be shown.

b) In Example 11-2, $N_f = 0.53w$ so $f = 0.53$ and $d = 2.46$ m. The general resul derived in part (a) gives $x = fd = (0.53)(2.46 \text{ m}) = 1.30$ m, the same as calculated in the example.

11-43 a) Seesaw set so Karen exerts maximum torque about the pivot implies pivot moved as far as possible from Karen, so pivot is 0.20 m to left of center of seesaw

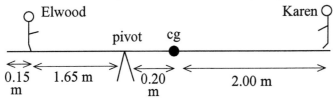

w_E is the weight of Elwood, w_K is the weight of Karen, and w_{ss} is the weight of the seesaw N is the normal force exerted by the pivot

Free-body diagram for the seesaw:

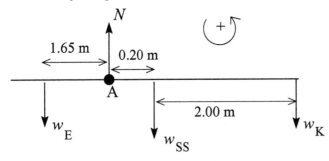

Apply $\sum \tau = 0$ with the axis at the pivot (point A) and counterclockwise torques positive.

$$w_E(1.65 \text{ m}) - w_{ss}(0.20 \text{ m}) - w_K(2.20 \text{ m}) = 0$$

$$w_E = \frac{w_K(2.20 \text{ m}) + w_{ss}(0.20 \text{ m})}{1.65 \text{ m}} = \frac{(420 \text{ N})(2.20 \text{ m}) + (240 \text{ N})(0.20 \text{ m})}{1.65 \text{ m}} = 589 \text{ N}$$

b) The center of gravity of each piece of the system (Elwood, Karen, and the seesaw is located some distance above the pivot. When the seesaw is horizontal the forc diagram that takes this account gives forces with lines of action and moment arm the same as in part (a):

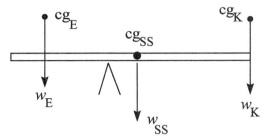

The center of gravity cg$_{\text{tot}}$ of the total system (seesaw, Elwood, Karen) is located a distance y_{cm} directly above the pivot P. When the seesaw is horizontal the line of action of the total weight Mg and the normal N coincide and there is zero net torque.

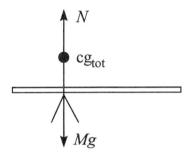

Now consider the force diagram when the system is rotated through an angle θ:

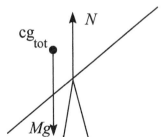

The line of action of the total weight force no longer passes through the pivot so now produces a counterclockwise torque about the pivot. The resultant torque in this direction rotates the system farther from equilibrium. (The same conclusion is reached if the seesaw is rotated clockwise from the horizontal.)

11-47 Free-body diagram for the bar:

N is the normal force exerted on the bar by the surface. There is no friction force at this surface.

H_h and H_v are the components of the force exerted on the bar by the hinge. The components of the force of the bar on the hinge will be equal in magnitude and opposite in direction.

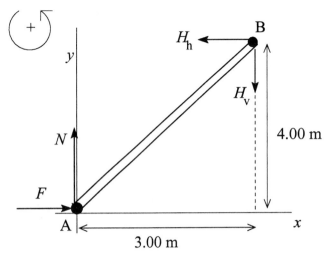

$$\sum F_x = ma_x$$
$$F = H_h = 120 \text{ N}$$

$$\sum F_y = ma_y$$
$$N - H_v = 0$$

$H_v = N$, but we don't know either of these forces.

$\sum \tau_B = 0$ gives $F(4.00 \text{ m}) - N(3.00 \text{ m}) = 0$

$N = (4.00 \text{ m}/3.00 \text{ m})F = \frac{4}{3}(120 \text{ N}) = 160 \text{ N}$ and then $H_v = 160 \text{ N}$

Force of bar on hinge:

horizontal component 120 N, to right

vertical component 160 N, upward

11-51 a)

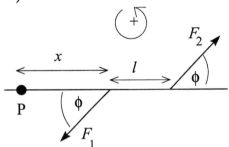

$F_1 = F_2 = F$

For an axis at P.
$$\tau_1 = -F_1 x \sin \phi = -F x \sin \phi$$
$$\tau_2 = +F_2(x + l) \sin \phi = +F(x + l) \sin \phi$$

$\tau_P = \tau_1 + \tau_2 = F \sin \phi (x + l - x) = F l \sin \phi$, which is independent of x. Therefor the resultant torque is the same for an axis at any point along the rod.

b)

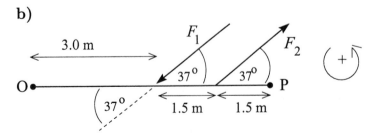

For an axis at O,
$$\tau_1 = -F_1(3.0 \text{ m}) \sin 37° = -(14.0 \text{ N})(3.0 \text{ m}) \sin 37° = -25.28 \text{ N} \cdot \text{m}$$
$$\tau_2 = +F_2(4.5 \text{ m}) \sin 37° = +(14.0 \text{ N})(4.5 \text{ m}) \sin 37° = +37.91 \text{ N} \cdot \text{m}$$

$$\sum \tau_O = \tau_1 + \tau_2 = -25.28 \ \text{N} \cdot \text{m} + 37.91 \ \text{N} \cdot \text{m} = +12.6 \ \text{N} \cdot \text{m}.$$

For an axis at P,

$$\tau_1 = +F_1(3.0 \ \text{m}) \sin 37° = +(14.0 \ \text{N})(3.0 \ \text{m}) \sin 37° = +25.28 \ \text{N} \cdot \text{m}$$

$$\tau_2 = -F_2(1.5 \ \text{m}) \sin 37° = -(14.0 \ \text{N})(1.5 \ \text{m}) \sin 37° = -12.64 \ \text{N} \cdot \text{m}$$

$$\sum \tau_P = \tau_1 + \tau_2 = +25.28 \ \text{N} \cdot \text{m} - 12.64 \ \text{N} \cdot \text{m} = +12.6 \ \text{N} \cdot \text{m}.$$

The general result derived in part (a) gives

$$\sum \tau = Fl \sin \phi = (14.0 \ \text{N})(1.5 \ \text{m}) \sin 37° = +12.6 \ \text{N} \cdot \text{m}.$$

The torque about P is the same as about O, and agrees with the general result derived in part (a).

11-61 Free-body diagram for the crate:

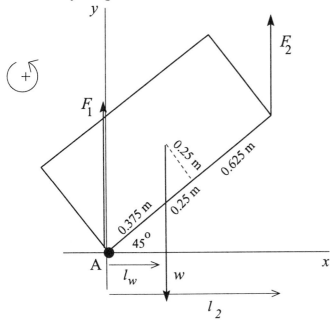

$l_w = (0.375 \ \text{m}) \cos 45°$
$l_2 = (1.25 \ \text{m}) \cos 45°$

Let $\vec{F}_1$ and $\vec{F}_2$ be the vertical forces exerted by you and your friend. Take the origin at the lower left-hand corner of the crate (point A).

$\sum F_y = ma_y$ gives $F_1 + F_2 - w = 0$

$F_1 + F_2 = w = (200 \ \text{kg})(9.80 \ \text{m/s}^2) = 1960 \ \text{N}$

$\sum \tau_A = 0$ gives $F_2 l_2 - w l_w = 0$

$$F_2 = w \left(\frac{l_w}{l_2} \right) = 1960 \ \text{N} \left(\frac{0.375 \ \text{m} \cos 45°}{1.25 \ \text{m} \cos 45°} \right) = 590 \ \text{N}$$

Then $F_1 = w - F_2 = 1960 \ \text{N} - 590 \ \text{N} = 1370 \ \text{N}.$

The person below (you) applies a force of 1370 N. The person above (your friend) applies a force of 590 N. It is better to be the person above.

11-63 Free-body diagram for the gate:

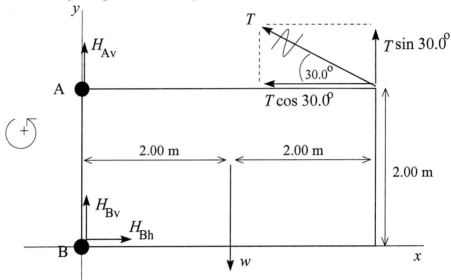

Use coordinates with the origin at B. Let $\vec{H}_A$ and $\vec{H}_B$ be the forces exerted by th
hinges at A and B. The problem states that $\vec{H}_A$ has no horizontal component.
Replace the tension $\vec{T}$ by its horizontal and vertical components.

a) $\sum \tau_B = 0$ gives $+(T \sin 30.0°)(4.00 \text{ m}) + (T \cos 30.0°)(2.00 \text{ m}) - w(2.00 \text{ m}) =$
$T(2 \sin 30.0° + \cos 30.0°) = w$

$$T = \frac{w}{2 \sin 30.0° + \cos 30.0°} = \frac{500 \text{ N}}{2 \sin 30.0° + \cos 30.0°} = 268 \text{ N}$$

b) $\sum F_x = ma_x$ says $H_{Bh} - T \cos 30.0° = 0$
$H_{Bh} = T \cos 30.0° = (268 \text{ N}) \cos 30.0° = 232 \text{ N}$

c) $\sum F_y = ma_y$ says $H_{Av} + H_{Bv} + T \sin 30.0° - w = 0$
$H_{Av} + H_{Bv} = w - T \sin 30.0° = 500 \text{ N} - (268 \text{ N}) \sin 30.0° = 366 \text{ N}$

11-65 a) Find the angle where the bale starts to tip:

When starts to tip only the lower left-hand corner of the bale makes contact wit
the conveyor belt. Therefore the line of action of the normal force N passes throug
the left-hand edge of the bale. Consider $\sum \tau_A = 0$ with point A at the lower lef
hand corner. Then $\tau_N = 0$ and $\tau_f = 0$, so it must be that $\tau_{mg} = 0$ also. Th
means that the line of action of the gravity must pass through point A. Thus th
free-body diagram must be as follows:

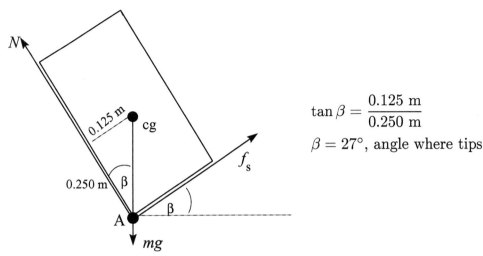

$$\tan \beta = \frac{0.125 \text{ m}}{0.250 \text{ m}}$$

$\beta = 27°$, angle where tips

At the angle where the bale is ready to slip down the incline f_s has its maximum possible value, $f_s = \mu_s N$. Free-body diagram for the bale, with the origin of coordinates at the cg:

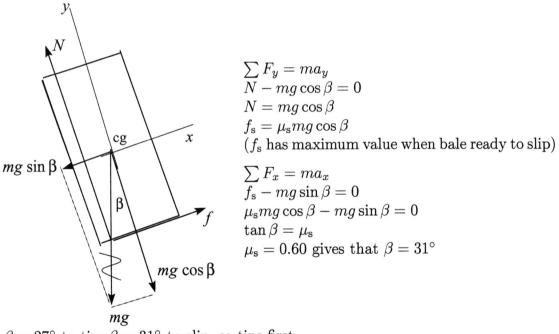

$$\sum F_y = ma_y$$
$$N - mg \cos \beta = 0$$
$$N = mg \cos \beta$$
$$f_s = \mu_s mg \cos \beta$$
(f_s has maximum value when bale ready to slip)

$$\sum F_x = ma_x$$
$$f_s - mg \sin \beta = 0$$
$$\mu_s mg \cos \beta - mg \sin \beta = 0$$
$$\tan \beta = \mu_s$$
$$\mu_s = 0.60 \text{ gives that } \beta = 31°$$

$\beta = 27°$ to tip; $\beta = 31°$ to slip, so tips first

b) The magnitude of the friction force didn't enter into the calculation of the tipping angle; still tips at $\beta = 27°$.

For $\mu_s = 0.40$ tips at $\beta = \arctan(0.40) = 22°$

Now the bale will start to slide down the incline before it tips.

11-67 a) Free-body diagram for the door:

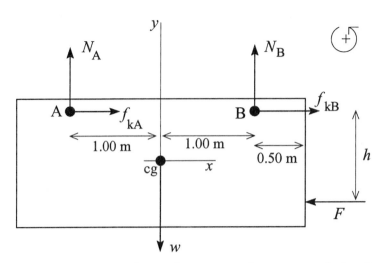

Take the origin of coordinates at the center of the door (at the cg). Let N_A f_{kA} N_B, and f_{kB} be the normal and friction forces exerted on the door at each wheel

$\sum F_y = ma_y$

$N_A + N_B - w = 0$

$N_A + N_B = w = 950 \text{ N}$

$\sum F_x = ma_x$

$f_{kA} + f_{kB} - F = 0$

$F = f_{kA} + f_{kB}$

$f_{kA} = \mu_k N_A, \quad f_{kB} = \mu_k N_B,$

so $F = \mu_k(N_A + N_B) = \mu_k w = (0.52)(950 \text{ N}) = 494 \text{ N}$

$\sum \tau_B = 0$

N_B, f_{kA}, and f_{kB} all have zero moment arms and hence zero torque about this point.

Thus $+w(1.00 \text{ m}) - N_A(2.00 \text{ m}) - F(h) = 0$

$$N_A = \frac{w(1.00 \text{ m}) - F(h)}{2.00 \text{ m}} = \frac{(950 \text{ N})(1.00 \text{ m}) - (494 \text{ N})(1.60 \text{ m})}{2.00 \text{ m}} = 80 \text{ N}$$

And then $N_B = 950 \text{ N} - N_A = 950 \text{ N} - 80 \text{ N} = 870 \text{ N}$.

b) If h is too large the torque of F will cause wheel A to leave the track. When wheel A just starts to lift off the track N_A and f_{kA} both go to zero.

The equations in part (a) still apply:

$N_A + N_B - w = 0$ gives $N_B = w = 950 \text{ N}$

Then $f_{kB} = \mu_k N_B = 0.52(950 \text{ N}) = 494 \text{ N}$

$F = f_{kA} + f_{kB} = 494 \text{ N}$

$$+w(1.00 \text{ m}) - N_A(2.00 \text{ m}) - F(h) = 0$$

$$h = \frac{w(1.00 \text{ m})}{F} = \frac{(950 \text{ N})(1.00 \text{ m})}{494 \text{ N}} = 1.92 \text{ m}$$

11-69 **a)** Free-body diagram for the pole:

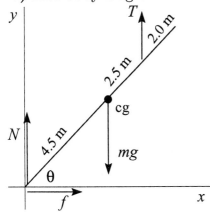

N and f are the vertical and horizontal components of the force the ground exerts on the pole.

$$\sum F_x = ma_x$$
$$f = 0$$

The force exerted by the ground has no horizontal component.

$$\sum \tau_A = 0$$
$$+T(7.0 \text{ m}) \cos\theta - mg(4.5 \text{ m}) \cos\theta = 0$$
$$T = mg(4.5 \text{ m}/7.0 \text{ m}) = (4.5/7.0)(5700 \text{ N}) = 3700 \text{ N}$$

$$\sum F_y = 0$$
$$N + T - mg = 0$$
$$N = mg - T = 5700 \text{ N} - 3700 \text{ N} = 2000 \text{ N}$$

The force exerted by the ground is vertical (upward) and has magnitude 2000 N.

b) In the $\sum \tau_A = 0$ equation the angle θ divided out. All forces on the pole are vertical and their moment arms are all proportional to $\cos\theta$.

11-71 Calculate the tension in the wire as the mass passes through the lowest point.

a) Free-body diagram for the mass:

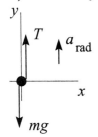

The mass moves in an arc of a circle with radius $R = 0.50$ m. It has acceleration $\vec{a}_{rad}$ directed in toward the center of the circle, so at this point $\vec{a}_{rad}$ is upward.

$$\sum F_y = ma_y$$
$$T - mg = mR\omega^2 \text{ so that } T = m(g + R\omega^2).$$

But ω must be in rad/s:

$\omega = (120 \text{ rev/min})(2\pi \text{ rad}/1 \text{ rev})(1 \text{ min}/60 \text{ s}) = 12.57 \text{ rad/s}.$

Then $T = (12.0 \text{ kg})(9.80 \text{ m/s}^2 + (0.50 \text{ m})(12.57 \text{ rad/s})^2) = 1066 \text{ N}.$

Now calculate the elongation Δl of the wire that this tensile force produces:

$$Y = \frac{F_\perp l_0}{A \Delta l} \text{ so } \Delta l = \frac{F_\perp l_0}{YA} = \frac{(1066 \text{ N})(0.50 \text{ m})}{(7.0 \times 10^{10} \text{ Pa})(0.014 \times 10^{-4} \text{ m}^2)} = 0.54 \text{ cm}.$$

b) The acceleration $\vec{a}_{\text{rad}}$ is directed in towards the center of the circular path, and at this point in the motion this direction is downward.

$\sum F_y = ma_y$

$mg + T = mR\omega^2$

$T = m(R\omega^2 - g)$

$T = (12.0 \text{ kg})((0.50 \text{ m})(12.57 \text{ rad/s})^2 - 9.80 \text{ m/s}^2) = 830 \text{ N}$

$$\Delta l = \frac{F_\perp l_0}{YA} = \frac{(830 \text{ N})(0.50 \text{ m})}{(7.0 \times 10^{10} \text{ Pa})(0.014 \times 10^{-4} \text{ m}^2)} = 0.42 \text{ cm}.$$

11-73 a) stress $= F_\perp/A$, so equal stress implies T/A same for each wire.

$T_A/2.00 \text{ mm}^2 = T_B/4.00 \text{ mm}^2$ so $T_B = 2.00T_A$

The question is where along the rod to hang the weight in order to produce this relation between the tensions in the two wires. Let the weight be suspended at point C, a distance x to the right of wire A. The free-body diagram for the rod is then

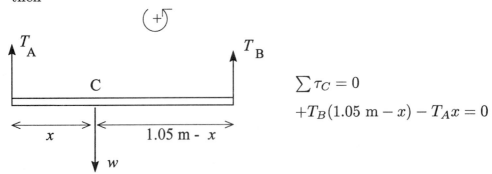

$\sum \tau_C = 0$

$+T_B(1.05 \text{ m} - x) - T_A x = 0$

But $T_B = 2.00T_A$ so $2.00T_A(1.05 \text{ m} - x) - T_A x = 0$

$2.10 \text{ m} - 2.00x = x$ and $x = 2.10 \text{ m}/3.00 = 0.70 \text{ m}$ (measured from A).

b) $Y = $ stress/strain gives that strain $= $ stress$/Y = F_\perp/AY$.

Equal strain thus implies

$$\frac{T_A}{(2.00 \text{ mm}^2)(1.80 \times 10^{11} \text{ Pa})} = \frac{T_B}{(4.00 \text{ mm}^2)(1.20 \times 10^{11} \text{ Pa})}$$

$$T_B = \left(\frac{4.00}{2.00}\right)\left(\frac{1.20}{1.80}\right) T_A = 1.333 T_A.$$

The $\sum \tau_C = 0$ equation still gives $T_B(1.05 \text{ m} - x) - T_A x = 0$.

But now $T_B = 1.333 T_A$ so $(1.333 T_A)(1.05 \text{ m} - x) - T_A x = 0$

$1.40 \text{ m} = 2.33x$ and $x = 1.40 \text{ m}/2.33 = 0.60 \text{ m}$ (measured from A).

1-75 Each piece of the composite rod is subjected to a tensile force of 4.00×10^4 N.

a) $Y = \dfrac{F_\perp l_0}{A \Delta l}$ so $\Delta l = \dfrac{F_\perp l_0}{YA}$

$\Delta l_b = \Delta l_n$ gives that $\dfrac{F_\perp l_{0,b}}{Y_b A_b} = \dfrac{F_\perp l_{0,n}}{Y_n A_n}$ (b for brass and n for nickel); $l_{0,n} = L$

But the $F_\perp$ is the same for both, so

$l_{0,n} = \dfrac{Y_n}{Y_b} \dfrac{A_n}{A_b} l_{0,b}$

$L = \left(\dfrac{21 \times 10^{10} \text{ Pa}}{9.0 \times 10^{10} \text{ Pa}}\right)\left(\dfrac{1.00 \text{ cm}^2}{2.00 \text{ cm}^2}\right)(1.40 \text{ m}) = 1.63 \text{ m}$

b) stress $= F_\perp/A = T/A$

brass: stress $= T/A = (4.00 \times 10^4 \text{ N})/(2.00 \times 10^{-4} \text{ m}^2) = 2.00 \times 10^8$ Pa

nickel: stress $= T/A = (4.00 \times 10^4 \text{ N})/(1.00 \times 10^{-4} \text{ m}^2) = 4.00 \times 10^8$ Pa

c) $Y = $ stress/strain and strain $= $ stress/Y

brass: strain $= (2.00 \times 10^8 \text{ Pa})/(9.0 \times 10^{10} \text{ Pa}) = 2.22 \times 10^{-3}$

nickel: strain $= (4.00 \times 10^8 \text{ Pa})/(21 \times 10^{10} \text{ Pa}) = 1.90 \times 10^{-3}$

11-77 $Y = F_\perp l_0/A \Delta l$ (Eq.11-10 holds since the problem states that the stress is proportional to the strain.)

Thus $\Delta l = F_\perp l_0/AY$

a) Change l_0 but $F_\perp$ (same floodlamp), A (same diameter wire), and Y (same material) all stay the same.

$\dfrac{\Delta l}{l_0} = \dfrac{F_\perp}{AY} = $ constant, so $\dfrac{\Delta l_1}{l_{01}} = \dfrac{\Delta l_2}{l_{02}}$

$\Delta l_2 = \Delta l_1 (l_{02}/l_{01}) = 2\Delta l_1 = 2(0.18 \text{ mm}) = 0.36 \text{ mm}$

b) $A = \pi(d/2)^2 = \frac{1}{4}\pi d^2$, so $\Delta l = \dfrac{F_\perp l_0}{\frac{1}{4}\pi d^2 Y}$

$F_\perp$, l_0, Y all stay the same, so $\Delta l(d^2) = F_\perp l_0/(\frac{1}{4}\pi Y) = \text{constant}$
$\Delta l_1(d_1^2) = \Delta l_2(d_2^2)$
$\Delta l_2 = \Delta l_1(d_1/d_2)^2 = (0.18 \text{ mm})(1/2)^2 = 0.045 \text{ mm}$

c) $F_\perp$, l_0, A all stay the same so $\Delta l\, Y = F_\perp l_0/A = \text{constant}$
$\Delta l_1 Y_1 = \Delta l_2 Y_2$
$\Delta l_2 = \Delta l_1(Y_1/Y_2) = (0.18 \text{ mm})(20 \times 10^{10} \text{ Pa}/11 \times 10^{10} \text{ Pa}) = 0.33 \text{ mm}$

11-79

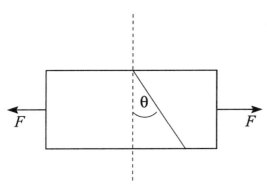

a)

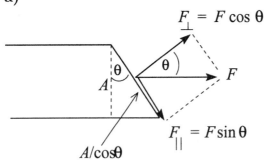

$F_\perp = F \cos \theta$

$F_\parallel = F \sin \theta$

$A/\cos\theta$

The area of the diagonal face is $A/\cos \theta$.

$\text{tensile stress} = \dfrac{F_\perp}{(A/\cos\theta)} = F\cos\theta/(A/\cos\theta) = \dfrac{F\cos^2\theta}{A}.$

b) $\text{shear stress} = \dfrac{F_\parallel}{(A/\cos\theta)} = F\sin\theta/(A/\cos\theta) = \dfrac{F\sin\theta\cos\theta}{A} = \dfrac{F\sin 2\theta}{2A}$ (usin
a trig identity).

c) From the result of part (a) the tensile stress is a maximum for $\cos\theta = 1$, s
$\theta = 0°$.

d) From the result of part (b) the shear stress is a maximum for $\sin 2\theta = 1$, so fo
$2\theta = 90°$ and thus $\theta = 45°$

xercises

2-1 $F_{\text{S on m}} = G\dfrac{m_S m_m}{r_{Sm}^2}$ (S = sun, m = moon); $F_{\text{E on m}} = G\dfrac{m_E m_m}{r_{Em}^2}$ (E = earth)

$$\frac{F_{\text{S on m}}}{F_{\text{E on m}}} = \left(G\frac{m_S m_m}{r_{Sm}^2}\right)\left(\frac{r_{Em}^2}{G m_E m_m}\right) = \frac{m_S}{m_E}\left(\frac{r_{Em}}{r_{Sm}}\right)^2$$

r_{Em}, the radius of the moon's orbit around the earth is given in Appendix F as 3.84×10^8 m. The moon is much closer to the earth than it is to the sun, so take the distance r_{Sm} of the moon from the sun to be r_{SE}, the radius of the earth's orbit around the sun.

$$\frac{F_{\text{S on m}}}{F_{\text{E on m}}} = \left(\frac{1.99 \times 10^{30} \text{ kg}}{5.98 \times 10^{24} \text{ kg}}\right)\left(\frac{3.84 \times 10^8 \text{ m}}{1.50 \times 10^{11} \text{ m}}\right)^2 = 2.18.$$

The force exerted by the sun is larger than the force exerted by the earth. The moon's motion is a combination of orbiting the sun and orbiting the earth.

2-3 $F_{12} = G\dfrac{m_1 m_2}{r_{12}^2}$

For the second set of spheres

$$F_{12}' = G\frac{(nm_1)(nm_2)}{(nr_{12})^2} = \frac{n^2}{n^2}G\frac{m_1 m_2}{r_{12}^2} = G\frac{m_1 m_2}{r_{12}^2}.$$

The force is the same as for the first set of spheres.

2-5 a)

The distance from the earth to the sun is $r = 1.50 \times 10^{11}$ m. Let the ship be a distance x from the earth; it is then a distance $r - x$ from the sun.

$F_E = F_S$ says that $Gmm_E/x^2 = Gmm_S/(r-x)^2$

$m_E/x^2 = m_S/(r-x)^2$ and $(r-x)^2 = x^2(m_S/m_E)$

$r - x = x\sqrt{m_S/m_E}$ and $r = x(1 + \sqrt{m_S/m_E})$

$$x = \frac{r}{1 + \sqrt{m_s/m_E}} = \frac{1.50 \times 10^{11} \text{ m}}{1 + \sqrt{1.99 \times 10^{30} \text{ kg}/5.97 \times 10^{24} \text{ kg}}} = 2.59 \times 10^8 \text{ m (from}$$

center of earth)

b) At the instant when the spaceship passes through this point its acceleration

zero.

12-11

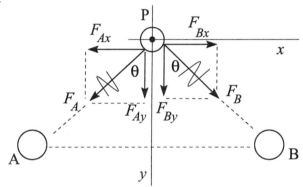

$\sin\theta = 0.80$

$\cos\theta = 0.60$

Take the origin of coordinates at point P.

$F_A = G\frac{m_A m}{r^2} = (6.673 \times 10^{-11} \text{ N} \cdot \text{m}^2/\text{kg}^2)\frac{(0.26 \text{ kg})(0.010 \text{ kg})}{(0.100 \text{ m})^2} = 1.735 \times 10^{-11}$ N

$F_B = G\frac{m_B m}{r^2} = 1.735 \times 10^{-11}$ N

$F_{Ax} = -F_A \sin\theta = -(1.735 \times 10^{-11} \text{ N})(0.80) = -1.39 \times 10^{-11}$ N

$F_{Ay} = -F_A \cos\theta = +(1.735 \times 10^{-11} \text{ N})(0.60) = +1.04 \times 10^{-11}$ N

$F_{Bx} = +F_B \sin\theta = +1.39 \times 10^{-11}$ N

$F_{By} = +F_B \sin\theta = +1.04 \times 10^{-11}$ N

$\sum F_x = ma_x$ gives $F_{Ax} + F_{Bx} = ma_x$

$0 = ma_x$ so $a_x = 0$

$\sum F_y = ma_y$ gives $F_{Ay} + F_{By} = ma_y$

$2(1.04 \times 10^{-11} \text{ N}) = (0.010 \text{ kg})a_y$

$a_y = 2.1 \times 10^{-9} \text{ m/s}^2$, directed downward midway between A and B

12-15 **a)** The acceleration due to gravity at the surface of Titania

is given by $g_T = Gm_T/R_T^2$, where m_T is its mass and R_T is its radius.

For the earth, $g_E = Gm_E/R_E^2$.

For Titania, $m_T = m_E/1700$ and $R_T = R_E/8$, so

$$g_T = \frac{Gm_T}{R_T^2} = \frac{G(m_E/1700)}{(R_E/8)^2} = \left(\frac{64}{1700}\right)\frac{Gm_E}{R_E^2} = 0.0377g_E.$$

Since $g_E = 9.80$ m/s^2, $g_T = (0.0377)(9.80$ m/s$^2) = 0.37$ m/s^2.

b) We know that the density of the earth is 5500 kg/m^3. For Titania,

$$\rho_T = \frac{m_T}{\frac{4}{3}\pi R_T^3} = \frac{m_E/1700}{\frac{4}{3}\pi(R_E/8)^3} = \frac{512}{1700}\rho_E = \frac{512}{1700}(5500 \text{ kg/m}^3) = 1700 \text{ kg/m}^3$$

2-17 Use the measured gravitational force to calculate the gravitational constant G:
$$F_g = G\frac{m_1 m_2}{r^2} \text{ so}$$

$$G = \frac{F_g r^2}{m_1 m_2} = \frac{(8.00 \times 10^{-10} \text{ N})(0.0100 \text{ m})^2}{(0.400 \text{ kg})(3.00 \times 10^{-3} \text{ kg})} = 6.667 \times 10^{-11} \text{ N} \cdot \text{m}^2/\text{kg}^2.$$

Then use $g = Gm_E/R_E^2$ (Eq.12-4) to calculate the mass of the earth:

$$g = \frac{Gm_E}{R_E^2} \text{ gives } m_E = \frac{R_E^2 g}{G} = \frac{(6.38 \times 10^6 \text{ m})^2(9.80 \text{ m/s}^2)}{6.667 \times 10^{-11} \text{ N} \cdot \text{m}^2/\text{kg}^2} = 5.98 \times 10^{24} \text{ kg}.$$

2-19 Example 12-5 gives the escape speed as $v_1 = \sqrt{2GM/R}$,

where M and R are the mass and radius of the astronomical object.

$$v_1 = \sqrt{2(6.673 \times 10^{-11} \text{ N} \cdot \text{m}^2/\text{kg}^2)(3.6 \times 10^{12} \text{ kg})/700 \text{ m}} = 0.83 \text{ m/s}.$$

At this speed a person can walk 100 m in 120 s; easily achieved for the average person.

12-23 a) Find the orbit radius r: $\dfrac{Gmm_E}{r^2} = m\dfrac{v^2}{r}$.

$r = Gm_E/v^2 = (6.673 \times 10^{-11} \text{ N} \cdot \text{m}^2/\text{kg}^2)(5.97 \times 10^{24} \text{ kg})/(6200 \text{ m/s})^2$

$= 1.036 \times 10^7$ m

The period (time for one revolution) is then given by

$T = 2\pi r/v = 2\pi(1.036 \times 10^7 \text{ m})/(6200 \text{ m/s}) = 1.05 \times 10^4 \text{ s} = 175 \text{ min}$

b) $a_{\text{rad}} = v^2/r = (6200 \text{ m/s})^2/1.036 \times 10^7 \text{ m} = 3.71 \text{ m/s}^2$.

12-25

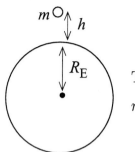

The radius of the orbit is $r = h + R_E$.

$r = 7.80 \times 10^5$ m $+ 6.38 \times 10^6$ m $= 7.16 \times 10^6$ m.

Free-body diagram for the satellite:

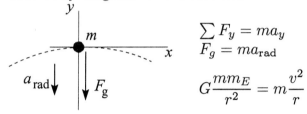

$$\sum F_y = ma_y$$
$$F_g = ma_{\text{rad}}$$

$$G\frac{mm_E}{r^2} = m\frac{v^2}{r}$$

$$v = \sqrt{\frac{Gm_E}{r}} = \sqrt{\frac{(6.673 \times 10^{-11} \text{ N} \cdot \text{m}^2/\text{kg}^2)(5.97 \times 10^{24} \text{ kg})}{7.16 \times 10^6 \text{ m}}} = 7.46 \times 10^3 \text{ m/s}$$

12-31 **a)** The gravitational force exerted on the spacecraft by the sun is

$F_g = Gm_Sm_H/r^2$, where m_S is the mass of the sun and m_H is the mass of th
Helios B spacecraft.

For a circular orbit, $a_{\text{rad}} = v^2/r$ and $\sum F = m_Hv^2/r$. If we neglect all forces o
the spacecraft except for the force exerted by the sun,

$F_g = \sum F = m_Hv^2/r$, so $Gm_Sm_H/r^2 = m_Hv^2/r$

$v = \sqrt{Gm_S/r} = \sqrt{(6.673 \times 10^{-11} \text{ N} \cdot \text{m}^2/\text{kg}^2)(1.99 \times 10^{30} \text{ kg})/43 \times 10^9 \text{ m}}$

$= 5.6 \times 10^4$ m/s $= 56$ km/s

The actual speed is 71 km/s, so the orbit cannot be circular.

b) The orbit is a circle or an ellispe if it is closed, a parabola or hyperbola if open
The orbit is closed if the total energy (kinetic + potential) is negative, so that th
object cannot reach $r \to \infty$.

For Helios B,

$K = \frac{1}{2}m_Hv^2 = \frac{1}{2}m_H(71 \times 10^3 \text{ m/s})^2 = (2.52 \times 10^9 \text{ m}^2/\text{s}^2)m_H$

$U = -Gm_Sm_H/r = m_H(-(6.673 \times 10^{-11} \text{ N} \cdot \text{m}^2/\text{kg}^2)(1.99 \times 10^{30} \text{ kg})/(43 \times$
$10^9 \text{ m})) = -(3.09 \times 10^9 \text{ m}^2/\text{s}^2)m_H$

$E = K + U = (2.52 \times 10^9 \text{ m}^2/\text{s}^2)m_H - (3.09 \times 10^9 \text{ m}^2/\text{s}^2)m_H$

$= -(5.7 \times 10^8 \text{ m}^2/\text{s}^2)m_H$

The total energy E is negative, so the orbit is closed. We know from part (a) that it is not circular, so it must be elliptical.

2-33 a)

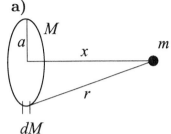

Divide the ring up into small segments dM.

The gravitational potential energy of dM and m is $dU = -Gm\,dM/r$.

The total gravitational potential energy of the ring and particle is $U = \int dU = -Gm \int dM/r$.

But $r = \sqrt{x^2 + a^2}$ is the same for all segments of the ring, so

$$U = -\frac{Gm}{r} \int dM = -\frac{GmM}{r} = -\frac{GmM}{\sqrt{x^2 + a^2}}$$

b) When $x \gg a$, $\sqrt{x^2 + a^2} \rightarrow \sqrt{x^2} = x$ and $U = -GmM/x$. This is the gravitational potential energy of two point masses separated by a distance x. This is the expected result.

c) $F_x = -\dfrac{dU}{dx} = -\dfrac{d}{dx}\left(-\dfrac{GmM}{\sqrt{x^2 + a^2}}\right)$

$F_x = +GmM\dfrac{d}{dx}(x^2 + a^2)^{-1/2} = GmM\left(-\dfrac{1}{2}(2x)(x^2 + a^2)^{-3/2}\right)$

$F_x = -GmMx/(x^2 + a^2)^{3/2}$; the minus sign means the force is attractive.

d) For $x \gg a$, $(x^2 + a^2)^{3/2} \rightarrow (x^2)^{3/2} = x^3$

Then $F_x = -GmMx/x^3 = -GmM/x^2$. This is the force between two point masses separated by a distance x and is the expected result.

e) For $x = 0$, $U = -GMm/a$. Each small segment of the ring is the same distance from the center and the potential is the same as that due to a point charge of mass M located at a distance a.

For $x = 0$, $F_x = 0$. When the particle is at the center of the ring, symmetrically placed segments of the ring exert equal and opposite forces and the total force exerted by the ring is zero.

12-35 a) $g_0 = Gm/R^2 = (6.673 \times 10^{-11} \text{ N} \cdot \text{m}^2/\text{kg}^2)(1.0 \times 10^{26} \text{ kg})/(2.5 \times 10^7 \text{ m})^2$

$= 10.7 \text{ m/s}^2$. This agrees with the value of g given in the problem.

$F = w_0 = mg_0 = (5.0 \text{ kg})(10.7 \text{ m/s}^2) = 53 \text{ N}$; this is the true weight of the object.

b) From Eq.(12-29), $w = w_0 - mv^2/R$

$$T = \frac{2\pi r}{v} \text{ gives } v = \frac{2\pi r}{T} = \frac{2\pi(2.5 \times 10^7 \text{ m})}{(16 \text{ h})(3600 \text{ s/1 h})} = 2.727 \times 10^3 \text{ m/s}$$

$v^2/R = (2.727 \times 10^3 \text{ s})^2/2.5 \times 10^7 \text{ m} = 0.297 \text{ m/s}^2$

Then $w = 53 \text{ N} - (5.0 \text{ kg})(0.297 \text{ m/s}^2) = 52 \text{ N}$

12-39 A black hole with the earth's mass M has the Schwarzschild radius R_S given by Eq.(12-32):

$R_S = 2Gm/c^2 = 2(6.673 \times 10^{-11} \text{ N} \cdot \text{m}^2/\text{kg}^2)(5.97 \times 10^{24} \text{ kg})/(2.998 \times 10^8 \text{ m/s})^2$
8.865×10^{-3} m

The ratio of R_S to the current radius R is $R_S/R = 8.865 \times 10^{-3} \text{ m}/6.38 \times 10^6 \text{ m}$
1.39×10^{-9}.

Problems

12-41 a)

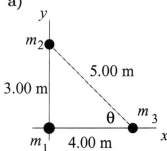

Section 12-7 proves that any two spherically symmetric masses interact as though they were point masses with all the mass concentrated at their centers.

Force diagram for m_3:

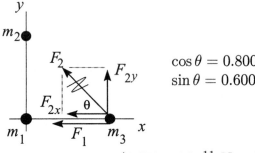

$\cos\theta = 0.800$
$\sin\theta = 0.600$

$$F_1 = G\frac{m_1 m_3}{r_{13}^2} = \frac{(6.673 \times 10^{-11} \text{ N} \cdot \text{m}^2/\text{kg}^2)(60.0 \text{ kg})(0.500 \text{ kg})}{(4.00 \text{ m})^2}$$
$= 1.251 \times 10^{-10}$ N

$$F_2 = G\frac{m_2 m_3}{r_{23}^2} = \frac{(6.673 \times 10^{-11} \text{ N} \cdot \text{m}^2/\text{kg}^2)(80.0 \text{ kg})(0.500 \text{ kg})}{(5.00 \text{ m})^2}$$
$= 1.068 \times 10^{-10}$ N

$F_{1x} = -1.251 \times 10^{-11}$ N, $F_{1y} = 0$

$F_{2x} = -F_2 \cos\theta = -(1.068 \times 10^{-10}$ N$)(0.800) = -8.544 \times 10^{-11}$ N

$F_{2y} = +F_2 \sin\theta = +(1.068 \times 10^{-10}$ N$)(0.600) = +6.408 \times 10^{-11}$ N

$F_x = F_{1x} + F_{2x} = -1.251 \times 10^{-10}$ N $- 8.544 \times 10^{-11}$ N $= -2.105 \times 10^{-10}$ N

$F_y = F_{1y} + F_{2y} = 0 + 6.408 \times 10^{-11}$ N $= +6.408 \times 10^{-11}$ N

$F = \sqrt{F_x^2 + F_y^2}$

$F = \sqrt{(-2.105 \times 10^{-10}\text{ N})^2 + (+6.408 \times 10^{-11}\text{ N})^2}$

$F = 2.20 \times 10^{-10}$ N

$\tan\theta = \dfrac{F_y}{F_x} = \dfrac{+6.408 \times 10^{-11}\text{ N}}{-2.105 \times 10^{-10}\text{ N}}; \quad \theta = 163°$

b) For the forces on it to be opposite in direction the third sphere must be on the y-axis and between the other two spheres.

$F_{\text{net}} = 0$ if $F_1 = F_2$

$G\dfrac{m_1 m_3}{y^2} = G\dfrac{m_2 m_3}{(3.00 \text{ m} - y)^2}$

$\dfrac{60.0}{y^2} = \dfrac{80.0}{(3.00 \text{ m} - y)^2}$

$\sqrt{80.0}\,y = \sqrt{60.0}(3.00 \text{ m} - y)$

$(\sqrt{80.0} + \sqrt{60.0})y = (3.00 \text{ m})\sqrt{60.0}$ and $y = 1.39$ m

Thus the sphere would have to be placed at thre point $x = 0$, $y = 1.39$ m

12-47 a) To stay above the same point on the surface of the earth the orbital period of the satellite must equal the orbital period of the earth:

$T = 1$ d$(24$ h$/$ 1 d$)(3600$ s$/$ 1 h$) = 8.64 \times 10^4$ s

Eq.(12-14) gives the relation between the orbit radius and the period:

$T = \dfrac{2\pi r^{3/2}}{\sqrt{Gm_E}}$ and $T^2 = \dfrac{4\pi^2 r^3}{Gm_E}$

$r = \left(\dfrac{T^2 G m_E}{4\pi^2}\right)^{1/3} =$

$\left(\dfrac{(8.64 \times 10^4 \text{ s})^2(6.673 \times 10^{-11}\text{ N}\cdot\text{m}^2/\text{kg}^2)(5.97 \times 10^{24}\text{ kg})}{4\pi^2}\right)^{1/3} = 4.23 \times 10^7$ m

This is the radius of the orbit; it is related to the height h above the earth's surface and the radius R_E of the earth by $r = h + R_E$. Thus $h = r - R_E = 4.23 \times 10^7$ m $- 6.38 \times 10^6$ m $= 3.59 \times 10^7$ m.

b)

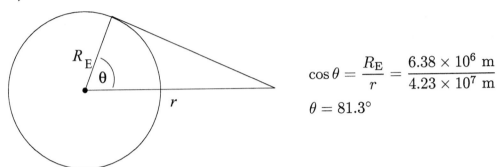

$$\cos\theta = \frac{R_E}{r} = \frac{6.38 \times 10^6 \text{ m}}{4.23 \times 10^7 \text{ m}}$$

$$\theta = 81.3°$$

A line from the satellite is tangent to a point on the earth that is at an angle of 81.3° above the equator. The sketch shows that points at higher latitudes are blocked by the earth from viewing the satellite.

12-51 $\underline{U = mgy}$ (Eq.(7-2))

Let $y = 0$ at the earth's surface. Then $U_1 = mgy_1 = 0$ and $U_2 = mgy_2 = mgh$.

$(\Delta U)_{\text{approx}} = U_2 - U_1 = mgh$

$\underline{U = -Gm_E m/r}$ (Eq.12-9)

This equation has built into it that $U \to 0$ as $r \to \infty$.

$U_1 = -Gm_E m/R_E$, $U_2 = -Gm_E m/(R_E + h)$

$$\Delta U = -(Gm_E m)\left(\frac{1}{R_E + h} - \frac{1}{R_E}\right) = -\frac{Gm_E m}{R_E}\left(\frac{1}{1 + h/R_E} - 1\right)$$

For only a 1% error in $\Delta U = mgh$ the value of h must be small compared to R_E, so use the binomial theorem to expand $(1 + h/R_E)^{-1}$ in powers of h/R_E:

$(1 + h/R_E)^{-1} = 1 - h/R_E + h^2/R_E^2 - \ldots$

$\Delta U = -(Gm_E m/R_E)(1 - h/R_E + h^2/R_E^2 - \ldots - 1) = (Gm_E m/R_E^2)h(1 - h/R_E)$

$g = Gm_E/R_E^2$, so $\Delta U = mgh(1 - h/R_E)$

Then $\Delta U - (\Delta U)_{\text{approx}} = mgh - mgh + mgh(h/R_E) = mgh(h/R_E)$

Want h such that $\dfrac{\Delta U - (\Delta U)_{\text{approx}}}{(\Delta U)_{\text{approx}}} = 0.010$, so $0.010 = \dfrac{mgh(h/R_E)}{mgh} = h/R_E$

$h = 0.010 R_E = 6.4 \times 10^4$ m $= 64$ km

12-53 First use the radius of the orbit to find the initial orbital speed:

$v = \sqrt{Gm/r}$ and $r = R_M + h = 1.74 \times 10^6$ m $+ 50.0 \times 10^3$ m $= 1.79 \times 10^6$ m

Thus $v = \sqrt{\dfrac{(6.673 \times 10^{-11} \text{ N} \cdot \text{m}^2/\text{kg}^2)(7.35 \times 10^{22} \text{ kg})}{1.79 \times 10^6 \text{ m}}} = 1.655 \times 10^3$ m/s

After the speed decreases by 20.0 m/s it becomes 1.655×10^3 m/s $- 20.0$ m/s $= 1.635 \times 10^3$ m/s.

Use conservation of energy to find the speed when the spacecraft reaches the lunar surface:

$K_1 + U_1 + W_{\text{other}} = K_2 + U_2$

Gravity is the only force that does work so $W_{\text{other}} = 0$ and $K_2 = K_1 + U_1 - U_2$

$U_1 = -Gm_{\text{m}}m/r; \quad U_2 = -Gm_{\text{m}}m/R_{\text{m}}$

$\frac{1}{2}mv_2^2 = \frac{1}{2}mv_1^2 + Gmm_{\text{m}}(1/R_{\text{m}} - 1/r)$

And the mass m divides out to give $v_2 = \sqrt{v_1^2 + 2Gm_{\text{m}}(1/R_{\text{m}} - 1/r)}$

$v_2 = 1.682 \times 10^3$ m/s(1 km/1000 m)(3600 s/1 h) $= 6060$ km/h

2-55

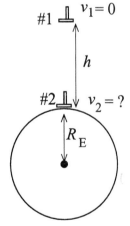

Take point 1 to be where the hammer is released and point 2 to be just above the surface of the earth, so $r_1 = R_E + h$ and $r_2 = R_E$.

$K_1 + U_1 + W_{\text{other}} = K_2 + U_2$

Only gravity does work, so $W_{\text{other}} = 0$.

$K_1 = 0, \quad K_2 = \frac{1}{2}mv_2^2$

$U_1 = -G\dfrac{mm_E}{r_1} = -\dfrac{Gmm_E}{h + R_E}, \quad U_2 = -G\dfrac{mm_E}{r_2} = -\dfrac{Gmm_E}{R_E}$

Thus $-G\dfrac{mm_E}{h + R_E} = \dfrac{1}{2}mv_2^2 - G\dfrac{mm_E}{R_E}$

$v_2^2 = 2Gm_E\left(\dfrac{1}{R_E} - \dfrac{1}{R_E + h}\right) = \dfrac{2Gm_E}{R_E(R_E + h)}(R_E + h - R_E) = \dfrac{2Gm_E h}{R_E(R_E + h)}$

$$v_2 = \sqrt{\frac{2Gm_{\mathrm{E}}h}{R_{\mathrm{E}}(R_{\mathrm{E}} + h)}}$$

12-61 a)

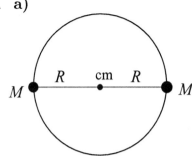

The two stars are separated by a distance $2R$, so

$$F_g = GM^2/(2R)^2 = GM^2/4R^2$$

b) $F_g = ma_{\mathrm{rad}}$

$GM^2/4R^2 = M(v^2/R)$ so $v = \sqrt{GM/4R}$

And $T = 2\pi R/v = 2\pi R\sqrt{4R/GM} = 4\pi\sqrt{R^3/GM}$

c) Apply $K_1 + U_1 + W_{\mathrm{other}} = K_2 + U_2$ to the system of the two stars.

Separate to infinity implies $K_2 = 0$ and $U_2 = 0$.

$K_1 = \frac{1}{2}Mv^2 + \frac{1}{2}Mv^2 = 2(\frac{1}{2}M)(GM/4R) = GM^2/4R$

$U_1 = -GM^2/2R$

Thus the energy required is $W_{\mathrm{other}} = -(K_1 + U_1) = -(GM^2/4R - GM^2/2R) = GM^2/4R$.

12-63 Use conservation of energy: $K_1 + U_1 + W_{\mathrm{other}} = K_2 + U_2$

The gravity force exerted by the sun is the only force that does work on the comet so $W_{\mathrm{other}} = 0$.

$K_1 = \frac{1}{2}mv_1^2, \quad v_1 = 2.0 \times 10^4$ m/s

$U_1 = -Gm_{\mathrm{S}}m/r_1, \quad r_1 = 2.5 \times 10^{11}$ m

$K_2 = \frac{1}{2}mv_2^2$

$U_2 = -Gm_{\mathrm{S}}m/r_2, \quad r_2 = 5.0 \times 10^{10}$ m

$\frac{1}{2}mv_1^2 - Gm_{\mathrm{S}}m/r_1 = \frac{1}{2}mv_2^2 - Gm_{\mathrm{S}}m/r_2$

$$v_2^2 = v_1^2 + 2Gm_{\mathrm{S}}\left(\frac{1}{r_2} - \frac{1}{r_1}\right) = v_1^2 + 2Gm_{\mathrm{S}}\left(\frac{r_1 - r_2}{r_1 r_2}\right)$$

$v_2 = 6.8 \times 10^4$ m/s (The comet has greater speed when it is closer to the sun.)

12-65 a) Find the value of a for the elliptical orbit:

$2a = r_a + r_p = R_{\mathrm{E}} + h_a + R_{\mathrm{E}} + h_p$, where h_a and h_p are the heights at apogee and perigee, respectively.

$a = R_{\text{E}} + (h_a + h_p)/2$

$a = 6.38 \times 10^6 \text{ m} + (400 \times 10^3 + 4000 \times 10^3 \text{ m})/2 = 8.58 \times 10^6 \text{ m}$

$$T = \frac{2\pi a^{3/2}}{\sqrt{GM_{\text{E}}}} = \frac{2\pi(8.58 \times 10^6 \text{ m})^{3/2}}{\sqrt{(6.673 \times 10^{-11} \text{ N} \cdot \text{m}^2/\text{kg}^2)(5.97 \times 10^{24} \text{ kg})}} = 7.91 \times 10^3 \text{ s}$$

b) Conservation of angular momentum gives $r_a v_a = r_p v_p$

$$\frac{v_p}{v_a} = \frac{r_a}{r_p} = \frac{6.38 \times 10^6 \text{ m} + 4.00 \times 10^6 \text{ m}}{6.38 \times 10^6 \text{ m} + 4.00 \times 10^5 \text{ m}} = 1.53$$

c) Conservation of energy applied to apogee and perigee gives $K_a + U_a = K_p + U_p$

$\frac{1}{2}mv_a^2 - Gm_{\text{E}}m/r_a = \frac{1}{2}mv_p^2 - Gm_{\text{E}}m/r_p$

$v_p^2 - v_a^2 = 2Gm_{\text{E}}(1/r_p - 1/r_a) = 2Gm_{\text{E}}(r_a - r_p)/r_a r_p$

But $v_p = 1.532 v_a$, so $1.347 v_a^2 = 2Gm_{\text{E}}(r_a - r_p)/r_a r_p$

$v_a = 5.51 \times 10^3 \text{ m/s}, \quad v_p = 8.43 \times 10^3 \text{ m/s}$

d) Need v so that $E = 0$, where $E = K + U$.

<u>at perigee:</u> $\frac{1}{2}mv^2 - Gm_{\text{E}}m/r_p = 0$

$v_p = \sqrt{2Gm_{\text{E}}/r_p} =$

$\sqrt{2(6.673 \times 10^{-11} \text{ N} \cdot \text{m}^2/\text{kg}^2)(5.97 \times 10^{24} \text{ kg})/6.78 \times 10^6 \text{ m}} = 1.084 \times 10^4 \text{ m/s}$

This means an increase of $1.084 \times 10^4 \text{ m/s} - 8.43 \times 10^3 \text{ m/s} = 2.41 \times 10^3 \text{ m/s}$.

<u>at apogee:</u> $v_a = \sqrt{2Gm_{\text{E}}/r_a} =$

$\sqrt{2(6.673 \times 10^{-11} \text{ N} \cdot \text{m}^2/\text{kg}^2)(5.97 \times 10^{24} \text{ kg})/1.038 \times 10^7 \text{ m}} = 8.761 \times 10^3 \text{ m/s}$

This means an increase of $8.761 \times 10^3 \text{ m/s} - 5.51 \times 10^3 \text{ m/s} = 3.25 \times 10^3 \text{ m/s}$.

Perigee is more efficient.

12-67 $K_1 + U_1 + W_{\text{other}} = K_2 + U_2$

$U_1 = -Gm_{\text{M}}m/r_1$, where m_{M} is the mass of Mars and $r_1 = R_{\text{M}} + h$, where R_{M} is the radius of Mars and $h = 2000 \times 10^3 \text{ m}$

$$U_1 = -(6.673 \times 10^{-11} \text{ N} \cdot \text{m}^2/\text{kg}^2)\frac{(6.42 \times 10^{23} \text{ kg})(3000 \text{ kg})}{3.40 \times 10^6 \text{ m} + 2000 \times 10^3 \text{ m}} = -2.380 \times 10^{10} \text{ J}$$

$U_2 = -Gm_{\text{m}}m/r_2$, where r_2 is the new orbit radius.

$$U_2 = -(6.673 \times 10^{-11} \text{ N} \cdot \text{m}^2/\text{kg}^2)\frac{(6.42 \times 10^{23} \text{ kg})(3000 \text{ kg})}{3.40 \times 10^6 \text{ m} + 4000 \times 10^3 \text{ m}} = -1.737 \times 10^{10} \text{ J}$$

For a circular orbit $v = \sqrt{Gm_M/r}$ (Eq.12-12, with the mass of Mars rather tha$_\cdot$ the mass of the earth).

Using this gives $K = \frac{1}{2}mv^2 = \frac{1}{2}m(Gm_M/r) = \frac{1}{2}Gm_Mm/r$, so $K = -\frac{1}{2}U$

$K_1 = -\frac{1}{2}U_1 = +1.190 \times 10^{10}$ J and $K_2 = -\frac{1}{2}U_2 = +8.685 \times 10^9$ J

Then $K_1 + U_1 + W_{\text{other}} = K_2 + U_2$ gives

$W_{\text{other}} = (K_2 - K_1) + (U_2 - U_1) = (8.685 \times 10^9 \text{ J} - 1.190 \times 10^{10} \text{ J}) + (-2.380$ 10^{10} J $+ 1.737 \times 10^{10}$ J$)$

$W_{\text{other}} = -3.215 \times 10^9 \text{ J} + 6.430 \times 10^9 \text{ J} = 3.22 \times 10^9$ J

(Note: When the orbit radius increases the kinetic energy decreases, the grav$_\cdot$ taitonal potential energy increases, and the total energy increases.)

12-69 $T = 30{,}000 \text{ y}(3.156 \times 10^7 \text{ s/1 y}) = 9.468 \times 10^{11}$ s

Eq.(12-19): $T = \dfrac{2\pi a^{3/2}}{\sqrt{Gm_S}}$, $\quad T^2 = \dfrac{4\pi^2 a^3}{Gm_S}$

$a = \left(\dfrac{Gm_S T^2}{4\pi^2}\right)^{1/3} = 1.4 \times 10^{14}$ m.

The orbit radius of Pluto is 5.9×10^{12} m (Appendix F); the semi-major axis f$_\cdot$ this comet is larger by a factor of 24.

4.3 light years = 4.3 light years$(9.461 \times 10^{15}$ m/1 light year$) = 4.1 \times 10^{16}$ m

The distance of Alpha Centauri is larger by a factor of 300.

12-71 Use a coordinate system with the origin at the left-hand end of the rod and the x'-axis along the rod. Divide the rod into small segments of length dx (Use x' for the coordinate so not to confuse with the distance x from the end $_\cdot$ the rod to the particle.)

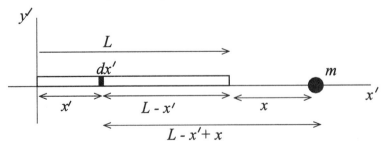

The mass of each segment is $dM = dx'(M/L)$. Each segment is a distance $L - x' +$ from mass m, so the force on the particle due to a segment is

$$dF = \frac{Gm\,dM}{(L - x' + x)^2} = \frac{GMm}{L}\frac{dx'}{(L - x' + x)^2}.$$

$$dF = \frac{Gm\,dM}{(L - x' + x)^2} = \frac{GMm}{L}\frac{dx'}{(L - x' + x)^2}.$$

$$F = \int_L^0 dF = \frac{GMm}{L}\int_L^0 \frac{dx'}{(L - x' + x)^2} = \frac{GMm}{L}\left(-\frac{1}{L - x' + x}\Big|_L^0\right)$$

$$F = \frac{GMm}{L}\left(\frac{1}{x} - \frac{1}{L + x}\right) = \frac{GMm}{L}\frac{(L + x - x)}{x(L + x)} = \frac{GMm}{x(L + x)}$$

For $x \gg L$ this result becomes $F = GMm/x^2$, the same as for a pair of point masses.

CHAPTER 13
PERIODIC MOTION

Exercises 1, 5, 9, 11, 13, 17, 19, 21, 25, 27, 31, 33, 35, 41, 43, 47
Problems 51, 53, 59, 61, 63, 71, 73, 75, 77, 79

Exercises

13-1 **a)** $f = 220$ Hz
$T = 1/f = 1/220$ Hz $= 4.54 \times 10^{-3}$ s
$\omega = 2\pi f = 2\pi(220$ Hz$) = 1380$ rad/s

b) $f = 4(220$ Hz$) = 880$ Hz
$T = 1/f = 1/880$ Hz $= 1.14 \times 10^{-3}$ s (smaller by a factor of 4)
$\omega = 2\pi f = 2\pi(880$ Hz$) = 5530$ rad/s (factor of 4 larger)

13-5 **a)** $T = 1/f = 1/6.00$ Hz $= 0.167$ s
b) $\omega = 2\pi f = 2\pi(6.00$ Hz$) = 37.7$ rad/s

c) $\omega = \sqrt{k/m}$ implies $m = k/\omega^2 = (120$ N/m$)/(37.7$ rad/s$)^2 = 0.0844$ kg

13-9 $f = 440$ Hz, $A = 3.0$ mm, $\phi = 0$
a) $x = A\cos(\omega t + \phi)$
$\omega = 2\pi f = 2\pi(440$ Hz$) = 2.76 \times 10^3$ rad/s
$x = (3.0 \times 10^{-3}$ m$)\cos((2.76 \times 10^3$ rad/s$)t)$

b) $v = -\omega A\sin(\omega t + \phi)$
$v_{max} = \omega A = (2.76 \times 10^3$ rad/s$)(3.0 \times 10^{-3}$ m$) = 8.3$ m/s (maximum magnitude o
velocity)

$a = -\omega^2 A\cos(\omega t + \phi)$
$a_{max} = \omega^2 A = (2.76 \times 10^3$ rad/s$)^2(3.0 \times 10^{-3}$ m$) = 2.3 \times 10^4$ m/s^2 (maximur
magnitude of acceleration)

c) $a = -\omega^2 A\cos\omega t$
$da/dt = +\omega^3 A\sin\omega t = [2\pi(440$ Hz$)]^3(3.0 \times 10^{-3}$ s$)\sin[(2.76 \times 10^3$ rad/s$]t)$
$= (6.3 \times 10^7$ m/s$^3)\sin[(2.76 \times 10^3$ rad/s$]t)$
Maximum magnitude of the jerk is $\omega^3 A = 6.3 \times 10^7$ m/s^3

13-11 **a)** Eq.(13-19): $A = \sqrt{x_0^2 + v_0^2/\omega^2} = \sqrt{x_0^2 + mv_0^2/k}$

$$A = \sqrt{(0.200 \text{ m})^2 + (2.00 \text{ kg})(-4.00 \text{ m/s})^2/(300 \text{ N/m})} = 0.383 \text{ m}$$

b) Eq.(13-18): $\phi = \arctan(-v_0/\omega x_0)$

$\omega = \sqrt{k/m} = \sqrt{(300 \text{ N/m})/2.00 \text{ kg}} = 12.25 \text{ rad/s}$

$$\phi = \arctan\left(-\frac{(-4.00 \text{ m/s})}{12.25 \text{ rad/s})(0.200 \text{ m})}\right) = \arctan(+1.633) = 58.5° \text{ (or 1.02 rad)}$$

c) $x = A\cos(\omega t + \phi)$ gives $x = (0.383 \text{ m})\cos([12.2 \text{ rad/s}]t + 1.02 \text{ rad})$

3-13 $x = 0$ at $t = 0$ implies that $\phi = \pm\pi/2$ rad

Thus $x = A\cos(\omega t \pm \pi/2)$.

$T = 2\pi/\omega$ so $\omega = 2\pi/T = 2\pi/1.20 \text{ s} = 5.236 \text{ rad/s}$

$x = (0.600 \text{ m})\cos([5.236 \text{ rad/s}][0.480 \text{ s}] \pm \pi/2) = \mp 0.353 \text{ m}$.

The distance of the object from the equilibrium position is 0.353 m.

3-17 **a)** $U + K = E$

$U = K$ says that $2U = E$

$2(\frac{1}{2}kx^2) = \frac{1}{2}kA^2$ and $x = \pm A/\sqrt{2}$; magnitude is $A/\sqrt{2}$

But can also say $U = K$ implies that $2K = E$

$2(\frac{1}{2}mv^2) = \frac{1}{2}kA^2$ and $v = \pm\sqrt{k/m}A/\sqrt{2} = \pm\omega A/\sqrt{2}$; magnitude is $\omega A/\sqrt{2}$.

b) In one cycle x goes from A to 0 to $-A$ to 0 to $+A$. Thus $x = +A\sqrt{2}$ twice and $x = -A/\sqrt{2}$ twice in each cycle. Therefore, $U = K$ four times each cycle.

The time between $U = K$ occurrences is the time Δt_a for $x_1 = +A/\sqrt{2}$ to $x_2 = -A\sqrt{2}$, time Δt_b for $x_1 = -A/\sqrt{2}$ to $x_2 = +A\sqrt{2}$, time Δt_c for $x_1 = +A/\sqrt{2}$ to $x_2 = +A\sqrt{2}$, or the time Δt_d for $x_1 = -A/\sqrt{2}$ to $x_2 = -A/\sqrt{2}$,

$\Delta t_a = \Delta t_b$

$\Delta t_c = \Delta t_d$

Calculation of Δt_a:

Specify x in $x = A\cos\omega t$ (choose $\phi = 0$ so $x = A$ at $t = 0$) and solve for t.

$x_1 = +A/\sqrt{2}$ implies $A/\sqrt{2} = A\cos(\omega t_1)$

$\cos\omega t_1 = 1\sqrt{2}$ so $\omega t_1 = \arccos(1/\sqrt{2}) = \pi/4$ rad

$t_1 = \pi/4\omega$

$x_2 = -A/\sqrt{2}$ implies $-A/\sqrt{2} = A\cos(\omega t_2)$

$\cos \omega t_2 = -1\sqrt{2}$ so $\omega t_1 = 3\pi/4$ rad

$t_2 = 3\pi/4\omega$

$\Delta t_a = t_2 - t_1 = 3\pi/4\omega - \pi/4\omega = \pi/2\omega$ (Note that this is $T/4$, one fourth period.

Calculation of Δt_d:

$x_1 = -A/\sqrt{2}$ implies $t_1 = 3\pi/4\omega$

$x_2 = -A/\sqrt{2}$, t_2 is the <u>next</u> time after t_1 that gives $\cos \omega t_2 = -1/\sqrt{2}$

Thus $\omega t_2 = \omega t_1 + \pi/2 = 5\pi/4$ and $t_2 = 5\pi/4\omega$.

$\Delta t_d = t_2 - t_1 = 5\pi/4\omega - 3\pi/4\omega = \pi/2\omega$, so is the same as Δt_a.

Therefore the occurrences of $K = U$ are equally spaced in time, with a time interva
between them of $\pi/2\omega$. (This is one-fourth T, as it must be if there are 4 equall
spaced occurrences each period.)

c) $x = A/2$ and $U + K = E$

$K = E - U = \frac{1}{2}kA^2 - \frac{1}{2}kx^2 = \frac{1}{2}kA^2 - \frac{1}{2}k(A/2)^2 = \frac{1}{2}kA^2 - \frac{1}{8}kA^2 = 3kA^2/8$

Then $\dfrac{K}{E} = \dfrac{3kA^2/8}{\frac{1}{2}kA^2} = \dfrac{3}{4}$ and $\dfrac{U}{E} = \dfrac{\frac{1}{8}kA^2}{\frac{1}{2}kA^2} = \dfrac{1}{4}$

13-19 a) $-kx = ma$ so $a = -(k/m)x$ (Eq.13-4)

But the maximum $|x|$ is A, so $a_{max} = (k/m)A = \omega^2 A$.

$f = 0.850$ Hz implies $\omega = \sqrt{k/m} = 2\pi f = 2\pi(0.850$ Hz$) = 5.34$ rad/s

$a_{max} = \omega^2 A = (5.34$ rad/s$)^2(0.180$ m$) = 5.13$ m/s^2.

$\frac{1}{2}mv^2 + \frac{1}{2}kx^2 = \frac{1}{2}kA^2$

$v = v_{max}$ when $x = 0$ so $\frac{1}{2}mv_{max}^2 = \frac{1}{2}kA^2$

$v_{max} = \sqrt{k/m}A = \omega A = (5.34$ rad/s$)(0.180$ m$) = 0.961$ m/s

b) $a = -(k/m)x = -\omega^2 x = -(5.34$ rad/s$)^2(0.090$ m$) = -2.57$ m/s^2

$\frac{1}{2}mv^2 + \frac{1}{2}kx^2 = \frac{1}{2}kA^2$ says that $v = \pm\sqrt{k/m}\sqrt{A^2 - x^2} = \pm\omega\sqrt{A^2 - x^2}$

$v = \pm(5.34$ rad/s$)\sqrt{(0.180$ m$)^2 - (0.090$ m$)^2} = \pm 0.832$ m/s

The speed is 0.832 m/s.

c) $x = A\cos(\omega t + \phi)$

Let $\phi = -\pi/2$ so that $x = 0$ at $t = 0$.

Then $x = A\cos(\omega t - \pi/2) = A\sin(\omega t)$ [Using the trig identity $\cos(a - \pi/2) = \sin$

Find the time t that gives $x = 0.120$ m.

$0.120 \text{ m} = (0.180 \text{ m})\sin(\omega t)$

$\sin\omega t = 0.6667$

$t = \arcsin(0.6667)/\omega = 0.7297 \text{ rad}/(5.34 \text{ rad/s}) = 0.137 \text{ s}$

Note: It takes one-fourth of a period for the object to go from $x = 0$ to $x = A = 0.180$ m. So the time we have calculated should be less than $T/4$. $T = 1/f = 1/0.850 \text{ Hz} = 1.18 \text{ s}$, $T/4 = 0.295$ s, and the time we calculated is less than this.

d) The conservation of energy equation relates v and x and $F = ma$ relates a and x. So the speed and acceleration can be found by energy methods but the time cannot.

3-21 **a)** $E = \frac{1}{2}mv^2 + \frac{1}{2}kA^2$

$E = \frac{1}{2}(0.150 \text{ kg})(0.300 \text{ m/s})^2 + \frac{1}{2}(300 \text{ N/m})(0.012 \text{ m})^2 = 0.0284 \text{ J}$

b) $E = \frac{1}{2}kA^2$ so $A = \sqrt{2E/k} = \sqrt{2(0.0284 \text{ J})/300 \text{ N/m}} = 0.014 \text{ m}$

c) $E = \frac{1}{2}mv_{\text{max}}^2$ so $v_{\text{max}} = \sqrt{2E/m} = \sqrt{2(0.0284 \text{ J})/0.150 \text{ kg}} = 0.615 \text{ m/s}$

3-25

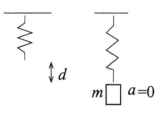

Let d be the distance the spring stretches when the cheese hangs at rest.

Free-body diagram for the cheese:

$\sum F_y = ma_y$

$mg - kd = 0$

$d = mg/k$

$T = 2\pi\sqrt{m/k}$ says that $m/k = (T/2\pi)^2$

Thus $d = (T/2\pi)^2 g = (0.400 \text{ s}/2\pi)^2(9.80 \text{ m/s}^2) = 0.0397 \text{ m}$.

3-27 Ticks four times each second implies 0.25 s per tick.

Each tick is half a period, so $T = 0.50$ s and $f = 1/T = 1/0.50 \text{ s} = 2.00$ Hz

a) Thin rim implies $I = MR^2$ (from Table 9-2)

$I = (0.900 \times 10^{-3} \text{ kg})(0.55 \times 10^{-2} \text{ m})^2 = 2.7 \times 10^{-8} \text{ kg} \cdot \text{m}^2$

b) $T = 2\pi\sqrt{I/\kappa}$ so $\kappa = I(2\pi/T)^2 = (2.7 \times 10^{-8} \text{ kg} \cdot \text{m}^2)(2\pi/0.50 \text{ s})^2$
$= 4.3 \times 10^{-6} \text{ N·m/rad}$

13-31 Eq.(13-25): $U = U_0[(R_0/r)^{12} - 2(R_0/r)^6]$
Let $r = R_0 + x$.

$$U = U_0\left[\left(\frac{R_0}{R_0 + x}\right)^{12} - 2\left(\frac{R_0}{R_0 + x}\right)^6\right] = U_0\left(\left(\frac{1}{1 + x/R_0}\right)^{12} - 2\left(\frac{1}{1 + x/R_0}\right)^6\right)$$

$$\left(\frac{1}{1 + x/R_0}\right)^{12} = (1 + x/R_0)^{-12}; \quad |x/R_0| << 1$$

Apply Eq.(13-28) with $n = -12$ and $u = +x/R_0$:

$$\left(\frac{1}{1 + x/R_0}\right)^{12} = 1 - 12x/R_0 + 66x^2/R_0^2 - \ldots$$

For $\left(\dfrac{1}{1 + x/R_0}\right)^6$ apply Eq.(13-28) with $n = -6$ and $u = +x/R_0$:

$$\left(\frac{1}{1 + x/R_0}\right)^6 = 1 - 6x/R_0 + 15x^2/R_0^2 - \ldots$$

Thus $U = U_0(1 - 12x/R_0 + 66x^2/R_0^2 - 2 + 12x/R_0 - 30x^2/R_0^2) = -U_0 + 36U_0x^2/R_0^2$
This is in the form $U = \frac{1}{2}kx^2 - U_0$ with $k = 72U_0/R_0^2$, which is the same as the force constant in Eq.(13-29).

13-33 Let the period on earth be $T_E = 2\pi\sqrt{L/g_E}$, where $g_E = 9.80 \text{ m/s}^2$, the value on earth.

Let the period on Mars be $T_M = 2\pi\sqrt{L/g_M}$, where $g_M = 3.71 \text{ m/s}^2$, the value of Mars.

We can eliminate L, which we don't know, by taking a ratio:

$$\frac{T_M}{T_E} = 2\pi\sqrt{\frac{L}{g_M}}\frac{1}{2\pi}\sqrt{\frac{g_E}{L}} = \sqrt{\frac{g_E}{g_M}}.$$

$$T_M = T_E\sqrt{\frac{g_E}{g_M}} = (1.60 \text{ s})\sqrt{\frac{9.80 \text{ m/s}^2}{3.71 \text{ m/s}^2}} = 2.60 \text{ s}.$$

13-35 vertical SHM: $f_b = \dfrac{1}{2\pi}\sqrt{\dfrac{k}{m}}$

pendulum motion (small amplitude): $f_p = \dfrac{1}{2\pi}\sqrt{\dfrac{g}{L}}$

The problem specifies that $f_p = \frac{1}{2}f_b$.

$$\frac{1}{2\pi}\sqrt{\frac{g}{L}} = \frac{1}{2}\frac{1}{2\pi}\sqrt{\frac{k}{m}}$$

$g/L = k/4m$ so $L = 4gm/k = 4w/k = 4(1.00\ \text{N})/1.50\ \text{N/m} = 2.67\ \text{m}$

But this is the <u>stretched</u> length of the spring, its length when the apple is hanging from it. (Note: Small angle of swing means v is small as the apple passes through the lowest point, so a_{rad} is small and the component of mg perpendicular to the spring is small. Thus the amount the spring is stretched changes very little as the apple swings back and forth.)

Calculate the distance the spring is stretched from its unstretched length when the apple hangs from it. Free-body diagram for the apple hanging at rest on the end of the spring:

$$\sum F_y = ma_y$$
$$k\,\Delta L - mg = 0$$
$$\Delta L = mg/k = w/k = 1.00\ \text{N}/1.50\ \text{N} = 0.667$$

Thus the unstretched length of the spring is $2.67\ \text{m} - 0.67\ \text{m} = 2.00\ \text{m}$.

13-41 The ornament is a physical pendulum: $T = 2\pi\sqrt{I/mgd}$ (Eq.13-39).

$I = 5MR^2/3$, the moment of inertia about an axis at the edge of the sphere.

d is the distance from the axis to the center of gravity, which is at the center of the sphere, so $d = R$.

Thus $T = 2\pi\sqrt{5/3}\sqrt{R/g} = 2\pi\sqrt{5/3}\sqrt{0.050\ \text{m}/(9.80\ \text{m/s}^2)} = 0.58\ \text{s}$.

13-43 **a)** Eq.(13-43) says $\omega' = \sqrt{(k/m) - (b^2/4m^2)}$

$$\omega' = \sqrt{\frac{2.50\ \text{N/m}}{0.300\ \text{kg}} - \frac{(0.900\ \text{kg/s})^2}{4(0.300\ \text{kg})^2}} = 24.7\ \text{rad/s}$$

$f' = \omega'/2\pi = (24.7\ \text{rad/s})/2\pi = 0.393\ \text{Hz}$

b) The condition for critical damping is $b = 2\sqrt{km}$ (Eq.13-44)

$b = 2\sqrt{(2.50\ \text{N/m})(0.300\ \text{kg})} = 1.73\ \text{kg/s}$

13-47 Eq.(13-46): $A = \dfrac{F_{max}}{\sqrt{(k - m\omega_d^2)^2 + b^2\omega_d^2}}$

a) Consider the special case where $k - m\omega_d^2 = 0$, so $A = F_{max}/b\omega_d$ and $b = F_{max}/A\omega_d$.

Units of $\dfrac{F_{max}}{A\omega_d}$ are $\dfrac{\text{kg} \cdot \text{m/s}^2}{(\text{m})(\text{s}^{-1})} = \text{kg/s}$.

For units consistency the units of b must be kg/s.

b) Units of $\sqrt{km}$: $[(\text{N/m})\text{kg}]^{1/2} = (\text{N kg/m})^{1/2} = [(\text{kg·m/s}^2)(\text{kg})/\text{m}]^{1/2} = (\text{kg}^2/\text{s}^2)^{1/2} = \text{kg/s}$, the same as the units for b.

c) For $\omega_d = \sqrt{k/m}$ (at resonance) $A = (F_{max}/b)\sqrt{m/k}$.

(i) $b = 0.2\sqrt{km}$

$A = F_{max}\sqrt{\dfrac{m}{k}}\dfrac{1}{0.2\sqrt{km}} = \dfrac{F_{max}}{0.2k} = 5.0\dfrac{F_{max}}{k}$.

(ii) $b = 0.4\sqrt{km}$

$A = F_{max}\sqrt{\dfrac{m}{k}}\dfrac{1}{0.4\sqrt{km}} = \dfrac{F_{max}}{0.4k} = 2.5\dfrac{F_{max}}{k}$.

Both these results agree with what is shown in Fig.13-24.

Problems

13-51 a) $T = 2\pi\sqrt{m/k}$; independent of A so period doesn't change

$f = 1/T$; doesn't change

$\omega = 2\pi f$; doesn't change

b) $E = \frac{1}{2}kA^2$ when $k = \pm A$. When A is halved E decreases by a factor of 4 $E_2 = E_1/4$.

c) $v_{max} = \omega A = 2\pi f A$

$v_{max,1} = 2\pi f A_1$ $v_{max,2} = 2\pi f A_2$ (f doesn't change)

Since $A_2 = \frac{1}{2}A_1$, $v_{max,2} = 2\pi f(\frac{1}{2}A_1) = \frac{1}{2}2\pi f A_1 = \frac{1}{2}v_{max,1}$; v_{max} is one-half as great

d) $v = \pm\sqrt{k/m}\sqrt{A^2 - x^2}$

$x = \pm A_1/4$ gives $v = \pm\sqrt{k/m}\sqrt{A^2 - A_1^2/16}$

With the original amplitude $v_1 = \pm\sqrt{k/m}\sqrt{A_1^2 - A_1^2/16} = \pm\sqrt{15/16}(\sqrt{k/m})A_1$

With the reduced amplitude $v_2 = \pm\sqrt{k/m}\sqrt{A_2^2 - A_1^2/16}$

$= \pm\sqrt{k/m}\sqrt{(A_1/2)^2 - A_1^2/16} = \pm\sqrt{3/16}(\sqrt{k/m})A_1$

$v_1/v_2 = \sqrt{15/3} = \sqrt{5}$, so $v_2 = v_1/\sqrt{5}$; the speed at this x is $1/\sqrt{5}$ times as great.

e) $U = \frac{1}{2}kx^2$; same x so same U.

$K = \frac{1}{2}mv^2$; $K_1 = \frac{1}{2}mv_1^2$

$K_2 = \frac{1}{2}mv_2^2 = \frac{1}{2}m(v_1/\sqrt{5})^2 = \frac{1}{5}(\frac{1}{2}mv_1^2) = K_1/5$; $1/5$ times as great.

13-53 a) $T = 2\pi\sqrt{m/k}$

We are given information about v at a particular x. The expression relating these two quantities comes from conservation of energy:

$\frac{1}{2}mv^2 + \frac{1}{2}kx^2 = \frac{1}{2}kA^2$

We can solve this equation for $\sqrt{m/k}$, and then use that result to calculate T.

$mv^2 = k(A^2 - x^2)$

$$\sqrt{\frac{m}{k}} = \frac{\sqrt{A^2 - x^2}}{v} = \frac{\sqrt{(0.100 \text{ m})^2 - (0.060 \text{ m})^2}}{0.300 \text{ s}} = 0.267 \text{ s}$$

Then $T = 2\pi\sqrt{m/k} = 2\pi(0.267 \text{ s}) = 1.68 \text{ s}$.

b) We are asked to relate x and v, so use conservation of energy equation:

$\frac{1}{2}mv^2 + \frac{1}{2}kx^2 = \frac{1}{2}kA^2$

$kx^2 = kA^2 - mv^2$

$x = \sqrt{A^2 - (m/k)v^2} = \sqrt{(0.100 \text{ m})^2 - (0.267 \text{ s})^2(0.160 \text{ m/s})^2} = 0.090 \text{ m}$

c) For the plate, $-kx = ma$ and $a = -(k/m)a$.

The maximum $|x|$ is A, so $a_{\max} = (k/m)A$.

If the carrot slice doesn't slip then the static friction force must be able to give it this much acceleration.

Free-body diagram for the carrot slice (mass m'):

$\sum F_y = ma_y$

$N - m'g = 0$

$N = m'g$

$\sum F_x = ma_x$

$\mu_s N = m'a$

$\mu_s m'g = m'a$ and $a = \mu_s g$

But we require that $a = a_{max} = (k/m)A$, so $(k/m)A = \mu_s g$ and

$$\mu_s = \frac{k}{m}\frac{A}{g} = \left(\frac{1}{0.267 \text{ s}}\right)^2 \left(\frac{0.100 \text{ m}}{9.80 \text{ m/s}^2}\right) = 0.143$$

13-59 Measure x from the equilibrium position of the object, where the gravity and spring forces balance. Let $+x$ be downward.

a) Conservation of energy:

$\frac{1}{2}mv^2 + \frac{1}{2}kx^2 = \frac{1}{2}kA^2$

For $x = 0$, $\frac{1}{2}mv^2 = \frac{1}{2}kA^2$ and $v = A\sqrt{k/m}$, just as for horizontal SHM.

We can use the period to calculate $\sqrt{k/m}$: $T = 2\pi\sqrt{m/k}$ implies $\sqrt{k/m} = 2\pi/T$

Thus $v = 2\pi A/T = 2\pi(0.100 \text{ m})/4.20 \text{ s} = 0.150 \text{ m/s}$.

b) $ma = -kx$ so $a = -(k/m)x$

$+x$-direction is downward, so here $x = -0.050$ m

$a = -(2\pi/T)^2(-0.050 \text{ m}) = +(2\pi/4.20 \text{ s})^2(0.050 \text{ m}) = 0.112 \text{ m/s}^2$ (positive, s direction is downward)

c) This is twice the time it takes to go from $x = 0$ to $x = +0.050$ m.

$x(t) = A\cos(\omega t + \phi)$

Let $\phi = -\pi/2$, so $x = 0$ at $t = 0$.

Then $x = A\cos(\omega t - \pi/2) = A\sin\omega t = A\sin(2\pi t/T)$.

Find the time t that gives $x = +0.050$ m:

$0.050 \text{ m} = (0.100 \text{ m})\sin(2\pi t/T)$

$2\pi t/T = \arcsin(0.50) = \pi/6$ and $t = T/12 = 4.20 \text{ s}/12 = 0.350 \text{ s}$

The time asked for in the problem is twice this, 0.700 s.

d) The problem is asking for the distance d that the spring stretches when th object hangs at rest from it.

Free-body diagram for the object:

$a = 0$

$\sum F_x = ma_x$

$mg - kd = 0$

$d = (m/k)g$

But $\sqrt{k/m} = 2\pi/T$ (part (a)) and $m/k = (T/2\pi)^2$

$$d = \left(\frac{T}{2\pi}\right)g = \left(\frac{4.20}{2\pi}\right)^2 (9.80 \text{ m/s}^2) = 4.38 \text{ m}.$$

3-61 **a)** First find the speed of the steak just before it strikes the pan.

Use a coordinate system with $+y$ downward.

$v_{0y} = 0$ (released from rest); $y - y_0 = 0.40$ m; $a_y = +9.80$ m/s^2; $v_y = ?$

$v_y^2 = v_{0y}^2 + 2a_y(y - y_0)$

$v_y = +\sqrt{2a_y(y - y_0)} = +\sqrt{2(9.80 \text{ m/s}^2)(0.40 \text{ m})} = +2.80 \text{ m/s}$

Apply conservation of momentum to the collision between the steak and the pan. After the collision the steak and the pan are moving together with common velocity v_2. Let A be the steak and B be the pan.

before   after

P_y conserved: $m_A v_{A1y} + m_B v_{B1y} = (m_A + m_B)v_{2y}$

$m_A v_{A1} = (m_A + m_B)v_2$

$$v_2 = \left(\frac{m_A}{m_A + m_B}\right)v_{A1} = \left(\frac{2.2 \text{ kg}}{2.2 \text{ kg} + 0.20 \text{ kg}}\right)(2.80 \text{ m/s}) = 2.57 \text{ m/s}$$

b) Conservation of energy applied to the SHM gives:

$\frac{1}{2}mv_0^2 + \frac{1}{2}kx_0^2 = \frac{1}{2}kA^2$ where v_0 and x_0 are the initial speed and displacement of the object and where the displacement is measured from the equilibrium position of the object.

The weight of the steak will stretch the spring an additional distance d given by

$$kd = mg \text{ so } d = \frac{mg}{k} = \frac{(2.2 \text{ kg})(9.80 \text{ m/s}^2)}{400 \text{ N/m}} = 0.0539 \text{ m}.$$ So just after the steak hits the pan, before the pan has had time to move, the steak plus pan is 0.0539 m above the equilibrium position of the combined object. Thus $x_0 = 0.0539$ m.

From part (a) $v_0 = 2.57$ m/s, the speed of the combined object just after the collision.

Then $\frac{1}{2}mv_0^2 + \frac{1}{2}kx_0^2 = \frac{1}{2}kA^2$ gives

$$A = \sqrt{\frac{mv_0^2 + kx_0^2}{k}} = \sqrt{\frac{2.4 \text{ kg}(2.57 \text{ m/s})^2 + (400 \text{ N/m})(0.0539 \text{ m})^2}{400 \text{ N/m}}} = 0.21 \text{ m}$$

c) $T = 2\pi\sqrt{m/k} = 2\pi\sqrt{\dfrac{2.4 \text{ kg}}{400 \text{ N/m}}} = 0.49 \text{ s}$

13-63

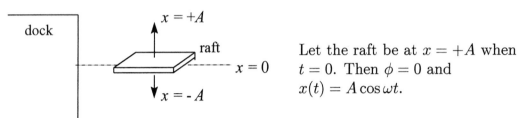

dock

$x = +A$

raft

$x = 0$

$x = -A$

Let the raft be at $x = +A$ when $t = 0$. Then $\phi = 0$ and $x(t) = A\cos\omega t$.

Calculate the time it takes the raft to move from $x = +A = +0.200$ m to x $A - 0.100$ m $= 0.100$ m.

Write the equation for $x(t)$ in terms of T rather than ω:

$\omega = 2\pi/T$ gives that $x(t) = A\cos(2\pi t/T)$

$x = A$ at $t = 0$

$x = 0.100$ m implies 0.100 m $= (0.200 \text{ m})\cos(2\pi t/T)$

$\cos(2\pi t/T) = 0.500$ so $2\pi t/T = \arccos(0.500) = 1.047$ rad

$t = (T/2\pi)(1.047 \text{ rad}) = (2.80 \text{ s}/2\pi)(1.047 \text{ rad}) = 0.4666$ s

This is the time for the raft to move down from $x = 0.200$ m to $x = 0.100$ m. B▪ people can also get off while the raft is moving up from $x = 0.100$ m to $x = 0.2($ m, so during each period of the motion the time the people have to get off $2t = 2(0.4666 \text{ s}) = 0.933$ s.

13-71 **a)** $U = A\left[\dfrac{1}{r} - \dfrac{1}{(r - 2R_0)}\right]$

$F_r = -\dfrac{dU}{dr} = -A\left[-\dfrac{1}{r^2} + \dfrac{1}{(r - 2R_0)^2}\right]$, as was to be shown.

b) At equilibirum $F_r = 0$ so $1/r^2 - 1/(r - 2R_0)^2 = 0$

$r^2 = (r - 2R_0)^2$ and $r = \pm(r - 2R_0)$

$r = +(r - 2R_0)$; no solution unless $R_0 = 0$

or $r = -(r - 2R_0)$, which says $r = R_0$, as was to be shown.

c) Write $r = R_0 + x$, where x/R_0 is small.

Note that $(r - 2R_0)^2 = (R_0 + x - 2R_0)^2 = (x - R_0)^2 = (R_0 - x)^2$, so

$$F_r = A\left[\dfrac{1}{(R_0 + x)^2} - \dfrac{1}{(R_0 - x)^2}\right] = \dfrac{A}{R_0^2}\left[\dfrac{1}{(1 + x/R_0)^2} - \dfrac{1}{(1 - x/R_o)^2}\right]$$

For $(1 + x/R_0)^{-2}$ apply Eq.(13-28) with $n = -2$ and $u = x/R_0$.

This gives $(1 + x/R_0)^{-2} \sim 1 - 2(x/R_0)$.

For $(1 - x/R_0)^{-2}$ apply Eq.(13-28) with $n = -2$ and $u = -x/R_0$. This gives $(1 - x/R_0)^{-2} \sim 1 + 2(x/R_0)$.

Then $F_r \sim (A/R_0^2)[1 - 2x/R_0 - (1 + 2x/R_0)] = -4Ax/R_0^3$.

$F_r = -kx$ gives $k = 4A/R_0^3$.

d) $f = (1/2\pi)\sqrt{k/m}$.

Using $k = \dfrac{4A}{R_0^3}$ from part (c) gives $f = \dfrac{1}{2\pi}\sqrt{\dfrac{4A}{mR_0^3}} = \dfrac{1}{\pi}\sqrt{\dfrac{A}{mR_0^3}}$.

13-73

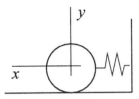

Let the origin of coordinates be at the center of the cylinders when they are at their equilibrium position.

Free-body diagram for the cylinders when they are displaced a distance x to the left:

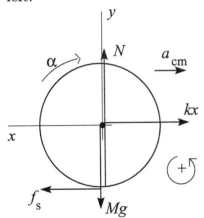

$\sum \tau = I_{cm}\alpha$

$f_s R = (\tfrac{1}{2}MR^2)\alpha$

$f_s = \tfrac{1}{2}MR\alpha$

But $R\alpha = a_{cm}$ so

$f_s = \tfrac{1}{2}Ma_{cm}$

$\sum F_x = ma_x$

$f_s - kx = -Ma_{cm}$

$\tfrac{1}{2}Ma_{cm} - kx = -Ma_{cm}$

$kx = \tfrac{3}{2}Ma_{cm}$

$(2k/3M)x = a_{cm}$

Eq.(13-8): $a = -\omega^2 x$ (The minus sign says that x and a have opposite directions, as our diagram shows.)

Our result for a_{cm} is of this form, with $\omega^2 = 2k/3M$ and $\omega = \sqrt{2k/3M}$.

Thus $T = 2\pi/\omega = 2\pi\sqrt{3M/2k}$.

13-75

The bell swings as a physical pendulum so its period of oscillation is given b

$T = 2\pi\sqrt{I/mgd} = 2\pi\sqrt{18 \text{ kg} \cdot \text{m}^2/(34.0 \text{ kg})(9.80 \text{ m/s}^2)(0.60 \text{ m})} = 1.885 \text{ s}$

The clapper is a simple pendulum so its period is given by $T = 2\pi\sqrt{L/g}$.

Thus $L = g(T/2\pi)^2 = (9.80 \text{ m/s}^2)(1.885 \text{ s}/2\pi)^2 = 0.88 \text{ m}$

13-77 **a)** $T = 2\pi\sqrt{L/g}$

$L = g(T/2\pi)^2 = (9.80 \text{ m/s}^2)(4.00 \text{ s}/2\pi)^2 = 3.97 \text{ m}$

b) Use a uniform slender rod of mass M and length $L = 0.50$ m. Pivot the rod about an axis that is a distance d abover the center of the rod. The rod will oscillat as a physical pendulum with period $T = 2\pi\sqrt{I/Mgd}$.

Choose d so that $T = 4.00$ s.

$I = I_{\text{cm}} + Md^2 = \frac{1}{12}ML^2 + Md^2 = M(\frac{1}{12}L^2 + d^2)$

$T = 2\pi\sqrt{\dfrac{I}{Mgd}} = 2\pi\sqrt{\dfrac{M(\frac{1}{12}L^2 + d^2)}{Mgd}} = 2\pi\sqrt{\dfrac{\frac{1}{12}L^2 + d^2}{gd}}.$

Note that $T \to \infty$ as $d \to 0$ (pivot at center of rod) and that if pivot is at top o rod then $d = L/2$ and

$T = 2\pi\sqrt{\dfrac{\frac{1}{12}L^2 + \frac{1}{4}L^2}{Lg/2}} = 2\pi\sqrt{\dfrac{L}{g}\dfrac{4}{6}} = 2\pi\sqrt{\dfrac{2L}{3g}} = 2\pi\sqrt{\dfrac{2(0.50 \text{ m})}{3(9.80 \text{ m/s}^2)}} = 1.16 \text{ s}, \text{ which}$

is less than the desired 4.00 s. Thus it is reasonable to expect that there is a value of d between 0 and $L/2$ for which $T = 4.00$ s.

$T = 2\pi\sqrt{\dfrac{\frac{1}{12}L^2 + d^2}{gd}}$; solve for d:

$gd(T/2\pi)^2 = \frac{1}{12}L^2 + d^2$

$d^2 - (T/2\pi)^2 gd + L^2/12 = 0$

$d^2 - (4.00 \text{ s}/2\pi)^2(9.80 \text{ m/s}^2)d + (0.50 \text{ m})^2/12 = 0$

$d^2 - 3.9718d + 0.020833 = 0$

The quadratic formula gives

$d = \frac{1}{2}[3.9718 \pm \sqrt{(3.9718)^2 - 4(0.020833)}] \text{ m}$

$d = (1.9859 \pm 1.9806) \text{ m}$ so $d = 3.97$ m or $d = 0.0053$ m.

The maximum value d can have is $L/2 = 0.25$ m, so the answer we want is $d = 0.0053 \text{ m} = 0.53 \text{ cm}$.

Therefore, take a slender rod of length 0.50 m and pivot it about an axis that is 0.53 cm above its center.

3-79 a)

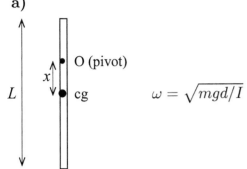

$$\omega = \sqrt{mgd/I}$$

$d = x$, the distance from the cg of the object (which is at its geometrical center) from the pivot

I is the moment of inertia about the axis of rotation through O. By the parallel axis theorem $I_0 = md^2 + I_{cm}$. $I_{cm} = \frac{1}{12}mL^2$ (Table 9-2), so $I_0 = mx^2 + \frac{1}{12}mL^2$.

$$\omega = \sqrt{\frac{mgx}{mx^2 + \frac{1}{12}mL^2}} = \sqrt{\frac{gx}{x^2 + L^2/12}}$$

b) The maximum ω as vary x occurs when $d\omega/dx = 0$.

$\dfrac{d\omega}{dx} = 0$ gives $\sqrt{g}\dfrac{d}{dx}\left(\dfrac{x^{1/2}}{(x^2 + L^2/12)^{1/2}}\right) = 0.$

$$\frac{\frac{1}{2}x^{-1/2}}{(x^2 + L^2/12)^{1/2}} - \frac{1}{2}\frac{2x}{(x^2 + L^2/12)^{3/2}}(x^{1/2}) = 0$$

$$x^{-1/2} - \frac{2x^{3/2}}{x^2 + L^2/12} = 0$$

$x^2 + L^2/12 = 2x^2$ so $x = L/\sqrt{12}$. Get maximum ω when the pivot is a distance $L/\sqrt{12}$ above the center of the rod.

c) To answer this question we need an expression for ω_{max}:

In $\omega = \sqrt{\dfrac{gx}{x^2 + L^2/12}}$ substitute $x = L/\sqrt{12}$.

$$\omega_{max} = \sqrt{\frac{g(L/\sqrt{12})}{L^2/12 + L^2/12}} = \frac{g^{1/2}(12)^{-1/4}}{(L/6)^{1/2}} = \sqrt{g/L}(12)^{-1/4}(6)^{1/2} = \sqrt{g/L}(3)^{1/4}$$

$\omega_{max}^2 = (g/L)\sqrt{3}$ and $L = g\sqrt{3}/\omega_{max}^2$

$\omega_{max} = 2\pi$ rad/s gives $L = \dfrac{(9.80\ \text{m/s}^2)\sqrt{3}}{(2\pi\ \text{rad/s})^2} = 0.430$ m.

CHAPTER 14
FLUID MECHANICS

Exercises 1, 7, 9, 13, 15, 19, 21, 25, 27, 29, 31, 35, 39, 41, 43
Problems 49, 51, 55, 57, 59, 63, 67, 69, 71, 73, 75, 77, 83, 85, 87, 91, 93

Exercises

14-1 $\rho = m/V$ so $m = \rho V$

From Table 14-1, $\rho = 7.8 \times 10^3$ kg/m^3.

For a cylinder of length L and radius R, $V = (\pi R^2)L = \pi(0.01425\ \text{m})^2(0.858\ \text{m}) = 5.474 \times 10^{-4}$ m^3.

Then $m = \rho V = (7.8 \times 10^3\ \text{kg/m}^3)(5.474 \times 10^{-4}\ \text{m}^3) = 4.27$ kg, and

$w = mg = (4.27\ \text{kg})(9.80\ \text{m/s}^2) = 41.8$ N (about 9.4 lbs). A cart is not needed.

14-7   **a)** gauge pressure $= p - p_0 = \rho g h$

From Table 14-1 the density of seawater is 1.03×10^3 kg/m^3, so

$p - p_0 = \rho g h = (1.03 \times 10^3\ \text{kg/m}^3)(9.80\ \text{m/s}^2)(250\ \text{m}) = 2.52 \times 10^6$ Pa

b) The force on each side of the window is $F = pA$. Inside the pressure is p_0 and outside in the water the pressure is $p = p_0 + \rho g h$.

inside bell | outside bell

$F_1 = P_0 A$ | $F_2 = (P_0 + \rho g h)A$

The net force is
$$F_2 - F_1 = (p_0 + \rho g h)A - p_0 A = (\rho g h)A$$
$$F_2 - F_1 = (2.52 \times 10^6\ \text{Pa})\pi(0.150\ \text{m})^2$$
$$F_2 - F_1 = 1.78 \times 10^5\ \text{N}$$

14-9 $p_a = 980$ millibar $= 9.80 \times 10^4$ Pa

a) Apply $p = p_0 + \rho g h$ to the right-hand tube. The top of this tube is open to the air so $p_0 = p_a$. The density of the liquid (mercury) is 13.6×10^3 kg/m^3.

Thus $p = 9.80 \times 10^4$ Pa $+ (13.6 \times 10^3\ \text{kg/m}^3)(9.80\ \text{m/s}^2)(0.0700\ \text{m}) = 1.07 \times 10^5$ Pa.

b) $p = p_0 + \rho g h = 9.80 \times 10^4$ Pa $+ (13.6 \times 10^3\ \text{kg/m}^3)(9.80\ \text{m/s}^2)(0.0400\ \text{m}) = 1.03 \times 10^5$ Pa.

c) Since $y_2 - y_1 = 4.00$ cm the pressure at the mercury surface in the left-hand end tube equals that calculated in part (b). Thus the absolute pressure of gas in the tank is 1.03×10^5 Pa.

d) $p - p_0 = \rho g h = (13.6 \times 10^3 \text{ kg/m}^3)(9.80 \text{ m/s}^2)(0.0400 \text{ m}) = 5.33 \times 10^3 \text{ Pa}.$

4-13

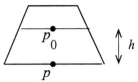

The pressure at the bottom of the tank is $p = \dfrac{F_\perp}{A} = \dfrac{16.4 \times 10^3 \text{ N}}{0.0700 \text{ m}^2} = 2.343 \times 10^5 \text{ Pa}$

The density of the kerosene is $\rho = \dfrac{m}{V} = \dfrac{205 \text{ kg}}{0.250 \text{ m}^3} = 820 \text{ kg/m}^3.$

$p = p_0 + \rho g h$ gives that $h = \dfrac{p - p_0}{\rho g} = \dfrac{2.343 \times 10^5 \text{ Pa} - 2.01 \times 10^5 \text{ Pa}}{(820 \text{ kg/m}^3)(9.80 \text{ m/s}^2)} = 4.14 \text{ m}$

4-15 The floating object is the slab of ice plus the woman; the buoyant force must support both. The volume of water displaced equals the volume V_{ice} of the ice.

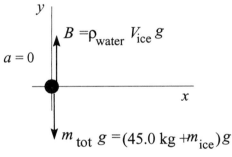

$\sum F_y = ma_y$
$B - m_{tot}g = 0$
$\rho_{water}V_{ice}g = (45.0 \text{ kg} + m_{ice})g$
But $\rho = m/V$ so $m_{ice} = \rho_{ice}V_{ice}$

$V_{ice} = \dfrac{45.0 \text{ kg}}{\rho_{water} - \rho_{ice}} = \dfrac{45.0 \text{ kg}}{1000 \text{ kg/m}^3 - 920 \text{ kg/m}^3} = 0.562 \text{ m}^3.$

4-19 a)

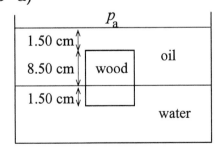

$p - p_0 = \rho g h$

The upper face is 1.50 cm below
the top of the oil, so
$p - p_0 = (790 \text{ kg/m}^3)(9.80 \text{ m/s}^2)(0.0150 \text{ m})$
$p - p_0 = 116 \text{ Pa}$

b) The pressure at the interface is $p_{interface} = p_a + \rho_{oil}g(0.100 \text{ m})$. The lower face of the block is 1.50 cm below the interface, so the pressure there is $p = p_{interface} + \rho_{water}g(0.0150 \text{ m})$. Combining these two equations gives
$p - p_a = \rho_{oil}g(0.100 \text{ m}) + \rho_{water}g(0.0150 \text{ m})$

$$p - p_a = [(790 \text{ kg/m}^3)(0.100 \text{ m}) + (1000 \text{ kg/m}^3)(0.0150 \text{ m})](9.80 \text{ m/s}^2)$$
$$p - p_a = 921 \text{ Pa}$$

c) Consider the forces on the block. The area of each face of the block is A = $(0.100 \text{ m})^2 = 0.0100 \text{ m}^2$. Let the absolute pressure at the top face be p_t and th[e] pressure at the bottom face be p_b. Then the free-body diagram for the block is:

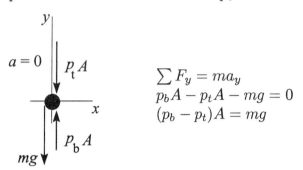

$$\sum F_y = ma_y$$
$$p_b A - p_t A - mg = 0$$
$$(p_b - p_t)A = mg$$

Note that $(p_b - p_t) = (p_b - p_a) - (p_t - p_a) = 921 \text{ Pa} - 116 \text{ Pa} = 805 \text{ Pa}$; th[e] difference in absolute pressures equals the difference in gauge pressures.

$$m = \frac{(p_b - p_t)A}{g} = \frac{(805 \text{ Pa})(0.0100 \text{ m}^2)}{9.80 \text{ m/s}^2} = 0.821 \text{ kg}.$$

And then $\rho = m/V = 0.821 \text{ kg}/(0.100 \text{ m})^3 = 821 \text{ kg/m}^3$.

14-21 Use Eq.(14-12): $p - p_a = 4\gamma/R = 4(25.0 \times 10^{-3} \text{ N/m})/1.50 \times 10^{-2} \text{ m} = 6.67 \text{ Pa}$

14-25 $w + T = 2\gamma l$
$T = 2\gamma l - w = 2\gamma l - mg$, where $m = 0.700 \text{ g} = 0.700 \times 10^{-3} \text{ kg}$.
$T = 2(25.0 \times 10^{-3} \text{ N/m})(0.220 \text{ m}) - (0.700 \times 10^{-3} \text{ kg})(9.80 \text{ m/s}^2) = 4.14 \times 10^{-3} \text{ N}$

14-27 **a)** $vA = 1.20 \text{ m}^3/\text{s}$
$$v = \frac{1.20 \text{ m}^3/\text{s}}{A} = \frac{1.20 \text{ m}^3/\text{s}}{\pi r^2} = \frac{1.20 \text{ m}^3/\text{s}}{\pi(0.150 \text{ m})^2} = 17.0 \text{ m/s}$$

b) $vA = 1.20 \text{ m}^3/\text{s}$
$v\pi r^2 = 1.20 \text{ m}^3/\text{s}$

$$r = \sqrt{\frac{1.20 \text{ m}^3/\text{s}}{v\pi}} = \sqrt{\frac{1.20 \text{ m}^3/\text{s}}{(3.80 \text{ m/s})\pi}} = 0.317 \text{ m}$$

14-29

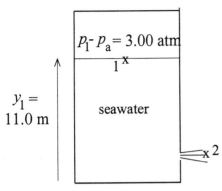

Apply Bernoulli's equation with points 1 and 2 chosen as shown in the sketch. Let $y = 0$ at the bottom of the tank so $y_1 = 11.0$ m and $y_2 = 0$.

$p_1 + \rho g y_1 + \frac{1}{2}\rho v_1^2 = p_2 + \rho g y_2 + \frac{1}{2}\rho v_2^2$

$A_1 v_1 = A_2 v_2$, so $v_1 = (A_2/A_1)v_2$. But the cross-section area of the tank (A_1) is much larger than the cross-section area of the hole (A_2), so $v_1 << v_2$ and the $\frac{1}{2}\rho v_1^2$ term can be neglected.

Thus $\frac{1}{2}\rho v_2^2 = (p_1 - p_2) + \rho g y_1$.

Use $p_2 = p_a$ and solve for v_2:

$$v_2 = \sqrt{2(p_1 - p_a)/\rho + 2gy_1} = \sqrt{\frac{2(3.039 \times 10^5 \text{ Pa})}{1010 \text{ kg/m}^3} + 2(9.80 \text{ m/s}^2)(11.0 \text{ m})}$$

$v_2 = 28.6$ m/s

14-31

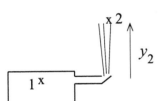

Apply Bernoulli's equation to points 1 and 2 as shown in the sketch. Point 1 is in the mains and point 2 is at the maximum height reached by the stream, so $v_2 = 0$.

$p_1 + \rho g y_1 + \frac{1}{2}\rho v_1^2 = p_2 + \rho g y_2 + \frac{1}{2}\rho v_2^2$

Let $y_1 = 0$, $y_2 = 15.0$ m. The mains have large diameter so $v_1 \approx 0$.

Thus $p_1 = p_2 + \rho g y_2$.

But $p_2 = p_a$, so $p_1 - p_a = \rho g y_2 = (1000 \text{ kg/m}^3)(9.80 \text{ m/s}^2)(15.0 \text{ m}) = 1.47 \times 10^5$ Pa.

14-35

$p_1 + \rho g y_1 + \frac{1}{2}\rho v_1^2 = p_2 + \rho g y_2 + \frac{1}{2}\rho v_2^2$

Horizontal pipe implies $y_1 = y_2$ so $p_1 + \frac{1}{2}\rho v_1^2 = p_2 + \frac{1}{2}\rho v_2^2$

If we solve for v_2 then we can use the discharge rate to calculate A_2 and hence r_2.

Also, $v_1 A_1 = 4.65 \times 10^{-4}$ m^3/s so $v_1 = \dfrac{4.65 \times 10^{-4} \text{ m}^3/\text{s}}{1.32 \times 10^{-3} \text{ m}^2} = 0.3523$ m/s.

Then $\frac{1}{2}\rho v_2^2 = \frac{1}{2}\rho v_1^2 + (p_1 - p_2)$ gives

$$v_2 = \sqrt{v_1^2 + \frac{2(p_1 - p_2)}{\rho}} = \sqrt{(0.3523 \text{ m/s})^2 + \frac{2(1.60 \times 10^5 \text{ Pa} - 1.20 \times 10^5 \text{ Pa})}{1000 \text{ kg/m}^3}}$$

$v_2 = 8.951$ m/s

Then $v_2 A_2 = v_2 \pi r_2^2 = 4.65 \times 10^{-4}$ m^3/s gives

$$r_2 = \sqrt{\frac{4.65 \times 10^{-4} \text{ m}^3/\text{s}}{v_2 \pi}} = \sqrt{\frac{4.65 \times 10^{-4} \text{ m}^3/\text{s}}{(8.951 \text{ m/s})\pi}} = 4.1 \times 10^{-3} \text{ m} = 0.41 \text{ cm}.$$

14-39 $\eta = 1.005$ centipoise $= 1.005 \times 10^{-3}$ N $\cdot$ s/m^2

a) The volume flow rate dV/dt is given by Eq.(14-26):

$$\frac{dV}{dt} = \frac{\pi}{8}\left(\frac{R^4}{\eta}\right)\left(\frac{p_1 - p_2}{L}\right)$$

The absolute pressure at the pump is p_1 and $p_2 = p_a$ is the pressure at the ope end of the pipe, so $p_1 - p_2 = p_1 - p_a$, the gauge pressure at the pump.

$$\frac{dV}{dt} = \left(\frac{\pi}{8}\right)\left(\frac{(4.50 \times 10^{-2} \text{ m})^4}{1.005 \times 10^{-3} \text{ N} \cdot \text{s/m}^2}\right)\left(\frac{1200 \text{ Pa}}{15.0 \text{ m}}\right) = 0.128 \text{ m}^3/\text{s}.$$

b) For the same volume flow rate $R^4 \Delta p$ must stay constant, where Δp is the gaug pressure maintained by the pump. Let $R_a = 4.50$ cm and $R_b = 1.50$ cm, so
$R_a^4 \Delta p_a = R_b^4 \Delta p_b$ gives $\Delta p_b = \Delta p_a (R_a/R_b)^4 = 1200 \text{ Pa}(4.50 \text{ cm}/1.50 \text{ cm})^4 =$
9.72×10^4 Pa.

c) Same R, Δp, and L so $\left(\dfrac{dV}{dt}\right)\eta = \left(\dfrac{\pi}{8}\right)R^4\left(\dfrac{p_1 - p_2}{L}\right) = $ constant.

Thus $\left(\dfrac{dV}{dt}\right)_b = \left(\dfrac{dV}{dt}\right)_a \dfrac{\eta_a}{\eta_b} = (0.128 \text{ m}^3/\text{s})\left(\dfrac{1.005 \text{ centipoise}}{0.469 \text{ centipoise}}\right) = 0.274 \text{ m}^3/\text{s}.$

14-41 The viscous drag force is given by Eq.(14-27): $F = 6\pi\eta r v$.

To compare this to the weight of the sphere, express the weight in terms of th density ρ and radius r of the sphere:
$m = \rho V = \rho(\frac{4}{3}\pi r^3)$ and $w = \rho V g = \rho g(\frac{4}{3}\pi r^3)$
Then $F = \frac{1}{4}w$ implies $6\pi\eta r v = \frac{1}{4}\rho g(\frac{4}{3}\pi r^3)$.

$$v = \frac{\rho r^2 g}{18\eta}$$

From Table 14-1. $\rho_{\text{aluminum}} = 2.70 \times 10^3 \text{ kg/m}^3$.

$$v = \frac{(2.70 \times 10^3 \text{ kg/m}^3)(2.00 \times 10^{-3} \text{ m})^2(9.80 \text{ m/s}^2)}{18(0.986 \text{ N} \cdot \text{s/m}^2)} = 5.96 \text{ mm/s}$$

4-43 Volume flow rate $\dfrac{dV}{dt} = \dfrac{\pi}{8}\left(\dfrac{R^4}{\eta}\right)\left(\dfrac{p_1 - p_2}{L}\right)$

a) $dV/dt \sim R^4$ so if double the diameter (so double the radius), dV/dt increases by a factor of $(2)^4 = 16$.

b) $dV/dt \sim \eta^{-1}$ so if double η, dV/dt changes by a factor of $\frac{1}{2}$.

c) $dV/dt \sim (p_1 - p_2)$ so if double $(p_1 - p_2)$, then dV/dt increases by a factor of 2.

d) $dV/dt \sim (p_1 - p_2)/L$ so if double $(p_1 - p_2)/L$ (the pressure gradient), then dV/dt increases by a factor of 2.

e) $dV/dt \sim L^{-1}$ so if double L, dV/dt changes by a factor of $\frac{1}{2}$.

Problems

14-49

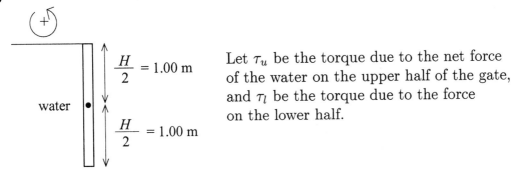

Let τ_u be the torque due to the net force of the water on the upper half of the gate, and τ_l be the torque due to the force on the lower half.

$\dfrac{H}{2} = 1.00 \text{ m}$

$\dfrac{H}{2} = 1.00 \text{ m}$

With the indicated sign convention, τ_l is positive and τ_u is negative, so the net torque about the hinge is $\tau = \tau_l - \tau_u$. Let H be the height of the gate.

Upper-half of gate:

Calculate the torque due to the force on a narrow strip of height dy located a distance y below the top of the gate. Then integrate to get the total torque.

The net force on the strip is $dF = p(y)\, dA$,
where $p(y) = \rho g y$ is the pressure at this depth and
$dA = W\, dy$ with $W = 4.00 \text{ m}$
$dF = \rho g y W\, dy$

The moment arm is $(H/2 - y)$, so $d\tau = \rho g W(H/2 - y)y\,dy$.

$\tau_u = \int_0^{H/2} d\tau = \rho g W \int_0^{H/2}(H/2 - y)y\,dy = \rho g W((H/4)y^2 - y^3/3)\,\big|_0^{H/2}$

$\tau_u = \rho g W(H^3/16 - H^3/24) = \rho g W(H^3/48)$

$\tau_u = (1000 \text{ kg/m}^3)(9.80 \text{ m/s}^2)(4.00 \text{ m})(2.00 \text{ m})^3/48 = 6.533 \times 10^3 \text{ N} \cdot \text{m}$

Lower-half of gate:

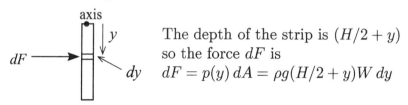

The depth of the strip is $(H/2 + y)$ so the force dF is

$dF = p(y)\,dA = \rho g(H/2 + y)W\,dy$

The moment arm is y, so $d\tau = \rho g W(H/2 + y)y\,dy$.

$\tau_l = \int_0^{H/2} d\tau = \rho g W \int_0^{H/2}(H/2 + y)y\,dy = \rho g W((H/4)y^2 + y^3/3)\,\big|_0^{H/2}$

$\tau_l = \rho g W(H^3/16 + H^3/24) = \rho g W(5H^3/48)$

$\tau_l = (1000 \text{ kg/m}^3)(9.80 \text{ m/s}^2)(4.00 \text{ m})5(2.00 \text{ m})^3/48 = 3.267 \times 10^4 \text{ N} \cdot \text{m}$

Then $\tau = \tau_l - \tau_u = 3.267 \times 10^4 \text{ N} \cdot \text{m} - 6.533 \times 10^3 \text{ N} \cdot \text{m} = 2.61 \times 10^4 \text{ N} \cdot \text{m}$.

14-51 $p - p_0 = \rho g d$

This expression gives that $g = (p - p_0)/\rho d = (p - p_0)V/md$.

But also $g = Gm_p/R^2$ (Eq.12-4 applied to the planet rather than to earth.)

Setting these two expressions for g equal gives $Gm_p/R^2 = (p - p_0)V/md$ and $m_p = (p - p_0)VR^2/Gmd$.

14-55 a)

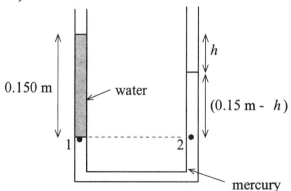

0.150 m

water

h

(0.15 m - h)

Apply $p = p_0 + \rho g h$ to the water in the left-hand arm of the tube.

mercury

$p_0 = p_a$, so the gauge pressure at the interface (point 1) is

$p - p_a = \rho g h = (1000 \text{ kg/m}^3)(9.80 \text{ m/s}^2)(0.150 \text{ m}) = 1470 \text{ Pa}$

b) The pressure at point 1 equals the pressure at point 2.

$p_1 = p_a + \rho_w g(0.150 \text{ m})$

$p_2 = p_a + \rho_{Hg} g(0.150 \text{ m} - h)$

$p_1 = p_2$ implies $\rho_w g(0.150 \text{ m}) = \rho_{Hg} g(0.150 \text{ m} - h)$

$$0.150 \text{ m} - h = \frac{\rho_w(0.150 \text{ m})}{\rho_{Hg}} = \frac{(1000 \text{ kg/m}^3)(0.150 \text{ m})}{13.6 \times 10^3 \text{ kg/m}^3} = 0.011 \text{ m}$$

$h = 0.150 \text{ m} - 0.011 \text{ m} = 0.139 \text{ m} = 13.9 \text{ cm}$

14-57 Free-body diagram for the barge plus coal:

$\sum F_y = ma_y$

$B - (m_{barge} + m_{coal})g = 0$

$\rho_w V_{barge} g = (m_{barge} + m_{coal})g$

$m_{coal} = \rho_w V_{barge} - m_{barge}$

$V_{barge} = (22 \text{ m})(12 \text{ m})(40 \text{ m}) = 1.056 \times 10^4 \text{ m}^3$

The mass of the barge is $m_{barge} = \rho_s V_s$, where s refers to steel.

From Table 14-1, $\rho_s = 7800 \text{ kg/m}^3$.

The volume V_s is 0.040 m times the total area of the five pieces of steel that make up the barge:

$V_s = (0.040 \text{ m})[2(22 \text{ m})(12 \text{ m}) + 2(40 \text{ m})(12 \text{ m}) + (22 \text{ m})(40 \text{ m})] = 94.7 \text{ m}^3$.

Therefore, $m_{barge} = \rho_s V_s = (7800 \text{ kg/m}^3)(94.7 \text{ m}^3) = 7.39 \times 10^5 \text{ kg}$.

Then $m_{coal} = \rho_w V_{barge} - m_{barge} = (1000 \text{ kg/m}^3)(1.056 \times 10^4 \text{ m}^3) - 7.39 \times 10^5 \text{ kg} = 9.8 \times 10^6 \text{ kg}$.

The volume of this mass of coal is $V_{coal} = m_{coal}/\rho_{coal} = 9.8 \times 10^6 \text{ kg}/1500 \text{ kg/m}^3 = 6500 \text{ m}^3$; this is less that V_{barge} so it will fit into the barge.

14-59 a) Free-body diagram for the floating car:

(V_{sub} is the volume that is submerged.)

$\sum F_y = ma_y$

$B - mg = 0$

$\rho_w V_{sub} g - mg = 0$

$V_{sub} = m/\rho_w = (900 \text{ kg})/(1000 \text{ kg/m}^3) = 0.900 \text{ m}^3$

$V_{sub}/V_{obj} = (0.900 \text{ m}^3)/(3.0 \text{ m}^3) = 0.30 = 30\%$

b) When the car starts to sink it is fully submerged and the buoyant force is equal to the weight of the car plus the water that is inside it.

When the car is full submerged $V_{sub} = V$, the volume of the car and

$B = \rho_w V g = (1000 \text{ kg/m}^3)(3.0 \text{ m}^3)(9.80 \text{ m/s}^2) = 2.94 \times 10^4 \text{ N}$

The weight of the car is $mg = (900 \text{ kg})(9.80 \text{ m/s}^2) = 8820 \text{ N}$.

Thus the weight of the water in the car when it sinks is the buoyant force minus the weight of the car itself:

$m_{water} = (2.94 \times 10^4 \text{ N} - 8820 \text{ N})/(9.80 \text{ m/s}^2) = 2.10 \times 10^3 \text{ kg}$

And $V_{water} = m_{water}/\rho_w = (2.10 \times 10^3 \text{ kg})/(1000 \text{ kg/m}^3) = 2.10 \text{ m}^3$

The fraction this is of the total interior volume is $(2.10 \text{ m}^3)/(3.00 \text{ m}^3) = 0.70 = 70\%$

14-63   a) Free-body diagram for the dirigible:

(The lift corresponds to a mass $m_{lift} = (120 \times 10^3 \text{ N})/(9.80 \text{ m/s}^2) = 1.224 \times 10^4$ kg. The mass m_{tot} is 1.224×10^4 kg plus the mass m_{gas} of the gas that fills the dirigible. B is the buoyant force exerted by the air.)

$B = \rho_{air} V g$

$a = 0$

$m_{tot} g = (9184 \text{ kg} + m_{gas}) g$

$\sum F_y = m a_y$

$B - m_{tot} g = 0$

$\rho_{air} V g = (1.224 \times 10^4 \text{ kg} + m_{gas}) g$

Write m_{gas} in terms of V: $m_{gas} = \rho_{gas} V$

And let g divide out; the equation becomes $\rho_{air} V = 1.224 \times 10^4 \text{ kg} + \rho_{gas} V$

$$V = \frac{1.224 \times 10^4 \text{ kg}}{1.20 \text{ kg/m}^3 - 0.0899 \text{ kg/m}^3} = 1.10 \times 10^4 \text{ m}^3$$

b) Let m_{lift} be the mass that could be lifted. From part (a),

$m_{lift} = (\rho_{air} - \rho_{gas})V = (1.20 \text{ kg/m}^3 - 0.166 \text{ kg/m}^3)(1.10 \times 10^4 \text{ m}^3) = 1.14 \times 10^4$ kg.

The lift force is $m_{lift} = (1.14 \times 10^4 \text{ kg})(9.80 \text{ m/s}^2) = 112 \text{ kN}$.

Hydrogen is not used because it is highly explosive in air.

14-67

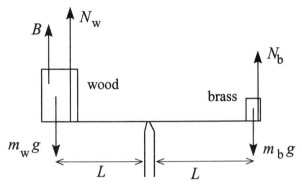

The buoyant force on the brass is negelcted, but we include the buoyant force B on the block of wood. N_w and N_b are the normal forces exerted by the balance arm on which the objects sit.

Free-body diagram for the balance arm:

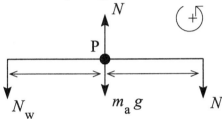

$$\tau_P = 0$$
$$N_w L - N_b L = 0$$
$$N_w = N_b$$

Free-body diagram for the brass mass:

$a = 0$

$$\sum F_y = ma_y$$
$$N_b - m_b g = 0$$
$$N_b = m_b g$$

Free-body diagram for the block of wood:

$a = 0$

$$\sum F_y = ma_y$$
$$N_w + B - m_w g = 0$$
$$N_w = m_w g - B$$

But $N_b = N_w$ implies $m_b g = m_w g - B$.

And $B = \rho_{air} V_w g = \rho_{air}(m_w / \rho_w)g$, so $m_b g = m_w g - \rho_{air}(m_w / \rho_w)g$.

$$m_w = \frac{m_b}{1 - \rho_{air}/\rho_w} = \frac{0.0950 \text{ kg}}{1 - ((1.20 \text{ kg/m}^3)/(150 \text{ kg/m}^3))} = 0.0958 \text{ kg}$$

The buoyancy in air of the brass can be neglected because the density of brass is much more than the density of air; the buoyancy force exerted on the brass by the

air is much less than the weight of the brass. The density of the balsa wood i
much less than the density of the brass, so the buoyancy force on the balsa woo
is not such a small fraction of its weight.

14-69 Free-body diagram for the piece of alloy:

y

$T = 39.0$ N

B

x

$m_{tot} \, g = 45.0$ N

$$\sum F_y = ma_y$$
$$B + T - m_{tot}g = 0$$
$$B = m_{tot}g - T$$
$$B = 45.0 \text{ N} - 39.0 \text{ N} = 6.0 \text{ N}$$

Also, $m_{tot}g = 45.0$ N so $m_{tot} = 45.0$ N$/(9.80$ m/s$^2) = 4.59$ kg.

We can use the known value of the buoyant force to calculate the volume of th
object:

$$B = \rho_w V_{obj}g = 6.0 \text{ N}$$

$$V_{obj} = \frac{6.0 \text{ N}}{\rho_w g} = \frac{6.0 \text{ N}}{(1000 \text{ kg/m}^3)(9.80 \text{ m/s}^2)} = 6.122 \times 10^{-4} \text{ m}^3$$

We know two things:

(1) The mass m_g of the gold plus the mass m_a of the aluminum must add to m_{tot}

$$m_g + m_a = m_{tot}$$

We write this in terms of the volumes V_g and V_a of the gold and aluminum:

$$\rho_g V_g + \rho_a V_a = m_{tot}$$

(2) The volumes V_a and V_g must add to give V_{obj}:

$V_a + V_g = V_{obj}$ so that $V_a = V_{obj} - V_g$

Use this in the equation in (1) to eliminate V_a:

$$\rho_g V_g + \rho_a(V_{obj} - V_g) = m_{tot}$$

$$V_g = \frac{m_{tot} - \rho_a V_{obj}}{\rho_g - \rho_a} = \frac{4.59 \text{ kg} - (2.7 \times 10^3 \text{ kg/m}^3)(6.122 \times 10^{-4} \text{ m}^3)}{19.3 \times 10^3 \text{ kg/m}^3 - 2.7 \times 10^3 \text{ kg/m}^3} =$$

1.769×10^{-4} m^3.

Then $m_g = \rho_g V_g = (19.3 \times 10^3 \text{ kg/m}^3)(1.769 \times 10^{-4} \text{ m}^3) = 3.41$ kg and the weigh
of gold is $w_g = m_g g = 33.4$ N.

14-71 a) Free-body diagram for the crown:

$$\sum F_y = ma_y$$
$$T + B - w = 0$$

$$T = fw$$
$$B = \rho_w V_c g, \text{ where } \rho_w = \text{density}$$
of water, V_c = volume of crown

Then $fw + \rho_w V_c g - w = 0$.

$(1 - f)w = \rho_w V_c g$

Use $w = \rho_c V_c g$, where ρ_c = density of crown.

$(1 - f)\rho_c V_c g = \rho_w V_c g$

$$\frac{\rho_c}{\rho_w} = \frac{1}{1 - f}, \text{ as was to be shown.}$$

$f \to 0$ gives $\rho_c/\rho_w = 1$ and $T = 0$. These values are consistent. If the density of the crown equals the density of the water, the crown just floats, fully submerged, and the tension should be zero.

When $f \to 1$, $\rho_c \gg \rho_w$ and $T = w$. If $\rho_c \gg \rho_w$ then B is negligible relative to the weight w of the crown and T should equal w.

b) "apparent weight" equals T in rope when the crown is immersed in water. $T = fw$, so need to compute f.

$\rho_c = 19.3 \times 10^3 \text{ kg/m}^3; \quad \rho_w = 1.00 \times 10^3 \text{ kg/m}^3$

$$\frac{\rho_c}{\rho_w} = \frac{1}{1 - f} \text{ gives } \frac{19.3 \times 10^3 \text{ kg/m}^3}{1.00 \times 10^3 \text{ kg/m}^3} = \frac{1}{1 - f}$$

$19.3 = 1/(1 - f)$ and $f = 0.9482$

Then $T = fw = (0.9482)(12.9 \text{ N}) = 12.2 \text{ N}$.

c) Now the density of the crown is very nearly the density of lead;

$\rho_c = 11.3 \times 10^3 \text{ kg/m}^3$.

$$\frac{\rho_c}{\rho_w} = \frac{1}{1 - f} \text{ gives } \frac{11.3 \times 10^3 \text{ kg/m}^3}{1.00 \times 10^3 \text{ kg/m}^3} = \frac{1}{1 - f}$$

$11.3 = 1/(1 - f)$ and $f = 0.9115$

Then $T = fw = (0.9115)(12.9 \text{ N}) = 11.8 \text{ N}$.

14-73 **a)** $B = \rho_{\text{water}} V_{\text{tot}} g$, where V_{tot} is the total volume of the object.

$V_{\text{tot}} = V_m + V_0$, where V_m is the volume of the metal.

$V_m = w/g\rho_m$ so $V_{tot} = w/g\rho_m + V_0$

This gives $B = \rho_{water} g(w/g\rho_m + V_0)$

Solving for V_0 gives $V_0 = B/(\rho_{water} g) - w/(\rho_m g)$, as was to be shown.

b) The expression derived in part (a) gives

$$V_0 = \frac{20\ N}{(1000\ \text{kg/m}^3)(9.80\ \text{m/s}^2)} - \frac{156\ N}{(8.9 \times 10^3\ \text{kg/m}^3)(9.80\ \text{m/s}^2)} = 2.52 \times 10^{-4}\ \text{m}^3$$

$$V_{tot} = \frac{B}{\rho_{water} g} = \frac{20\ N}{(1000\ \text{kg/m}^3)(9.80\ \text{m/s}^2)} = 2.04 \times 10^{-3}\ \text{m}^3 \text{ and}$$

$V_0/V_{tot} = (2.52 \times 10^{-4}\ \text{m}^3)/(2.04 \times 10^{-3}\ \text{m}^3) = 0.124.$

14-75 In both cases the total buoyant force must equal the weight of the barge plus the weight of the anchor. Thus the total amount of water must be the same when the anchor is in the boat as when it is over the side. When the anchor is in the water the barge displaces less water, less by the amount the anchor displaces. Thus the barge rises in the water.

The volume of the anchor is $V_{anchor} = m/\rho = (35.0\ \text{kg})/(7860\ \text{kg/m}^3) = 4.453 \times 10^{-3}\ \text{m}^3.$

The barge rises in the water a vertical distance h given by $hA = 4.453 \times 19^{-3}\ \text{m}^3$ where A is the area of the bottom of the barge.

$h = (4.453 \times 10^{-3}\ \text{m}^3)/(8.00\ \text{m}^2) = 5.57 \times 10^{-4}\ \text{m}.$

14-77 a) Free-body diagram for the block:

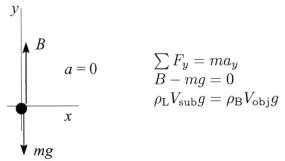

$$\sum F_y = ma_y$$
$$B - mg = 0$$
$$\rho_L V_{sub} g = \rho_B V_{obj} g$$

The fraction of the volume that is submerged is $V_{sub}/V_{obj} = \rho_B/\rho_L.$

Thus the fraction that is *above* the surface is $V_{above}/V_{obj} = 1 - \rho_B/\rho_L.$

b) Let the water layer have depth d.

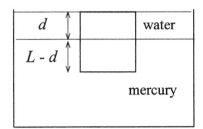

$p = p_0 + \rho_w g d + \rho_L g (L - d)$

Applying $\sum F_y = m a_y$ to the block gives
$(p - p_0) A - mg = 0$.

$[\rho_w g d + \rho_L g (L - d)] A = \rho_B L A g$

A and g divide out and $\rho_w d + \rho_L (L - d) = \rho_B L$

$d(\rho_w - \rho_L) = (\rho_B - \rho_L) L$

$d = \left(\dfrac{\rho_L - \rho_B}{\rho_L - \rho_w} \right) L$

c) $d = \left(\dfrac{13.6 \times 10^3 \text{ kg/m}^3 - 7.8 \times 10^3 \text{ kg/m}^3}{13.6 \times 10^3 \text{ kg/m}^3 - 1000 \text{ kg/m}^3} \right) (0.100 \text{ m}) = 0.0460 \text{ m} = 4.60 \text{ cm}$

14-83

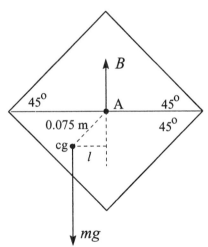

The resultant buoyant force acts at the geometrical center of the submerged portion of the object. The weight of the object acts at the center of gravity of the object. These two points are displaced from each other as in Fig.14-41b, and this gives rise to a restoring torque about point A.

For an axis at point A:

$\tau_B = 0$

$\tau_{mg} = mgl = mg(0.075 \text{ m}) \cos 45° = (0.053 \text{ m}) mg$

The block is floating, so $B = mg$. Calculate B in order to find the weight mg of the block. Half of the volume of the block is submerged, so

$V_{\text{sub}} = \frac{1}{2}(0.30 \text{ m})^3 = 0.0135 \text{ m}^3$.

Then $B = \rho_w V_{\text{sub}} g = (1000 \text{ kg/m}^3)(0.0135 \text{ m}^3)(9.80 \text{ m/s}^2) = 132.3 \text{ N}$.

Therefore, $\sum \tau_A = \tau_{mg} = (0.053 \text{ m})(132.3 \text{ N}) = 7.0 \text{ N} \cdot \text{m}$.

14-85 The water level in the vessel will rise until the volume flow rate into the vessel, 2.40×10^{-4} m^3/s, equals the volume flow rate out the hole in the bottom.

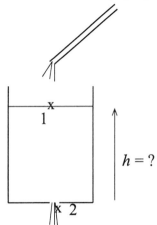

Let points 1 and 2 be chosen as in the sketch.

Bernoulli's equation: $p_1 + \rho g y_1 + \frac{1}{2}\rho v_1^2 = p_2 + \rho g y_2 + \frac{1}{2}\rho v_2^2$

Volume flow rate out of hole equals volume flow rate from tube gives that

$$v_2 A_2 = 2.40 \times 10^{-4} \text{ m}^3/\text{s and } v_2 = \frac{2.40 \times 10^{-4} \text{ m}^3/\text{s}}{1.50 \times 10^{-4} \text{ m}^2} = 1.60 \text{ m/s}$$

$A_1 >> A_2$ and $v_1 A_1 = v_2 A_2$ says that $\frac{1}{2}\rho v_1^2 << \frac{1}{2}\rho v_2^2$; neglect the $\frac{1}{2}\rho v_1^2$ term.

Measure y from the botom of the bucket, so $y_2 = 0$ and $y_1 = h$.

$p_1 = p_2 = p_a$ (air pressure)

Then $p_a + \rho g h = p_a + \frac{1}{2}\rho v_2^2$ and

$h = v_2^2/2g = (1.60 \text{ m/s})^2/2(9.80 \text{ m/s}^2) = 0.131 \text{ m} = 13.1 \text{ cm}$

14-87

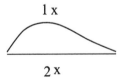

Apply Bernoulli's equation to points 1 and 2, where point 1 is just above the wing and point 2 is just below the wing.

$p_1 + \rho g y_1 + \frac{1}{2}\rho v_1^2 = p_2 + \rho g y_2 + \frac{1}{2}\rho v_2^2$

A "lift" of 2000 N/m^2 means that $p_2 - p_1 = 2000$ Pa.

Solving for v_1 gives $v_1 = \sqrt{2(p_2 - p_1)/\rho + v_2^2 - 2g(y_1 - y_2)}$

Note that $2(p_2 - p_1)/\rho = 2(2000 \text{ Pa})/(1.20 \text{ kg/m}^3) = 3333$ m^2/s^2. We aren't give a value for $y_1 - y_2$, but it must be 1 m or so. For $y_1 - y_2 = 1$ m, $2g(y_1 - y_2) = 19.$ m^2/s^2.

Thus $2g(y_1 - y_2) << 2(p_2 - p_1)/\rho$ and can be neglected.

Thus $v_1 = \sqrt{2(p_2 - p_1)/\rho + v_2^2} = \sqrt{3333 \text{ m}^2/\text{s}^2 + (120 \text{ m/s})^2} = 133$ m/s.

14-91

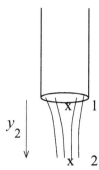

Let point 1 be at the end of the pipe and let point 2 be in the stream of liquid at a distance y_2 below the end of the tube.

Consider the free-fall of the liquid. Take $+y$ to be downward.

Free-fall implies $a = g$.

Then $v_2^2 = v_1^2 + 2a(y - y_0)$ gives $v_2^2 = v_1^2 + 2gy_2$ and $v_2 = \sqrt{v_1^2 + 2gy_2}$.

Equation of continuity says $v_1 A_1 = v_2 A_2$

And since $A = \pi r^2$ this becomes $v_1 \pi r_1^2 = v_2 \pi r_2^2$ and $v_2 = v_1 (r_1/r_2)^2$.

Use this in the above to eliminate v_2: $v_1 (r_1^2/r_2^2) = \sqrt{v_1^2 + 2gy_2}$

$r_2 = r_1 \sqrt{v_1}/(v_1^2 + 2gy_2)^{1/4}$

Note that this equation says that r_2 decreases with distance below the end of the pipe.

b) $v_1 = 1.20$ m/s

We want the value of y_2 that gives $r_2 = \frac{1}{2}r_1$, or $r_1 = 2r_2$

The result obtained in part (a) says $r_2^4 (v_1^2 + 2gy_2) = r_1^4 v_1^2$

Solving for y_2 gives $y_2 = \dfrac{[(r_1/r_2)^4 - 1]v_1^2}{2g} = \dfrac{(16 - 1)(1.20 \text{ m/s})^2}{2(9.80 \text{ m/s}^2)} = 1.10$ m

14-93 a) Free-body diagram for the bubble:

(Note that the viscous drag force F is downward, since the bubble is traveling upward.)

$a = 0$
(at the terminal speed)

$\sum F_y = ma_y$
$B - mg - F = 0$
$B = mg + F$

$B = \rho' V g = \frac{4}{3}\pi r^3 \rho' g$, where ρ' is the density of the liquid

$mg = \rho V g = \frac{4}{3}\pi r^3 \rho g$, where ρ is the density of the air in the bubble

$$F = 6\pi\eta r v_t$$

Thus $B = mg + F$ implies $\frac{4}{3}\pi r^3 \rho' g = \frac{4}{3}\pi r^3 \rho g + 6\pi\eta r v_t$.

$$v_t = \frac{2r^2 g}{9\eta}(\rho' - \rho) = \frac{2(1.00 \times 10^{-3} \text{ m})^2(9.80 \text{ m/s}^2)}{9(0.150 \text{ N} \cdot \text{s/m}^2)}(900 \text{ kg/m}^3 - 1.20 \text{ kg/m}^3)$$

$v_t = 1.30 \times 10^{-2}$ m/s $= 1.30$ cm/s.

(Note that the precise value of $\rho = \rho_{\text{air}}$ that is used is unimportant since $\rho' \gg \rho$

b) $v_t = \dfrac{2r^2 g}{9\eta}(\rho' - \rho) = \dfrac{2(1.00 \times 10^{-3} \text{ m})^2(9.80 \text{ m/s}^2)}{9(1.005 \times 10^{-3} \text{ N} \cdot \text{s/m}^2)}(1000 \text{ kg/m}^3 - 1.20 \text{ kg/m}^3$

$v_t = 2.16$ m/s

CHAPTER 15
TEMPERATURE AND HEAT

Exercises

15-3 $T_F = \frac{9}{5}T_c + 32°$

$T_F = T_c = T$ says $T = \frac{9}{5}T + 32°$

$\frac{4}{5}T = -32°$ and $T = -40°$

$-40 °C = -40 °F$; must be winter

15-13   If the pressure varies linearly with temperature, then

$p_2 = p_1 + \gamma(T_2 - T_1)$.

$\gamma = \dfrac{p_2 - p_1}{T_2 - T_1} = \dfrac{6.50 \times 10^4 \text{ Pa} - 4.80 \times 10^4 \text{ Pa}}{100°C - 0.01°C} = 170.0 \text{ Pa/C}°$

Apply $p = p_1 + \gamma(T - T_1)$ with $T_1 = 0.01°C$ and $p = 0$ to solve for T.

$0 = p_1 + \gamma(T - T_1)$

$T = T_1 - \dfrac{p_1}{\gamma} = 0.01°C - \dfrac{4.80 \times 10^4 \text{ Pa}}{170.0 \text{ Pa/C}°} = -282°C.$

b) Eq.(15-4) says $T_2/T_1 = p_2/p_1$, where T is in kelvins.

$p_2 = p_1 \left(\dfrac{T_2}{T_1}\right) = 6.50 \times 10^4 \text{ Pa} \left(\dfrac{0.01 + 273.15}{100 + 273.15}\right) = 4.76 \times 10^4 \text{ Pa}$; this differs from
the 4.80×10^4 Pa that was measured so Eq.(15-4) is not precisely obeyed.

15-19   Let $L_0 = 40.125$ cm; $T_0 = 20.0°C$

$\Delta T = 45.0°C - 20.0°C = 25.0 \text{ C}°$ gives $\Delta L = 0.023$ cm

Thus $\Delta L = \alpha L_0 \, \Delta T$ implies

$\alpha = \dfrac{\Delta L}{L_0 \, \Delta T} = \dfrac{0.023 \text{ cm}}{(40.125 \text{ cm})(25.0 \text{ C}°)} = 2.3 \times 10^{-5} \text{ (C}°)^{-1}.$

15-21   Consider ΔV for the ethanol.

From Table 15-2, β for ethanol is $75 \times 10^{-5} \text{ K}^{-1}$

$\Delta T = 10.0°C - 19.0°C = -9.0$ K.

Then $\Delta V = \beta V_0 \Delta T = (75 \times 10^{-5} \text{ K}^{-1})(1700 \text{ L})(-9.0 \text{ K}) = -11$ L.
The volume of the air space will be 11 L $= 0.011 \text{ m}^3$.

15-25   **a)** $A_0 = \pi r_0^2 = \pi(1.350 \text{ cm}/2)^2 = 1.431 \text{ cm}^2$

b) Exercise 15-24 says $\Delta A = 2\alpha A_0 \Delta T$, so
$\Delta A = 2(1.2 \times 10^{-5} \text{ C}°^{-1})(1.431 \text{ cm})(175°C - 25°C) = 5.15 \times 10^{-3} \text{ cm}^2$
$A = A_0 + \Delta A = 1.436 \text{ cm}^2$

15-27   **a)** $\Delta L = \alpha L_0 \Delta T$

$$\alpha = \frac{\Delta L}{L_0 \Delta T} = \frac{1.9 \times 10^{-2} \text{ m}}{(1.50 \text{ m})(420°C - 20°C)} = 3.2 \times 10^{-5} \text{ (C}°)^{-1}$$

b) Eq.(15-12): stress $F/A = -Y\alpha \Delta T$
$\Delta T = 20°C - 420°C = -400$ C° (ΔT always means final temperature minus initia
temperature)
$F/A = -(2.0 \times 10^{11} \text{ Pa})(3.2 \times 10^{-5}(\text{C}°)^{-1})(-400°C) = +2.6 \times 10^9$ Pa
F/A is positive means that the stress is a tensile (stretching) stress.

15-29   **a)** $Q = mc \Delta T$
$m = \frac{1}{2}(1.3 \times 10^{-3} \text{ kg}) = 0.65 \times 10^{-3}$ kg
$Q = (0.65 \times 10^{-3} \text{ kg})(1020 \text{ J/kg} \cdot \text{K})(37°C - (-20°C)) = 38$ J

b) 20 breaths/min (60 min/1 h) $= 1200$ breaths/h
So $Q = (1200)(38 \text{ J}) = 4.6 \times 10^4$ J.

15-31   The change in mechanical energy equals the decrease in
gravitational potential energy, $\Delta U = -mgh$; $\quad | \Delta U | = mgh$.
$Q = | \Delta U | = mgh$ implies $mc\Delta T = mgh$
$\Delta T = gh/c = (9.80 \text{ m/s}^2)(225 \text{ m})/(4190 \text{ J/kg} \cdot \text{K}) = 0.526 \text{ K} = 0.526$ C°

Note that the answer is independent of the mass of the object. Note also the sma
change in temperature that corresponds to this large change in height!

15-35   <u>kettle</u>
$Q = mc \Delta T$, $c = 910$ J/kg · K (from Table 15-3)
$Q = (1.50 \text{ kg})(910 \text{ J/kg} \cdot \text{K})(85.0°C - 20.0°C) = 8.873 \times 10^4$ J

water

$Q = mc\,\Delta T$, $c = 4190$ J/kg $\cdot$ K (from Table 15-3)

$Q = (1.80$ kg$)(4190$ J/kg $\cdot$ K$)(85.0°$C $- 20.0°$C$) = 4.902 \times 10^5$ J

Total $Q = 8.873 \times 10^4$ J $+ 4.902 \times 10^5$ J $= 5.79 \times 10^5$ J

15-37   **a)** $P = Q/t$, so the total heat transferred to the liquid is

$Q = Pt = (65.0$ W$)(120$ s$) = 7800$ J

Then $Q = mc\,\Delta T$ gives

$$c = \frac{Q}{m\,\Delta T} = \frac{7800 \text{ K}}{0.780 \text{ kg}(22.54°\text{C} - 18.55°\text{C})} = 2.51 \times 10^3 \text{ J/kg} \cdot \text{K}$$

b) Then the actual Q transferred to the liquid is less than 7800 J so the actual c is less than our calculated value; our result in part (a) is an overestimate.

15-41   Heat must be added to do the following:

ice at $-10.0°$C $\rightarrow$ ice at $0°$C

$Q_{\text{ice}} = mc_{\text{ice}}\,\Delta T = (12.0 \times 10^{-3}$ kg$)(2100$ J/kg $\cdot$ K$)(0°$C $- (-10.0°$C$)) = 252$ J

phase transition ice $(0°$C$) \rightarrow$ liquid water $(0°$C$)$ (melting)

$Q_{\text{melt}} = +mL_f = (12.0 \times 10^{-3}$ kg$)(334 \times 10^3$ J/kg$) = 4.008 \times 10^3$ J

water at $0°$C (from melted ice $\rightarrow$ water at $100°$C

$Q_{\text{water}} = mc_{\text{water}}\,\Delta T = (12.0 \times 10^{-3}$ kg$)(4190$ J/kg $\cdot$ K$)(100°$C $- 0°$C$) =$ 5.028×10^3 J

phase transition water $(100°$C$) \rightarrow$ steam $(100°$C$)$ (boiling)

$Q_{\text{boil}} = +mL_v = (12.0 \times 10^{-3}$ kg$)(2256 \times 10^3$ J/kg$) = 2.707 \times 10^4$ J

The total Q is $Q = 252$ J $+ 4.008 \times 10^3$ K $+ 5.028 \times 10^3$ J $+ 2.707 \times 10^4$ J $=$ 3.64×10^4 J

$(3.64 \times 10^4$ J$)(1$ cal$/4.186$ J$) = 8.70 \times 10^3$ cal

$(3.64 \times 10^4$ J$)(1$ Btu$/1055$ J$) = 34.5$ Btu

15-43   "two-ton air conditioner" means 2 tons (4000 lbs) of ice can be frozen from water at $0°$C in 24 h.

Find the mass m that corresponds to 4000 lb (weight of water):

$m = (4000$ lb$)(1$ kg$/2.205$ lb$) = 1814$ kg (The kg to lb equivalence from Appendix E has been used.)

The heat that must be removed from the water to freeze it is

$$Q = -mL_f = -(1814 \text{ kg})(334 \times 10^3 \text{ J/kg}) = -6.06 \times 10^8 \text{ J}.$$

The power required if this is to be done in 24 hours is

$$P = \frac{|Q|}{t} = \frac{6.06 \times 10^8 \text{ J}}{(24 \text{ h})(3600 \text{ s/1 h})} = 7010 \text{ W}$$

or $P = (7010 \text{ W})((1 \text{ Btu/h})/(0.293 \text{ W})) = 2.39 \times 10^4 \text{ Btu/h}.$

15-45   The heat that must be added to a lead bullet of mass m to melt it is

$Q = mc\,\Delta T + mL_f$ ($mc\,\Delta T$ is the heat required to raise the temperature from $25°C$ to the melting point of $327.3°C$; mL_f is the heat required to make the solid $\to$ liquid phase change.)

The kinetic energy of the bullet if its speed is v is $K = \frac{1}{2}mv^2$.

Then $K = Q$ says $\frac{1}{2}mv^2 = mc\,\Delta T + mL_f$.

$$v = \sqrt{2(c\Delta T + L_f)}$$
$$v = \sqrt{2[(130 \text{ J/kg} \cdot \text{K})(327.3°C - 25°C) + 24.5 \times 10^3 \text{ J/kg}]} = 357 \text{ m/s}$$

15-49 $Q_{\text{system}} = 0$

Calculate Q for each component of the system:

copper pot
$Q_{\text{can}} = mc\,\Delta T = (0.500 \text{ kg})(390 \text{ J/kg} \cdot \text{K})(T - 20.0°C) = (195 \text{ J/K})T - 3900 \text{ J}$

water
$Q_{\text{water}} = mc\,\Delta T = (0.170 \text{ kg})(4190 \text{ J/kg} \cdot \text{K})(T - 20.0°C) = (712.3 \text{ J/K})T$
$1.425 \times 10^4 \text{ J}$

iron
$Q_{\text{iron}} = mc\,\Delta T = (0.250 \text{ kg})(470 \text{ J/kg} \cdot \text{K})(T - 85.0°C) =$
$(117.5 \text{ J/K})T - 9987.5 \text{ J}$

$Q_{\text{system}} = 0$ gives that $Q_{\text{can}} + Q_{\text{water}} + Q_{\text{iron}} = 0$
$(195 \text{ J/K})T - 3900 \text{ J} + (712.3 \text{ J/K})T - 1.425 \times 10^4 \text{ J} + (117.5 \text{ J/K})T - 9987.5 \text{ J} =$
$(1024.8 \text{ J/K})T = 2.814 \times 10^4 \text{ J}$

$$T = \frac{2.814 \times 10^4 \text{ J}}{1024.8 \text{ J/K}} = 27.5°C$$

15-51 $Q_{\text{system}} = 0$

Calculate Q for each component of the system:

(Beaker has small mass says that $Q = mc\,\Delta T$ for beaker can be neglected.)

0.250 kg of water (cools from 75.0°C to 30.0°C)
$Q_{\text{water}} = mc\,\Delta T = (0.250 \text{ kg})(4190 \text{ J/kg} \cdot \text{K})(30.0°\text{C} - 75.0°\text{C}) = -4.714 \times 10^4 \text{ J}$

ice (warms to 0°C; melts; water from melted ice warms to 30.0°C)
$Q_{\text{ice}} = mc_{\text{ice}}\Delta T + mL_{\text{f}} + mc_{\text{water}}\Delta T$
$Q_{\text{ice}} = m[(2100 \text{ J/kg} \cdot \text{K})(0°\text{C} - (-20.0°\text{C})) + 334 \times 10^3 \text{ J/kg}+$
$(4190 \text{ J/kg} \cdot \text{K})(30.0°\text{C} - 0°\text{C})]$
$Q_{\text{ice}} = (5.017 \times 10^5 \text{ J/kg})m$

$Q_{\text{system}} = 0$ says $Q_{\text{water}} + Q_{\text{ice}} = 0$
$-4.714 \times 10^4 \text{ J} + (5.017 \times 10^5 \text{ J/kg})m = 0$

$m = \dfrac{4.714 \times 10^4 \text{ J}}{5.017 \times 10^5 \text{ J/kg}} = 0.0940 \text{ kg}$

15-53 $Q_{\text{system}} = 0$
Large block of ice implies that ice is left, so $T_2 = 0°\text{C}$.
Calculate Q for each component of the system:
ingot
$Q_{\text{ingot}} = mc\,\Delta T = (4.00 \text{ kg})(234 \text{ J/kg} \cdot \text{K})(0°\text{C} - 750°\text{C}) = -7.02 \times 10^5 \text{ J}$
ice
$Q_{\text{ice}} = +mL_{\text{f}}$, where m is the mass of the ice that changes phase (melts)

$Q_{\text{system}} = 0$ says $Q_{\text{ingot}} + Q_{\text{ice}} = 0$
$-7.02 \times 10^5 \text{ J} + m(334 \times 10^3 \text{ J/kg}) = 0$
$m = \dfrac{7.02 \times 10^5 \text{ J}}{334 \times 10^3 \text{ J/kg}} = 2.10 \text{ kg}$

15-57   a) temperature gradient $= (T_{\text{H}} - T_{\text{C}})/L =$
$(100.0°\text{C} - 0.0°\text{C})/0.450 \text{ m} = 222 \text{ C}°/\text{m} = 222 \text{ K/m}$

b) $H = kA(T_{\text{H}} - T_{\text{C}})/L$. From Table 15-5, $k = 385 \text{ W/m} \cdot \text{K}$,
so $H = (385 \text{ W/m} \cdot \text{K})(1.25 \times 10^{-4} \text{ m}^2)(222 \text{ K/m}) = 10.7 \text{ W}$

c) $H = 10.7 \text{ W}$ for all sections of the rod.

Apply $H = kA\,\Delta T/L$ to the 12.0 cm section:

$T_{\mathrm{H}} - T = LH/kA$ and $T = T_{\mathrm{H}} - LH/Ak =$

$$100.0^{\circ}\mathrm{C} - \frac{(0.120 \text{ m})(10.7 \text{ W})}{(1.25 \times 10^{-4} \text{ m}^2)(385 \text{ W/m} \cdot \text{K})} = 73.3^{\circ}\mathrm{C}$$

15-59   Call the temperature at the interface between the wood and the styrofoam T. The heat current in each material is given by $H = kA(T_{\mathrm{H}} - T_{\mathrm{C}})/L$.

Heat current through the wood:
$H_w = k_w A(T - T_1)/L_w$

Heat current through the stryofoam:
$H_s = k_s A(T_2 - T)/L_s$

In steady-state heat does not accumulate in either material. The same heat ha to pass through both materials in succession, so $H_w = H_s$.

This implies $k_w A(T - T_1)/L_w = k_s A(T_2 - T)/L_s$

$k_w L_s(T - T_1) = k_s L_w(T_2 - T)$

$$T = \frac{k_w L_s T_1 + k_s L_w T_2}{k_w L_s + k_s L_w} = \frac{-0.0176 \text{ W} \cdot^{\circ}\text{C/K} + 0.0057 \text{ W} \cdot^{\circ}\text{C/K}}{0.00206 \text{ W/K}} = -5.8^{\circ}\mathrm{C}$$

b) Heat flow per square meter is $\dfrac{H}{A} = k\left(\dfrac{T_{\mathrm{H}} - T_{\mathrm{C}}}{L}\right)$. We can calculate this eith for the wood or for the styrofoam; the results must be the same.

wood

$$\frac{H_w}{A} = k_w \frac{T - T_1}{L_w} = (0.080 \text{ W/m} \cdot \text{K})\frac{(-5.8^{\circ}\mathrm{C} - (-10.0^{\circ}\mathrm{C}))}{0.030 \text{ m}} = 11 \text{ W/m}^2.$$

styrofoam

$$\frac{H_s}{A} = k_s \frac{T_2 - T}{L_s} = (0.010 \text{ W/m} \cdot \text{K})\frac{(19.0^{\circ}\mathrm{C} - (-5.8^{\circ}\mathrm{C}))}{0.022 \text{ m}} = 11 \text{ W/m}^2.$$

15-63   The heat conducted through the bottom of the pot goes into the water at $100^{\circ}\mathrm{C}$ to convert it to steam at $100^{\circ}\mathrm{C}$. We can calculate the amount of heat flo from the mass of material that changes phase.

$Q = mL_{\mathrm{v}} = (0.390 \text{ kg})(2256 \times 10^3 \text{ J/kg}) = 8.798 \times 10^5 \text{ J}$

$H = Q/t = 8.798 \times 10^5 \text{ J}/180 \text{ s} = 4.888 \times 10^3 \text{ J/s}$

Then $H = kA(T_\text{H} - T_\text{C})/L$ says that

$$T_\text{H} - T_\text{C} = \frac{HL}{kA} = \frac{(4.888 \times 10^3 \text{ J/s})(8.50 \times 10^{-3} \text{ m})}{(50.2 \text{ W/m} \cdot \text{K})(0.150 \text{ m}^2)} = 5.52 \text{ C}°$$

$T_\text{H} = T_\text{C} + 5.52 \text{ C}° = 100°\text{C} + 5.52 \text{ C}° = 105.5°\text{C}$

15-65 $H_\text{net} = Ae\sigma(T^4 - T_\text{s}^4)$ (Eq.15-26; T must be in kelvins)

Example 15-17 gives $A = 1.2 \text{ m}^2$, $e = 1.0$, and $T = 30°\text{C} = 303 \text{ K}$ (body surface temperature)

$T_\text{s} = 5.0°\text{C} = 278 \text{ K}$

$H_\text{net} = 573.5 \text{ W} - 406.4 \text{ W} = 167 \text{ W}$

(Note that this is larger than H_net calculated in Example 15-17. The lower temperature of the surroundings increases the rate of heat loss by radiation.)

15-67 $H = Ae\sigma T^4$ so $A = H/e\sigma T^4$

150-W and all electrical energy consumed is radiated says $H = 150 \text{ W}$

$$A = \frac{150 \text{ W}}{(0.35)(5.67 \times 10^{-8} \text{ W/m}^2 \cdot \text{K}^4)(2450 \text{ K})^4} =$$
$2.1 \times 10^{-4} \text{ m}^2(1 \times 10^4 \text{ cm}^2/1 \text{ m}^2) = 2.1 \text{ cm}^2$

15-71   Eq.(15-27): $H = (T_\text{ic} - T_\text{amb})/r_\text{th}$

$T_\text{amb} = T_\text{ic} - Hr_\text{th}$

$T_\text{amb} = 120°\text{C} - (28 \text{ W})(4.5 \text{ K/W}) = 120°\text{C} - 126°\text{C} = -6°\text{C}$

Problems

15-77   Call the metals A and B. Use the data given to calculate α for each metal.

$\Delta L = L_0 \alpha \, \Delta T$ so $\alpha = \Delta L/(L_0 \, \Delta T)$

metal A: $\alpha_A = \dfrac{\Delta L}{L_0 \, \Delta T} = \dfrac{0.0650 \text{ cm}}{(30.0 \text{ cm})(100 \text{ C}°)} = 2.167 \times 10^{-5} \text{ (C}°)^{-1}$

metal B: $\alpha_B = \dfrac{\Delta L}{L_0 \, \Delta T} = \dfrac{0.0350 \text{ cm}}{(30.0 \text{ cm})(100 \text{ C}°)} = 1.167 \times 10^{-5} \text{ (C}°)^{-1}$

Now consider the composite rod. Let L_A be the length of metal A in the composite rod.

$$\Delta L = \Delta L_A + \Delta L_B = (\alpha_A L_A + \alpha_B L_B)\Delta T$$
$$\Delta L/\Delta T = \alpha_A L_A + \alpha_B (0.300 \text{ m} - L_A)$$
$$L_A = \frac{\Delta L/\Delta T - (0.300 \text{ m})\alpha_B}{\alpha_A - \alpha_B} =$$
$$\frac{(0.058 \times 10^{-2} \text{ m}/100 \text{ C}°) - (0.300 \text{ m})(1.167 \times 10^{-5}(\text{C}°)^{-1})}{1.00 \times 10^{-5} \ (\text{C}°)^{-1}}$$
$$L_A = \left(\frac{5.80 \times 10^{-6} - 3.50 \times 10^{-6}}{1.00 \times 10^{-5}}\right) \text{ m} = 0.230 \text{ m} = 23.0 \text{ cm}$$

$$L_B = 30.0 \text{ cm} - L_A = 30.0 \text{ cm} - 23.0 \text{ m} = 7.0 \text{ cm}$$

15-81 **a)** Heat the ring to make its diameter equal to 2.5020 in. The diameter of the ring undergoes linear expansion.

$$\Delta L = \alpha L_0 \, \Delta T \text{ so } \Delta T = \frac{\Delta L}{L_0 \alpha} = \frac{0.0020 \text{ in.}}{(2.5000 \text{ in.})(1.2 \times 10^{-5}(\text{C}°)^{-1})} = 66.7 \text{ C}°$$

$$T = T_0 + \Delta T = 20.0°\text{C} + 66.7 \text{ C}° = 87°\text{C}$$

b) $L = L_0(1 + \alpha \, \Delta T)$
Want L_s (steel) $= L_b$ (brass) for the same ΔT for both materials:
$$L_{0s}(1 + \alpha_s \, \Delta T) = L_{0b}(1 + \alpha_b \, \Delta T)$$
$$L_{0s} + L_{0s}\alpha_s \, \Delta T = L_{0b} + L_{0b}\alpha_b \, \Delta T$$

$$\Delta T = \frac{L_{0b} - L_{0s}}{L_{0s}\alpha_s - L_{0b}\alpha_b} =$$
$$\frac{2.5020 \text{ in.} - 2.5000 \text{ in.}}{(2.5000 \text{ in.}(1.2 \times 10^{-5}(\text{C}°)^{-1}) - (2.5020 \text{ in.}(2.0 \times 10^{-5}(\text{C}°)^{-1})}$$

$$\Delta T = \frac{0.0020}{3.00 \times 10^{-5} - 5.00 \times 10^{-5}} \text{ C}° = -100 \text{ C}°$$

$$T = T_0 + \Delta T = 20.0°\text{C} - 100 \text{ C}° = -80°\text{C}$$

15-83 Let β_l and β_m be the coefficients of volume expansion for the liquid and for the metal. Let ΔT be the (negative) change in temperature when the system is cooled to the new temperature.

Change in volume of cylinder when cool: $\Delta V_m = \beta_m V_0 \, \Delta T$ (negative)
Change in volume of liquid when cool: $\Delta V_l = \beta_l V_0 \, \Delta T$ (negative)

The difference $\Delta V_l - \Delta V_m$ must be equal to the negative volume change due to the increase in pressure, which is $-\Delta p\, V_0/B = -k\,\Delta p\, V_0$.

Thus $\Delta V_l - \Delta V_m = -k\,\Delta p\, V_0$.

$$\Delta T = -\frac{k\,\Delta p}{\beta_l - \beta_m}$$

$$\Delta T = -\frac{(8.50 \times 10^{-10}\ \text{Pa}^{-1})(50.0\ \text{atm})(1.013 \times 10^5\ \text{Pa}/1\ \text{atm})}{4.80 \times 10^{-4}\ \text{K}^{-1} - 3.90 \times 10^{-5}\ \text{K}^{-1}} = -9.8\ \text{C}°$$

$T = T_0 + \Delta T = 30.0°\text{C} - 9.8\ \text{C}° = 20.2°\text{C}.$

15-85   **a)** The kinetic energy is $K = \frac{1}{2}mv^2$.

The heat required to raise its temperature by 600 C° (but not to melt it) is

$Q = mc\,\Delta T.$

The ratio is $\dfrac{K}{Q} = \dfrac{\frac{1}{2}mv^2}{mc\,\Delta T} = \dfrac{v^2}{2c\,\Delta T} = \dfrac{(7700\ \text{m/s})^2}{2(910\ \text{J/kg} \cdot \text{K})(600\ \text{C}°)} = 54.3.$

b) The heat generated when friction work (due to friction force exerted by the air) removes the kinetic energy of the satellite during reentry is very large, and could melt the satellite. Manned space vehicles must have heat shields made of very high melting temperature materials, and reentry must be made slowly.

15-87   **a)** Eq.(15-14) says $dQ = mc\,dT$

But the problem gives the molar heat capacity $C = Mc$; $dQ = nC\,dT$ (Eq.15-19).

$$Q = n\int_{T_1}^{T_2} C\,dT = n\int_{T_1}^{T_2} k(T^3/\Theta^3)\,dT = (nk/\Theta^3)\int_{T_1}^{T_2} T^3\,dT = (nk/\Theta^3)(\tfrac{1}{4}T^4\,\big|_{T_1}^{T_2})$$

$$Q = \frac{nk}{4\Theta^3}(T_2^4 - T_1^4) = \frac{(1.50\ \text{mol})(1940\ \text{J/mol} \cdot \text{K})}{4(281\ \text{K})^3}((40.0\ \text{K})^4 - (10.0\ \text{K})^4) = 83.6\ \text{J}$$

b) $C_{\text{av}} = \dfrac{1}{n}\dfrac{\Delta Q}{\Delta T} = \dfrac{1}{1.50\ \text{mol}}\left(\dfrac{83.6\ \text{J}}{40.0\ \text{K} - 10.0\ \text{K}}\right) = 1.86\ \text{J/mol} \cdot \text{K}$

c) $C = k(t/\Theta)^3 = (1940\ \text{J/mol} \cdot \text{K})(40.0\ \text{K}/281\ \text{K})^3 = 5.60\ \text{J/mol} \cdot \text{K}$

(C is increasing with T, so C at the upper end of the temperature interval is larger than its average value over the interval.)

15-91   **a)** Heat comes out of the 0.0°C water and into the ice cube to warm it to 0.0°C. The heat that comes out of the water causes a phase change liquid → solid.

Heat that goes into the ice cube:

$$Q_{\text{ice}} = mc_{\text{ice}}\,\Delta T = (0.075\text{ kg})(2100\text{ J/kg}\cdot\text{K})(0.0^\circ\text{C} - (-10.0^\circ\text{C})) = 1575\text{ J}$$

Heat that comes out of the water when mass m freezes:
$$Q_{\text{water}} = -mL_{\text{f}}.$$

$Q_{\text{system}} = 0$ implies $Q_{\text{ice}} + Q_{\text{water}} = 0.$

$$m = \frac{1575\text{ J}}{L_{\text{f}}} = \frac{1575\text{ J}}{334\times10^3\text{ J/kg}} = 4.7\times10^{-3}\text{ kg} = 4.7\text{ g}$$

b) Yes, just need for the heat that is absorbed by the ice when it warms to 0.0°C
to equal the heat that must be removed from the water to cause it to freeze.

15-93   **a)** Assume that all the ice melts and that all the steam condenses. If we
calculate a final temperature T that is outside the range 0°C to 100°C then we
know that this assumption is incorrect. Calculate Q for each piece of the system
and then set the total $Q_{\text{system}} = 0.$

copper can (changes temperature from 0.0°C to T; no phase change)
$$Q_{\text{can}} = mc\,\Delta T = (0.446\text{ kg})(390\text{ J/kg}\cdot\text{K})(T - 0.0^\circ\text{C}) = (173.9\text{ J/K})T$$

ice (melting phase change and then the water produced warms to T)
$$Q_{\text{ice}} = +mL_{\text{f}} + mc\,\Delta T =$$
$$(0.0950\text{ kg})(334\times10^3\text{ J/kg}) + (0.0950\text{ kg})(4190\text{ J/kg}\cdot\text{K})(T - 0.0^\circ\text{C})$$
$$Q_{\text{ice}} = 3.173\times10^4\text{ J} + (398.0\text{ J/K})T.$$

steam (condenses to liquid and then water produced cools to T)
$$Q_{\text{steam}} = -mL_{\text{v}} + mc\,\Delta T =$$
$$-(0.0350\text{ kg})(2256\times10^3\text{ J/kg}) + (0.0350\text{ kg})(4190\text{ J/kg}\cdot\text{K})(T - 100.0^\circ\text{C})$$
$$Q_{\text{steam}} = -7.896\times10^4\text{ J} + (146.6\text{ J/K})T - 1.466\times10^4\text{ J} = -9.362\times10^4\text{ J} + (146.6\text{ J/K})T$$

$Q_{\text{system}} = 0$ implies $Q_{\text{can}} + Q_{\text{ice}} + Q_{\text{steam}} = 0.$
$$(173.9\text{ J/K})T + 3.173\times10^4\text{ J} + (398.0\text{ J/K})T - 9.362\times10^4\text{ J} + (146.6\text{ J/K})T = 0$$
$$(718.5\text{ J/K})T = 6.189\times10^4\text{ J}$$

$$T = \frac{6.189\times10^4\text{ J}}{718.5\text{ J/K}} = 86.1^\circ\text{C}.$$

This is between 0°C and 100°C so our assumptions about the phase changes being
complete were correct.

b) No ice, no steam, $0.0950\text{ kg} + 0.0350\text{ kg} = 0.130\text{ kg}$ of liquid water.

5-95   **a)** Heat that must be removed from steam if all of it condenses:

$$Q = -mL_v = -(0.0400 \text{ kg})(2256 \times 10^3 \text{ J/kg}) = -9.02 \times 10^4 \text{ J}$$

Heat absorbed by the water if it heats all the way to the boiling point of 100°C:

$$Q = mc\,\Delta T = (0.200 \text{ kg})(4190 \text{ J/kg} \cdot \text{K})(50.0 \text{ C°}) = 4.19 \times 10^4 \text{ J}$$

Therefore, the water can't absorb enough heat for all the steam to condense. Steam is left and the final temperature then must be 100°C.

b) Mass of steam that condenses is $m = Q/L_v = 4.19 \times 10^4 \text{ J}/2256 \times 10^3 \text{ J/kg} = 0.0186$ kg

Thus there is 0.0400 kg − 0.0186 kg = 0.0214 kg of steam left.

The amount of liquid water is 0.0186 kg + 0.200 kg = 0.219 kg.

15-97   **a)** Heat that must go into ice for all of it to melt:

$$Q = +mL_f = +(0.150 \text{ kg})(334 \times 10^3 \text{ J/kg}) = 5.01 \times 10^4 \text{ J}.$$

Heat that must come out of the steam for all of it to condense:

$$Q = -mL_v = -(0.0950 \text{ kg})(2256 \times 10^3 \text{ J/kg}) = -2.143 \times 10^5 \text{ J}$$

Heat that goes into the 0.200 kg of water if it warms from 50.0°C to 100.0°C:

$$Q = mc\,\Delta T = (0.200 \text{ kg})(4190 \text{ J/kg} \cdot \text{K})(50.0 \text{ C°}) = +4.19 \times 10^4 \text{ J}$$

Heat that goes into water produced from melting the ice if it warms from 0.0°C (the temperature at which it is produced) to 100.0°C:

$$Q = mc\,\Delta T = (0.150 \text{ kg})(4190 \text{ J/kg} \cdot \text{K})(100.0 \text{ C°}) = +6.285 \times 10^4 \text{ J}$$

Thus the total heat that the rest of the system can absorb from the steam is

$$Q = -(5.01 \times 10^4 \text{ J} + 4.19 \times 10^4 \text{ J} + 6.285 \times 10^4 \text{ J}) = -1.548 \times 10^5 \text{ J}$$

This is less than the 2.143×10^5 J that must be removed from the steam for all of it to condense, so some steam is left and this means the final temperature is 100°C

b) Mass of steam that condenses is $m = (1.548 \times 10^5 \text{ J})/(2256 \times 10^3 \text{ J/kg}) = 0.0686$ kg

Mass of steam left: 0.0950 kg − 0.0686 kg = 0.0264 kg

Total mass of liquid water at the end is the mass of liquid water that started with plus the mass of the ice (all melts) plus the mass of steam that condenses: 0.200 kg + 0.150 kg + 0.0684 kg = 0.418 kg

There is no ice left.

c) Heat that must go into ice for all of it to melt:

$$Q = +mL_f = +(0.350 \text{ kg})(334 \times 10^3 \text{ J/kg}) = 1.169 \times 10^5 \text{ J}.$$

Heat that must come out of the steam for all of it to condense:
$$Q = -mL_v = -(0.0120 \text{ kg})(2256 \times 10^3 \text{ J/kg}) = -2.707 \times 10^4 \text{ J}$$

Heat that comes out of the 0.200 kg of water if it cools from $40.0°$C to $0.0°$C:
$$Q = mc\,\Delta T = (0.200 \text{ kg})(4190 \text{ J/kg} \cdot \text{K})(0.0°\text{C} - 40.0 \text{ C}°) = -3.352 \times 10^4 \text{ J}$$

Heat that comes out of water produced from condensing the steam if it cools from 100.0°C (the temperature at which it is produced) to 0.0°C:
$$Q = mc\,\Delta T = (0.012 \text{ kg})(4190 \text{ J/kg} \cdot \text{K})(0.0°\text{C} - 100.0°\text{C}) = -5028 \text{ J}$$

The total heat that the rest of the system can give to the ice to melt it is Q -2.707×10^4 J -3.352×10^4 J -5028 J $= -6.562 \times 10^4$ J.

This is less than the 1.169×10^5 J that it takes to melt all the ice so not all th ice melts and then the final temperature must be $0.0°$C.

The mass of ice that melts is $m = Q/L_f = (6.562 \times 10^4 \text{ J})/(334 \times 10^3 \text{ J/kg})$ 0.196 kg.

The mass of ice left is 0.350 kg $-$ 0.196 kg $=$ 0.154 kg.

The mass of liquid water at the end is the mass of liquid water at the start plu the mass of the steam (all condenses) plus the mass of the ice that melts: 0.20 kg + 0.012 kg + 0.196 kg = 0.408 kg.

No steam is left.

15-101 Use H written in terms of the thermal resistance R:
$$H = A\,\Delta T/R, \text{ where } R = 1/k \text{ and } R = R_1 + R_2 + \dots \text{ (additive)}.$$

<u>single pane</u>

$R_s = R_{\text{glass}} + R_{\text{film}}$, where $R_{\text{film}} = 0.15$ m$^2 \cdot$K/W is the combined thermal resistanc of the air films on the room and outdoor surfaces of the window.

$R_{\text{glass}} = L/k = (4.2 \times 10^{-3} \text{ m})/(0.80 \text{ W/m} \cdot \text{K}) = 0.00525 \text{ m}^2 \cdot \text{K/W}$

Thus $R_s = 0.00525$ m$^2 \cdot$ K/W $+ 0.15$ m$^2 \cdot$ K/W $= 0.1553$ m$^2 \cdot$ K/W.

<u>double pane</u>

$R_d = 2R_{\text{glass}} + R_{\text{air}} + R_{\text{film}}$, where R_{air} is the thermal resistance of the air spac between the panes.

$R_{\text{air}} = L/k = (7.0 \times 10^{-3} \text{ m})/(0.024 \text{ W/m} \cdot \text{K}) = 0.2917 \text{ m}^2 \cdot \text{K/W}$

Thus $R_d = 2(0.00525$ m$^2 \cdot$K/W$) + 0.2917$ m$^2 \cdot$K/W$ + 0.15$ m$^2 \cdot$K/W$ = 0.4522$ m^2 K/W

$H_s = A\,\Delta T/R_s$, $H_d = A\,\Delta T/R_d$, so $H_s/H_d = R_d/R_s$ (since A and ΔT are sam for both)

$H_s/H_d = (0.4522 \text{ m}^2 \cdot \text{K/W})/(0.1553 \text{ m}^2 \cdot \text{K/W}) = 2.9$

15-103 a) Heat must be conducted from the water to cool it to $0°C$ and to cause the phase transition. The entire volume of water is not at the phase transition temperature, just the upper surface that is in contact with the ice sheet.

b) Consider a section of ice that has area A. At time t let the thickness be h. Consider a short time interval t to $t + dt$. Let the thickness that freezes in this time be dh. The mass of the section that freezes in the time interval dt is $dm = \rho \, dV = \rho A \, dh$. The heat that must be conducted away from this mass of water to freeze it is

$dQ = dm \, L_f = (\rho A L_f) \, dh.$

$H = dQ/dt = kA(\Delta T/h)$, so the heat dQ conducted in time dt through the thickness h that is already there is

$$dQ = kA \left(\frac{T_H - T_C}{h} \right) dt.$$

Equate these expressions for dQ:

$$\rho A L_f \, dh = kA \left(\frac{T_H - T_C}{h} \right) dt$$

$$h \, dh = \left(\frac{k(T_H - T_C)}{\rho L_f} \right) dt$$

Integrate from $t = 0$ to time t. At $t = 0$ the thickness h is zero.

$\int_0^h h \, dh = [k(T_H - T_C)/\rho L_f] \int_0^t dt$

$\frac{1}{2}h^2 = \frac{k(T_H - T_C)}{\rho L_f} t$ and $h = \sqrt{\frac{2k(T_H - T_C)}{\rho L_f}} \sqrt{t}$

The thickness after time t is proportional to $\sqrt{t}$.

c) The expression in part (b) gives

$t = \dfrac{h^2 \rho L_f}{2k(T_H - T_C)} = \dfrac{(0.25 \text{ m})^2 (920 \text{ kg/m}^3)(334 \times 10^3 \text{ J/kg})}{2(1.6 \text{ W/m} \cdot \text{K})(0°C - (-10°C))} = 6.0 \times 10^5 \text{ s}$

$t = 170 \text{ h}.$

d) Find t for $h = 40$ m. t is proportional to h^2, so
$t = (40 \text{ m}/0.25 \text{ m})^2 (6.00 \times 10^5 \text{ s}) = 1.5 \times 10^{10} \text{ s}$. This is about 500 years. With our current climate this will not happen.

15-105 Work with a 1.00 m^2 area.

The heat current into a 1.00 m^2 area of ice due to the absorbed solar radiation i▮
$H = (0.70)(600 \text{ W/m}^2)(1.00 \text{ m}^2) = 420$ W.

The heat required to melt a $h = 2.50$ cm thick layer of ice that is initially at 0°C
is

$Q = mL_f = \rho h A L_f = (920 \text{ kg/m}^3)(0.0250 \text{ m})(1.00 \text{ m}^2)(334 \times 10^3 \text{ J/kg}) = $
7.68×10^6 J.

$H = Q/t$, so the time t it takes the heat current from solar radiation to input thi▮
amount of heat into the ice is

$t = Q/H = 7.68 \times 10^6 \text{ J}/420 \text{ W} = 1.83 \times 10^4 \text{ s} = 305$ min.

15-107 Calculate the net rate of radiation of heat from the can:

$H_{\text{net}} = Ae\sigma(T^4 - T_s^4).$

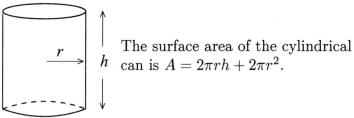

The surface area of the cylindrical
can is $A = 2\pi rh + 2\pi r^2$.

$A = 2\pi r(h + r) = 2\pi(0.045 \text{ m})(0.250 \text{ m} + 0.045 \text{ m}) = 0.08341 \text{ m}^2.$

$H_{\text{net}} = (0.08341 \text{ m}^2)(0.200)(5.67 \times 10^{-8} \text{ W/m}^2 \cdot \text{K}^4)(4.22 \text{ K})^4 - (77.3 \text{ K})^4)$
$H_{\text{net}} = -0.0338$ W (the minus sign says that the net heat current is into the can)▮

The heat that is put into the can by radiation in one hour is
$Q = -(H_{\text{net}})t = (0.0338 \text{ W})(3600 \text{ s}) = 121.7$ J.

This heat boils a mass m of helium according to the equation $Q = mL_f$, so

$$m = \frac{Q}{L_f} = \frac{121.7 \text{ J}}{2.09 \times 10^4 \text{ J/kg}} = 5.82 \times 10^{-3} \text{ kg} = 5.82 \text{ g}.$$

CHAPTER 16

THERMAL PROPERTIES OF MATTER

Exercises 1, 5, 9, 13, 15, 17, 21, 25, 27, 29, 31, 35, 37
Problems 43, 45, 49, 51, 53, 57, 59, 61, 65, 69, 73

Exercises

6-1 $n = \dfrac{m_{tot}}{M} = \dfrac{0.225 \text{ kg}}{4.00 \times 10^{-3} \text{ kg/mol}} = 56.2 \text{ mol}$

b) $pV = nRT$ implies $p = nRT/V$

T must be in kelvins; $T = (18 + 273) \text{ K} = 291 \text{ K}$

$p = \dfrac{(56.2 \text{ mol})(8.3145 \text{ J/mol} \cdot \text{K})(291 \text{ K})}{20.0 \times 10^{-3} \text{ m}^3} = 6.80 \times 10^6 \text{ Pa}$

$p = (6.80 \times 10^6 \text{ Pa})(1.00 \text{ atm}/1.013 \times 10^5 \text{ Pa}) = 67.1 \text{ atm}$

6-5 $pV = nRT$

n, R constant implies $pV/T = nR = $ constant and $p_1 V_1/T_1 = p_2 V_2/T_2$

$T_1 = (27 + 273) \text{ K} = 300 \text{ K}$

$p_1 = 1.01 \times 10^5 \text{ Pa}$

$p_2 = 2.72 \times 10^6 \text{ Pa} + 1.01 \times 10^5 \text{ Pa} = 2.82 \times 10^6 \text{ Pa}$ (in the ideal gas equation the pressures must be absolute, not gauge, pressures)

$T_2 = T_1 \left(\dfrac{p_2}{p_1}\right)\left(\dfrac{V_2}{V_1}\right) = 300 \text{ K} \left(\dfrac{2.82 \times 10^6 \text{ Pa}}{1.01 \times 10^5 \text{ Pa}}\right)\left(\dfrac{46.2 \text{ cm}^3}{499 \text{ cm}^3}\right) = 776 \text{ K}$

$T_2 = (776 - 273)°\text{C} = 503°\text{C}$

(Note that the units cancel in the V_2/V_1 volume ratio, so it was not necessary to convert the volumes in cm^3 to m^3.)

6-9 $pV = nRT$

n, R, p are constant so $V/T = nR/p = $ constant and $V_1/T_1 = V_2/T_2$

$T_1 = (19 + 273) \text{ K} = 292 \text{ K}$ (T must be in kelvins)

$V_2 = V_1(T_2/T_1) = (0.600 \text{ L})(77.3 \text{ K}/292 \text{ K}) = 0.159 \text{ L}$

6-13 a) $pV = nRT$

Find the initial pressure p_1:

$$p_1 = \frac{nRT_1}{V} = \frac{(11.0 \text{ mol})(8.3145 \text{ J/mol} \cdot \text{K})((23.0 + 273.13) \text{ K})}{3.10 \times 10^{-3} \text{ m}^3} = 8.737 \times 10^6 \text{ Pa}$$

$$p_2 = 100 \text{ atm}(1.013 \times 10^5 \text{ Pa}/1 \text{ atm}) = 1.013 \times 10^7 \text{ Pa}$$

$$p/T = nR/V = \text{constant, so } p_1/T_1 = p_2/T_2$$

$$T_2 = T_1 \left(\frac{p_2}{p_1}\right) = (296.15 \text{ K}) \left(\frac{1.013 \times 10^7 \text{ Pa}}{8.737 \times 10^6 \text{ Pa}}\right) = 343.4 \text{ K} = 70.2°\text{C}$$

b) The coefficient of volume expansion for a gas is much larger than for a solid, s•
the expansion of the tank is negligible.

16-15 Eq.(16-5): $\rho = pM/RT$ and $p = RT\rho/M$

$T = (-56.5 + 273.15) \text{ K} = 216.6 \text{ K}$

For air $M = 28.8 \times 10^{-3}$ kg/mol (Example 16-3)

$$p = \frac{(8.3145 \text{ J/mol} \cdot \text{K})(216.6 \text{ K})(0.364 \text{ kg/m}^3)}{28.8 \times 10^{-3} \text{ kg/mol}} = 2.28 \times 10^4 \text{ Pa}$$

16-17 **a)** $pV = nRT$, $n = N/N_A$ so $pV = (N/N_A)RT$

$$p = \left(\frac{N}{V}\right)\left(\frac{R}{N_A}\right)T$$

$$p = \left(\frac{80 \text{ molecules}}{1 \times 10^{-6} \text{ m}^3}\right)\left(\frac{8.3145 \text{ J/mol} \cdot \text{K}}{6.022 \times 10^{23} \text{ molecules/mol}}\right)(7500 \text{ K}) = 8.28 \times 10^{-12} \text{ Pa}$$

$p = 8.2 \times 10^{-17}$ atm. This is much lower than the laboratory pressure of $1 \times$
10^{-13} atm in Exercise 16-16.

b) The Lagoon Nebula is a very rarefied low pressure gas. The gas would exer•
very little force on an object passing through it.

16-21 **a)** Use the density and the mass of 5.00 mol to calculate the volume.

$\rho = m/V$ implies $V = m/\rho$, where $m = m_{tot}$, the mass of 5.00 mol of water.

$m_{tot} = nM = (5.00 \text{ mol})(18.0 \times 10^{-3} \text{ kg/mol}) = 0.0900 \text{ kg}$

Then $V = \dfrac{m}{\rho} = \dfrac{0.0900 \text{ kg}}{1000 \text{ kg/m}^3} = 9.00 \times 10^{-5} \text{ m}^3$

b) One mole contains $N_A = 6.022 \times 10^{23}$ molecules, so the volume occuppied b•
one molecule is

$$\frac{9.00 \times 10^{-5} \text{ m}^3/\text{mol}}{(5.00 \text{ mol})(6.022 \times 10^{23} \text{ molecules/mol})} = 2.989 \times 10^{-29} \text{ m}^3/\text{molecule}$$

$V = a^3$, where a is the length of each side of the cube occuppied by a molecule. $a^3 = 2.989 \times 10^{-29}$ m^3, so $a = 3.1 \times 10^{-10}$ m.

c) Atoms and molecules are on the order of 10^{-10} m in diameter, in agreement with the above estimates.

16-25 a) $\frac{1}{2}m(v^2)_{av} = \frac{3}{2}kT = \frac{3}{2}(1.38 \times 10^{-23}$ J/molecule $\cdot$ K$)(300$ K$) = 6.21 \times 10^{-21}$ J

b) We need the mass m of one atom:

$$m = \frac{M}{N_A} = \frac{32.0 \times 10^{-3} \text{ kg/mol}}{6.022 \times 10^{23} \text{ molecules/mol}} = 5.314 \times 10^{-26} \text{ kg/molecule}$$

Then $\frac{1}{2}m(v^2)_{av} = 6.21 \times 10^{-21}$ J (from part (a)) gives

$$(v^2)_{av} = \frac{2(6.21 \times 10^{-21} \text{ J})}{m} = \frac{2(6.21 \times 10^{-21} \text{ J})}{5.314 \times 10^{-26} \text{ kg}} = 2.34 \times 10^5 \text{ m}^2/\text{s}^2$$

c) $v_{rms} = \sqrt{(v^2)_{rms}} = \sqrt{2.34 \times 10^4 \text{ m}^2/\text{s}^2} = 484$ m/s

d) $p = mv_{rms} = (5.314 \times 10^{-26}$ kg$)(484$ m/s$) = 2.57 \times 10^{-23}$ kg $\cdot$ m/s

e) Time betwen collisions with one wall is

$$t = \frac{0.20 \text{ m}}{v_{rms}} = \frac{0.20 \text{ m}}{484 \text{ m/s}} = 4.13 \times 10^{-4} \text{ s}$$

In a collision $\vec{v}$ changes direction, so $\Delta p = 2mv_{rms} = 2(2.57 \times 10^{-23}$ kg $\cdot$ m/s$) = 5.14 \times 10^{-23}$ kg $\cdot$ m/s

$$F = \frac{dp}{dt} \text{ so } F_{av} = \frac{\Delta p}{\Delta t} = \frac{5.14 \times 10^{-23} \text{ kg} \cdot \text{m/s}}{4.13 \times 10^{-4} \text{ s}} = 1.24 \times 10^{-19} \text{ N}$$

f) pressure $= F/A = 1.24 \times 10^{-19}$ N$/(0.10$ m$)^2 = 1.24 \times 10^{-17}$ Pa (due to one atom)

g) pressure $= 1$ atm $= 1.013 \times 10^5$ Pa
Number of atoms needed is 1.013×10^5 Pa$/(1.24 \times 10^{-17}$ Pa/atom$) = 8.17 \times 10^{21}$ atoms

h) $pV = NkT$ (Eq.16-18), so

$$N = \frac{pV}{kT} = \frac{(1.013 \times 10^5 \text{ Pa})(0.10 \text{ m})^3}{(1.381 \times 10^{-23} \text{ J/molecule} \cdot \text{K})(300 \text{ K})} = 2.45 \times 10^{22} \text{ atoms}$$

i) From the factor of $\frac{1}{3}$ in $(v_x^2)_{av} = \frac{1}{3}(v^2)_{av}$.

16-27 $v_{\text{rms}} = \sqrt{3RT/M}$ (Eq.16-19), so

$v_{\text{rms}}^2/3R = T/M$, where T must be in kelvins.

Same v_{rms} so same T/M for the two gases and $T_{N_2}/M_{N_2} = T_{H_2}/M_{H_2}$.

$$T_{N_2} = T_{H_2}\left(\frac{M_{N_2}}{M_{H_2}}\right) = ((20 + 273)\text{ K})\left(\frac{28.014\text{ g/mol}}{2.016\text{ g/mol}}\right) = 4.071 \times 10^3\text{ K}$$

$$T_{N_2} = (4071 - 273)°\text{C} = 3800°\text{C}$$

16-29 **a)** $\frac{1}{2}R$ contribution to C_V for each degree of freedom, so

$C_V = 6(\frac{1}{2}R) = 3R = 3(8.3145\text{ J/mol}\cdot\text{K}) = 24.9\text{ J/mol}\cdot\text{K}$

b) For water vapor the specific heat capacity is $c = 2000\text{ J/kg}\cdot\text{K}$. The molar heat capacity is $C = Mc = (18.0 \times 10^{-3}\text{ kg/mol})(2000\text{ J/kg}\cdot\text{K}) = 36.0\text{ J/mol}\cdot\text{K}$.

The difference is $36.0\text{ J/mol}\cdot\text{K} - 24.9\text{ J/mol}\cdot\text{K} = 11.1\text{ J/mol}\cdot\text{K}$, which is about $2.7(\frac{1}{2}R)$; the vibrational degrees of freedom make a significant contribution.

16-31 **a)** $Q = nC_V\,\Delta T = n(\frac{5}{2}R)\,\Delta T$

$Q = (2.5\text{ mol})(\frac{5}{2})(8.3145\text{ J/mol}\cdot\text{K})(30.0\text{ K}) = 1560\text{ J}$

$Q = nC_V\,\Delta T = n(\frac{3}{2}R)\,\Delta T$

$Q = (2.5\text{ mol})(\frac{3}{2})(8.3145\text{ J/mol}\cdot\text{K})(30.0\text{ K}) = 935\text{ J}$

16-35 Eq.(16-33): $f(v) = \dfrac{8\pi}{m}\left(\dfrac{m}{2\pi kT}\right)^{3/2}\epsilon e^{-\epsilon/kT}$

At the maximum of $f(\epsilon)$, $\dfrac{df}{d\epsilon} = 0$

$\dfrac{df}{d\epsilon} = \dfrac{8\pi}{m}\left(\dfrac{m}{2\pi kT}\right)^{3/2}\dfrac{d}{d\epsilon}\left(\epsilon e^{-\epsilon/kT}\right) = 0$

This requires that $\dfrac{d}{d\epsilon}\left(\epsilon e^{-\epsilon/kT}\right) = 0$.

$e^{-\epsilon/kT} - (\epsilon/kT)e^{-\epsilon/kT} = 0$

$(1 - \epsilon/kT)e^{-\epsilon/kT} = 0$

This requires that $1 - \epsilon/kT = 0$ so $\epsilon = kT$, as was to be shown.

And then since $\epsilon = \frac{1}{2}mv^2$, this gives $\frac{1}{2}mv_{\text{mp}}^2 = kT$ and $v_{\text{mp}} = \sqrt{2kT/m}$, which is Eq.(16-34).

16-37 If the temperature at altitude y is below the freezing point only cirrus clouds can form. Use $T = T_0 - \alpha y$ to find the y that gives $T = 0.0°\text{C}$.

$$y = \frac{T_0 - T}{\alpha} = \frac{15.0°\text{C} - 0.0°\text{C}}{6.0\text{C}°/\text{km}} = 2.5\text{ km}$$

(Note: The solid-liquid phase transition occurs at $0°C$ only for $p = 1.01 \times 10^5$ Pa. Use the results of Example 16-4 to estimate the pressure at an altitude of 2.5 km.

$p_2 = p_1 e^{Mg(y_2-y_1)/RT}$

$Mg(y_2-y_1)/RT = 1.10(2500 \text{ m}/8863 \text{ m}) = 0.310$ (using the calculation in Example 16-4)

Then $p_2 = (1.01 \times 10^5 \text{ Pa})e^{-0.31} = 0.74 \times 10^5$ Pa.

This pressure is well above the triple point pressure for water. Figure 16-18 shows that the fusion curve has large slope and it takes a large change in pressure to change the phase transition temperature very much. Using $0.0°C$ introduces little error.)

Problems

16-43 $pV = nRT$ can be written $pV = (m/M)RT$

T, V, M, R are all constant, so $p/m = RT/MV = $ constant.

So $p_1/m_1 = p_2/m_2$, where m is the mass of the gas in the tank.

$p_1 = 1.30 \times 10^6$ Pa $+ 1.01 \times 10^5$ Pa $= 1.40 \times 10^6$ Pa

$p_2 = 2.50 \times 10^5$ Pa $+ 1.01 \times 10^5$ Pa $= 3.51 \times 10^5$ Pa

$m_1 = p_1 V M/RT; \quad V = hA = h\pi r^2 = (1.00 \text{ m})\pi(0.060 \text{ m})^2 = 0.01131 \text{ m}^3$

$$m_1 = \frac{(1.40 \times 10^6 \text{ Pa})(0.01131 \text{ m}^3)(44.1 \times 10^{-3} \text{ kg/mol})}{(8.3145 \text{ J/mol} \cdot \text{K})((22.0 + 273.15) \text{ K})} = 0.2845 \text{ kg}$$

Then $m_2 = m_1 \left(\dfrac{p_2}{p_1}\right) = (0.2845 \text{ kg}) \left(\dfrac{3.51 \times 10^5 \text{ Pa}}{1.40 \times 10^6 \text{ Pa})}\right) = 0.0713$ kg.

m_2 is the mass that remains in the tank. The mass that has been used is $m_1 - m_2 = 0.2848 \text{ kg} - 0.0713 \text{ kg} = 0.213$ kg.

16-45

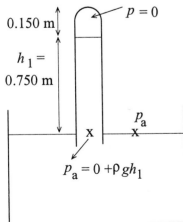

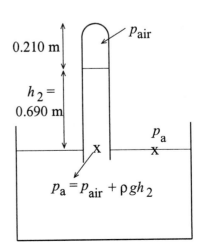

The pressure at points which are level with the surface of the mercury outside the tube is equal to atmospheric pressure, p_a. In the first sketch this gives $p_a = \rho g h$ and in the second sketch it gives $p_a = p_{air} + \rho g h_2$, where p_{air} is the pressure of the air in the space in the tube above the mercury.

Equating these two expressions for p_a gives $\rho g h_1 = p_{air} + \rho g h_2$.

$p_{air} = \rho g(h_1 - h_2) = (13.6 \times 10^3 \text{ kg/m}^3)(9.80 \text{ m/s}^2)(0.750 \text{ m} - 0.690 \text{ m}) = 7.997 \times 10^3 \text{ Pa}$

We know p, V, and T; calculate the mass m_{tot} of the air:

$$n = \frac{pV}{RT} = \frac{(7.997 \times 10^3 \text{ Pa})(0.210 \text{ m})(0.620 \times 10^{-4} \text{ m}^2)}{(8.3145 \text{ J/mol} \cdot \text{K})(293.15 \text{ K})} = 4.272 \times 10^{-5} \text{ mol}$$

$$m_{tot} = nM = (4.272 \times 10^{-5} \text{ mol})(28.8 \times 10^{-3} \text{ kg/mol}) = 1.23 \times 10^{-6} \text{ kg}$$

16-49 **a)** Consider the gas in one cylinder. Calculate the volume to which this volume of gas expands when the pressure is decreased from $(1.20 \times 10^6 \text{ Pa} + 1.01 \times 10^5 \text{ Pa}) = 1.30 \times 10^6 \text{ Pa}$ to $1.01 \times 10^5 \text{ Pa}$.

$pV = nRT$

n, R, T constant implies $pV = nRT = \text{constant}$, so $p_1 V_1 = p_2 V_2$.

$$V_2 = V_1(p_1/p_2) = (1.90 \text{ m}^3)\left(\frac{1.30 \times 10^6 \text{ Pa}}{1.01 \times 10^5 \text{ Pa}}\right) = 24.46 \text{ m}^3$$

The number of cylinders required to fill a 750 m³ balloon is 750 m³/24.46 m³ = 30.7 cylinders.

b) The upward force on the balloon is given by Archimedes' principle (Chapter 14):

$B = \text{weight of air displaced by balloon} = \rho_{air} V g$.

Free-body diagram for the balloon:

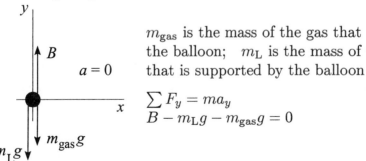

m_{gas} is the mass of the gas that is inside the balloon; m_L is the mass of the load that is supported by the balloon

$\sum F_y = ma_y$

$B - m_L g - m_{gas} g = 0$

$\rho_{air} V g - m_L g - m_{gas} g = 0$

$m_L = \rho_{air} V - m_{gas}$

Calculate m_{gas}, the mass of hydrogen that occupies 750 m³ at 15°C and $p = 1.01 \times 10^5$ Pa.

$pV = nRT = (m_{\text{gas}}/M)RT$ gives

$$m_{\text{gas}} = pVM/RT = \frac{(1.01 \times 10^5 \text{ Pa})(750 \text{ m}^3)(2.02 \times 10^{-3} \text{ kg/mol})}{(8.3145 \text{ J/mol} \cdot \text{K})(288 \text{ K})} = 63.9 \text{ kg}$$

Then $m_L = (1.23 \text{ kg/m}^3)(750 \text{ m}^3) - 63.9 \text{ kg} = 859 \text{ kg}$, and the weight that can be supported is $w_L = m_L g = (859 \text{ kg})(9.80 \text{ m/s}^2) = 8420 \text{ N}$.

c) $m_L = \rho_{\text{air}} V - m_{\text{gas}}$

$m_{\text{gas}} = pVM/RT = (63.9 \text{ kg})((4.00 \text{ g/mol})/(2.02 \text{ g/mol})) = 126.5 \text{ kg}$ (using the results of part (b)).

Then $m_L = (1.23 \text{ kg/m}^3)(750 \text{ m}^3) - 126.5 \text{ kg} = 796 \text{ kg}$.

$w_L = m_L g = (796 \text{ kg})(9.80 \text{ m/s}^2) = 7800 \text{ N}$.

16-51 a)

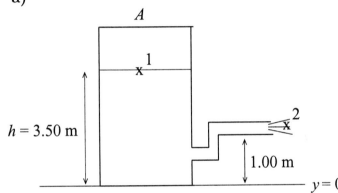

$p_1 = 4.20 \times 10^5$ Pa
$p_2 = p_{\text{air}} = 1.00 \times 10^5$ Pa
large tank implies $v_1 \approx 0$

$p_1 + \rho g y_1 + \frac{1}{2}\rho v_1^2 = p_2 + \rho g y_2 + \frac{1}{2}\rho v_2^2$
$\frac{1}{2}\rho v_2^2 = p_1 - p_2 + \rho g(y_1 - y_2)$
$v_2 = \sqrt{(2/\rho)(p_1 - p_2) + 2g(y_1 - y_2)}$
$v_2 = 26.2$ m/s

b) $h = 3.00$ m

The volume of the air in the tank increases so its pressure decreases.

$pV = nRT =$ constant, so $pV = p_0 V_0$ (p_0 is the pressure for $h_0 = 3.50$ m and p is the pressure for $h = 3.00$ m)

$p(4.00 \text{ m} - h)A = p_0(4.00 \text{ m} - h_0)A$

$$p = p_0\left(\frac{4.00 \text{ m} - h_0}{4.00 \text{ m} - h}\right) = (4.20 \times 10^5 \text{ Pa})\left(\frac{4.00 \text{ m} - 3.50 \text{ m}}{4.00 \text{ m} - 3.00 \text{ m}}\right) = 2.10 \times 10^5 \text{ Pa}$$

Repeat the calculation of part (a), but now $p_1 = 2.10 \times 10^5$ Pa and $y_1 = 3.00$ m.

$$v_2 = \sqrt{(2/\rho)(p_1 - p_2) + 2g(y_1 - y_2)}$$
$$v_2 = 16.1 \text{ m/s}$$

$\underline{h = 2.00 \text{ m}}$

$$p = p_0 \left(\frac{4.00 \text{ m} - h_0}{4.00 \text{ m} - h} \right) = (4.20 \times 10^5 \text{ Pa}) \left(\frac{4.00 \text{ m} - 3.50 \text{ m}}{4.00 \text{ m} - 2.00 \text{ m}} \right) = 1.05 \times 10^5 \text{ Pa}$$
$$v_2 = \sqrt{(2/\rho)(p_1 - p_2) + 2g(y_1 - y_2)}$$
$$v_2 = 5.44 \text{ m/s}$$

c) $v_2 = 0$ means $(2/\rho)(p_1 - p_2) + 2g(y_1 - y_2) = 0$

$$p_1 - p_2 = -\rho g(y_1 - y_2)$$
$$y_1 - y_2 = h - 1.00 \text{ m}$$

$$p = p_0 \left(\frac{0.50 \text{ m}}{4.00 \text{ m} - h} \right) = (4.20 \times 10^5 \text{ Pa}) \left(\frac{0.50 \text{ m}}{4.00 \text{ m} - h} \right)$$

$$(4.20 \times 10^5 \text{ Pa}) \left(\frac{0.50 \text{ m}}{4.00 \text{ m} - h} \right) - 1.00 \times 10^5 \text{ Pa} = (9.80 \text{ m/s}^2)(1000 \text{ kg/m}^3)(1.00 \text{ m} - h)$$

$(210/(4.00 - h)) - 100 = 9.80 - 9.80h$, with h in meters.

$210 = (4.00 - h)(109.8 - 9.80h)$

$9.80h^2 - 149h + 229.2 = 0$ and $h^2 - 15.20h + 23.39 = 0$

quadratic formula: $h = \frac{1}{2}(15.20 \pm \sqrt{(15.20)^2 - 4(23.39)}) = (7.60 \pm 5.86) \text{ m}$

h must be less than 4.00 m, so the only acceptable value is

$h = 7.60 \text{ m} - 5.86 \text{ m} = 1.74 \text{ m}$

16-53 Calculate the number of water molecules N:

$$n = \frac{m_{tot}}{M} = \frac{50 \text{ kg}}{18.0 \times 10^{-3} \text{ kg/mol}} = 2.778 \times 10^3 \text{ mol}$$

$N = nN_A = (2.778 \times 10^3 \text{ mol})(6.022 \times 10^{23} \text{ molecules/mol}) = 1.7 \times 10^{27}$ molecule

Each water molecule has three atoms, so the number of atoms is

$3(1.7 \times 10^{27}) = 5.1 \times 10^{27}$ atoms

16-57 a) $U(r) = U_0[(R_0/r)^{12} - 2(R_0/r)^6]$

Eq.(13.26): $F(r) = 12(U_0/R_0)[(R_0/r)^{13} - (R_0/r)^7]$

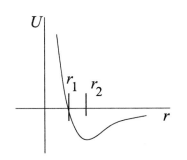
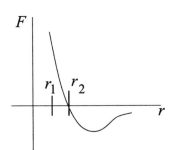

b) equilibrium requires $F = 0$; occurs at point r_2. r_2 is where U is a minimum (stable equilibirum).

c) $U = 0$ implies $[(R_0/r)^{12} - 2(R_0/r)^6] = 0$
$(r_1/R_0)^6 = 1/2$ and $r_1 = R_0/(2)^{1/6}$

$F = 0$ implies $[(R_0/r)^{13} - (R_0/r)^7] = 0$
$(r_2/R_0)^6 = 1$ and $r_2 = R_0$

Then $r_1/r_2 = (R_0/2^{1/6})/R_0 = 2^{-1/6}$

d) $W_{\text{other}} = \Delta U$
At $r \to \infty$, $U = 0$, so $W = -U(R_0) = -U_0[(R_0/R_0)^{12} - 2(R_0/R_0)^6] = +U_0$

6-59 **a)** $v_{\text{rms}} = \sqrt{3RT/M}$

$$v_{\text{rms}} = \sqrt{\frac{3(8.3145 \text{ J/mol} \cdot \text{K})(300 \text{ K})}{28.0 \times 10^{-3} \text{ kg/mol}}} = 517 \text{ m/s}$$

b) $(v_x^2)_{\text{av}} = \frac{1}{3}(v^2)_{\text{av}}$ so $\sqrt{(v_x^2)_{\text{av}}} = (1/\sqrt{3})\sqrt{(v^2)_{\text{av}}} = (1/\sqrt{3})v_{\text{rms}} =$
$(1/\sqrt{3})(517 \text{ m/s}) = 298 \text{ m/s}$

6-61 **a)** Apply conservation of energy $K_1 + U_1 + W_{\text{other}} = K_2 + U_2$,

where $U = -Gmm_p/r$. Let point 1 be at the surface of the planet, where the projectile is launched, and let point 2 be far from the earth. Just barely escapes says $v_2 = 0$.
Only gravity does work says $W_{\text{other}} = 0$.
$U_1 = -Gmm_p/R_p$; $r_2 \to \infty$ so $U_2 = 0$; $v_2 = 0$ so $K_2 = 0$.

The conservation of energy equation becomes
$K_1 - Gmm_p/R_p = 0$ and $K_1 = Gmm_p/R_p$.
But $g = Gm_p/R_p^2$ so $Gm_p/R_p = R_p g$ and $K_1 = mgR_p$, as was to be shown.

b) $\frac{1}{2}m(v^2)_{\text{av}} = mgR_p$ (from part (a))
But also, $\frac{1}{2}m(v^2)_{\text{av}} = \frac{3}{2}kT$, so $mgR_p = \frac{3}{2}kT$

$$T = \frac{2mgR_p}{3k}$$

nitrogen

$m_{N_2} = (28.0 \times 10^{-3} \text{ kg/mol})/(6.022 \times 10^{23} \text{ molecules/mol}) = 4.65 \times 10^{-26}$ kg/molecule

$$T = \frac{2mgR_p}{3k} = \frac{2(4.65 \times 10^{-26} \text{ kg/molecule})(9.80 \text{ m/s}^2)(6.38 \times 10^6 \text{ m})}{3(1.381 \times 10^{-23} \text{ J/molecule} \cdot \text{K})} =$$

1.40×10^5 K

hydrogen

$m_{H_2} = (2.02 \times 10^{-3} \text{ kg/mol})/(6.022 \times 10^{23} \text{ molecules/mol}) = 3.354 \times 10^{-27}$ kg/molecule

$$T = \frac{2mgR_p}{3k} = \frac{2(3.354 \times 10^{-27} \text{ kg/molecule})(9.80 \text{ m/s}^2)(6.38 \times 10^6 \text{ m})}{3(1.381 \times 10^{-23} \text{ J/molecule} \cdot \text{K})} =$$

1.01×10^4 K

c) $T = \dfrac{2mgR_p}{3k}$

nitrogen

$$T = \frac{2(4.65 \times 10^{-26} \text{ kg/molecule})(1.63 \text{ m/s}^2)(1.74 \times 10^6 \text{ m})}{3(1.381 \times 10^{-23} \text{ J/molecule} \cdot \text{K})} = 6370 \text{ K}$$

hydrogen

$$T = \frac{2(3.354 \times 10^{-27} \text{ kg/molecule})(1.63 \text{ m/s}^2)(1.74 \times 10^6 \text{ m})}{3(1.381 \times 10^{-23} \text{ J/molecule} \cdot \text{K})} = 459 \text{ K}$$

d) The "escape temperatures" are much less for the moon than for the earth. Fo the moon a larger fraction of the molecules at a given temperature will have speec in the Maxwell-Boltzmann distribution larger than the escape speed. After a lon time most of the molecules will have escaped from the moon.

16-65 **a)** The atoms have two degrees of freedom and hence each atom has an average kinetic energy $2(\frac{1}{2}kT) = kT$. Each atom also has potential energy du to the Hooke's law force that binds it at the surface to its position parallel to th surface. The average potential energy for this two-dimensional harmonic oscillato equals the average kinetic energy, just as for a one-dimensional or three-dimension: oscillator. The average total energy per atom is $kT + kT = 2kT$.

For n moles of atoms the average total energy is $2nRT$ and

$C_V = 2R = 2(8.3145 \text{ J/mol} \cdot \text{K}) = 16.6 \text{ J/mol} \cdot \text{K}.$

b) At low temperatures the average energy per atom is less than $2kT$ because most atoms remain in their lowest energy state. Thus the molar heat capacity will be less than the value found in part (a).

6-69 $\int_0^\infty v^2 f(v)\, dv = 4\pi(m/2\pi kT)^{3/2} \int_0^\infty v^4 e^{-mv^2/2kT}\, dv$

The integral formula with $n = 2$ gives $\int_0^\infty v^4 e^{-av^2}\, dv = (3/8a^2)\sqrt{\pi/a}$

Apply with $a = m/2kT$,

$\int_0^\infty v^2 f(v)\, dv = 4\pi(m/2\pi kT)^{3/2}(3/8)(2kT/m)^2\sqrt{2\pi kT/m} =$
$(3/2)(2kT/m) = 3kT/m$

Equation (16-16) says $\frac{1}{2}m(v^2)_{av} = 3kT/2$, so $(v^2)_{av} = 3kT/m$, in agreement with our calculation.

6-73 relative humidity $= \dfrac{\text{partial pressure of water vapor at temperature } T}{\text{vapor pressure of water at temperature } T}$

The experiment shows that the dew point is 16.0°C, so the partial pressure of water vapor at 30.0°C is equal to the vapor pressure at 16.0°C, which is 1.81×10^3 Pa.

Thus the relative humidity $= \dfrac{1.81 \times 10^3 \text{ Pa}}{4.25 \times 10^3 \text{ Pa}} = 0.426 = 42.6\%$.

CHAPTER 17
THE FIRST LAW OF THERMODYNAMICS

Exercises 1, 5, 9, 13, 15, 19, 23, 25, 27
Problems 31, 35, 39, 43, 45, 47, 51,

Exercises

17-1 **a)** The pressure is constant and the volume increases.

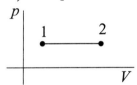

b) $W = \int_{V_1}^{V_2} p\, dV$

Constant: $W = p \int_{V_1}^{V_2} dV = p(V_2 - V_1)$

The problem gives T rather than p and V, so use the ideal gas law to rewrite the expression for W:

$pV = nRT$ so $p_1 V_1 = nRT_1$, $p_2 V_2 = nRT_2$; subtracting the two equations gives
$p(V_2 - V_1) = nR(T_2 - T_1)$

Thus $W = nR(T_2 - T_1)$ is an alternative expression for the work in a constant pressure process for an ideal gas.

Then $W = nR(T_2 - T_1) = (2.00 \text{ mol})(8.3145 \text{ J/mol} \cdot \text{K})(107°\text{C} - 27°\text{C}) = +1330$.

The gas expands when heated and does positive work.

17-5 **a)** pV-diagram

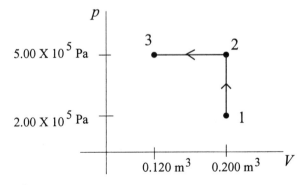

b) $1 \rightarrow 2$

$\Delta V = 0$, so $W = 0$

$2 \rightarrow 3$

p is constant, so $W = p\Delta V = (5.00 \times 10^5 \text{ Pa})(0.120 \text{ m}^3 - 0.200 \text{ m}^3) =$
-4.00×10^4 J (W is negative since the volume decreases in the process.)
$W_{\text{tot}} = W_{1 \rightarrow 2} + W_{2 \rightarrow 3} = -4.00 \times 10^4$ J

17-9   **a)** $W = \int_{V_1}^{V_2} p \, dV = p(V_2 - V_1)$ for this constant pressure process.

$W = (2.3 \times 10^5 \text{ Pa})(1.20 \text{ m}^3 - 1.70 \text{ m}^3) = -1.15 \times 10^5$ J (The volume decreases in
the process, so W is negative.)

b) $\Delta U = Q - W$

$Q = \Delta U + W = -1.40 \times 10^5 \text{ J} + (-1.15 \times 10^5 \text{ J}) = -2.55 \times 10^5$ J

Q negative means heat flows out of the gas.

c) $W = \int_{V_1}^{V_2} p \, dV = p(V_2 - V_1)$ (constant pressure) and $\Delta U = Q - W$ apply to
__any__ system, not just to an ideal gas. We did not use the ideal gas equation, either
directly or indirectly, in any of the calculations, so the results are the same whether
the gas is ideal or not.

17-13   **a)** For one cycle, $\Delta U = 0$.

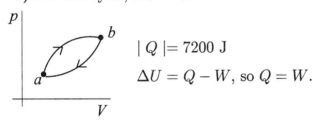

$|Q| = 7200$ J

$\Delta U = Q - W$, so $Q = W$.

The magnitude of the positive work done from a to b is larger than the magnitude
of the negative work done from b to a, so the net work done in one cycle is positive.
Then from $Q = W$, Q must be positive.

$Q = +7200$ J; the system absorbs heat.

b) $W = Q = +7200$ J

c)

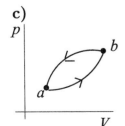

Now the net work done is negative.

Since $Q = W$, Q is negative; system liberates heat.

$|W|$ is the same as in part (b) and $Q = W$, so $Q = -7200$ J.

17-15 a) For the water $\Delta T > 0$, so by $Q = mc\,\Delta T$ heat has been added to the water. Thus heat energy comes from the burning fuel-oxygen mixture, and Q for th system (fuel and oxygen) is negative.

b) Constant volume implies $W = 0$.

c) The 1st law (Eq.17-4) says $\Delta U = Q - W$.

$Q < 0$, $W = 0$ so by the 1st law $\Delta U < 0$. The internal energy of the fuel-oxyge mixture decreased. In this process internal energy from the fuel-oxygen mixtur was transferred to the water, raising its temperature.

17-19 a) $W = \int_{V_1}^{V_2} p\,dV$

$pV = nRT$ so $p = nRT/V$

$W = \int_{V_1}^{V_2} (nRT/V)\,dV = nRT \int_{V_1}^{V_2} dV/V = nRT\ln(V_2/V_1)$ (work done during a isothermal process).

$W = (0.150 \text{ mol})(8.3145 \text{ J/mol} \cdot \text{K})(350 \text{ K})\ln(0.25V_1/V_1) = (436.5 \text{ J})\ln(0.25)$ = -605 J.

(W for the gas is negative, since the volume decreases.)

b) $\Delta U = nC_V\,\Delta T$ for any ideal gas process.
$\Delta T = 0$ (isothermal) so $\Delta U = 0$.

c) $\Delta U = Q - W$
$\Delta U = 0$ so $Q = W = -605$ J. (Q is negative; the gas liberates 605 J of heat to th surroundings.)

Note: $Q = nC_V\,\Delta T$ is only for a constant volume process so doesn't apply here.
$Q = nC_p\,\Delta T$ is only for a constant pressure process so doesn't apply here.

17-23 a) $Q = nC_p\,\Delta T$
Need to calculate C_p from $\gamma = C_p/C_V$ and $C_p = C_V + R$.
Combining these equations gives $\gamma = (C_V + R)/C_V = 1 + R/C_V$
$C_V = R/(\gamma - 1) = (8.3145 \text{ J/mol} \cdot \text{K})/(1.220 - 1) = 37.79 \text{ J/mol} \cdot \text{K}$
$C_p = C_V + R = 37.79 \text{ J/mol} \cdot \text{K} + 8.3145 \text{ J/mol} \cdot \text{K} = 46.10 \text{ J/mol} \cdot \text{K}$
Then $Q = nC_p\,\Delta T = 2.40(46.10 \text{ J/mol} \cdot \text{K})(25.0°\text{C} - 20.0°\text{C}) = +553$ J

b) For a constant pressure process,
$W = p\,\Delta V = nR\Delta T = (2.40 \text{ mol})(8.3145 \text{ J/mol} \cdot \text{K})(25.0°\text{C} - 20.0°\text{C}) = +99.7$ J

Then $\Delta U = Q - W = 553 \text{ J} - 99.7 \text{ J} = 453$ J.

17-25 a) In the process the pressure increases and the volume decreases.

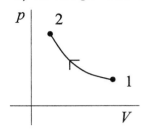

b) For an adiabatic process for an ideal gas

$T_1V_1^{\gamma-1} = T_2V_2^{\gamma-1}$, $\quad p_1V_1^{\gamma} = p_2V_2^{\gamma}$, $\quad$ and $pV = nRT$

From the first equation, $T_2 = T_1(V_1/V_2)^{\gamma-1} = (293\ \text{K})(V_1/0.0900V_1)^{1.4-1}$

$T_2 = (293\ \text{K})(11.11)^{0.4} = 768\ \text{K} = 495°\text{C}$

(Note: In the equation $T_1V_1^{\gamma-1} = T_2V_2^{\gamma-1}$ the temperature <u>must</u> be in kelvins.)

$p_1V_1^{\gamma} = p_2V_2^{\gamma}$ implies $p_2 = p_1(V_1/V_2)^{\gamma} = (1.00\ \text{atm})(V_1/0.0900V_1)^{1.4}$

$p_2 = (1.00\ \text{atm})(11.11)^{1.4} = 29.1\ \text{atm}$

Alternatively, we can use $pV = nRT$ to calculate p_2:

n, R constant implies $pV/T = nR = $ constant so $p_1V_1/T_1 = p_2V_2/T_2$

$p_2 = p_1(V_1/V_2)(T_2/T_1) = (1.00\ \text{atm})(V_1/0.0900V_1)(768\ \text{K}/293\ \text{K}) = 29.1\ \text{atm}$, which checks.

17-27 a) In the expansion the pressure decreases and the volume increases.

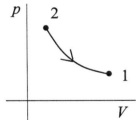

b) Adiabatic means $Q = 0$.

Then $\Delta U = Q - W$ gives $W = -\Delta U = -nC_V\,\Delta T = nC_V(T_1 - T_2)$ (Eq.17-25).

$C_V = 12.47\ \text{J/mol}\cdot\text{K}$ (Table 17-1)

$W = (0.450\ \text{mol})(12.47\ \text{J/mol}\cdot\text{K})(50.0°\text{C} - 10.0°\text{C}) = +224\ \text{J}$

W positive for $\Delta V > 0$ (expansion)

c) $Q = 0$ so there is no heat flow.

d Adiabatic process means $Q = 0$.

$\Delta U = Q - W$

$Q = 0$ so $\Delta U = -W = -224\ \text{J}$.

(In an adiabatic expansion of an ideal gas T decreases and U decreases.)

Problems

17-31

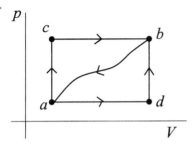

$Q_{acb} = +90.0$ J (positive since heat flows in)

$W_{acb} = +60.0$ J (positive since $\Delta V > 0$

a) $\Delta U = Q - W$

ΔU is path independent; Q and W depend on the path.

$\Delta U = U_b - U_a$

This can be calculated for any path from a to b, in particular for path acb:

$\Delta U_{a\rightarrow b} = Q_{acb} - W_{acb} = 90.0$ J $- 60.0$ J $= 30.0$ J.

Now apply $\Delta U = Q - W$ to path adb; $\Delta U = 30.0$ J for this path also.

$W_{adb} = +15.0$ J (positive since $\Delta V > 0$)

$\Delta U_{a\rightarrow b} = Q_{adb} - W_{adb}$ so $Q_{adb} = \Delta U_{a\rightarrow b} + W_{adb} = 30.0$ J $+ 15.0$ J $= +45.0$ J

b) Apply $\Delta U = Q - W$ to path ba:

$\Delta U_{b\rightarrow a} = Q_{ba} - W_{ba}$

$W_{ba} = -35.0$ J (negative since $\Delta V < 0$)

$\Delta U_{b\rightarrow a} = U_a - U_b = -(U_b - U_a) = -\Delta U_{a\rightarrow b} = -30.0$ J

Then $Q_{ba} = \Delta U_{b\rightarrow a} + W_{ba} = -30.0$ J $- 35.0$ J $= -65.0$ J.

($Q_{ba} < 0$; the system liberates heat.)

c) $U_a = 0$, $U_d = 8.0$ J

$\Delta U_{a\rightarrow b} = U_b - U_a = +30.0$ J, so $U_b = +30.0$ J.

process $a \rightarrow d$

$\Delta U_{a\rightarrow d} = Q_{ad} - W_{ad}$

$\Delta U_{a\rightarrow d} = U_d - U_a = +8.0$ J

$W_{adb} = +15.0$ J and $W_{adb} = W_{ad} + W_{db}$. But the work W_{db} for the process $d \rightarrow$ is zero since $\Delta V = 0$ for that process. Therefore $W_{ad} = W_{adb} = +15.0$ J.

Then $Q_{ad} = \Delta U_{a\rightarrow d} + W_{ad} = +8.0$ J $+ 15.0$ J $= +23.0$ J (positive implies hea absorbed).

process $d \to b$

$\Delta U_{d \to b} = Q_{db} - W_{db}$

$W_{db} = 0$, as already noted.

$\Delta U_{d \to b} = U_b - U_d = 30.0 \text{ J} - 8.0 \text{ J} = +22.0 \text{ J}$.

Then $Q_{db} = \Delta U_{d \to b} + W_{db} = +22.0 \text{ J}$ (positive; heat absorbed).

7-35 $\Delta U = Q - W$

$Q = -2.15 \times 10^5 \text{ J}$ (negative since heat energy goes out of the system)

$\Delta U = 0$ so $W = Q = -2.15 \times 10^5 \text{ J}$

Constant pressure, so $W = \int_{V_1}^{V_2} p \, dV = p(V_2 - V_1) = p \Delta V$.

Then $\Delta V = \dfrac{W}{p} = \dfrac{-2.15 \times 10^5 \text{ J}}{9.50 \times 10^5 \text{ J}} = -0.226 \text{ m}^3$.

7-39 The heat produced from the reaction is $Q_{\text{reaction}} = m L_{\text{reaction}}$,

where L_{reaction} is the heat of reaction of the chemicals.

$Q_{\text{reaction}} = W + \Delta U_{\text{spray}}$

For a mass m of spray, $W = \frac{1}{2}mv^2 = \frac{1}{2}m(19 \text{ m/s})^2 = (180.5 \text{ J/kg})m$ and

$\Delta U_{\text{spray}} = Q_{\text{spray}} = mc \, \Delta T = m(4190 \text{ J/kg·K})(100°\text{C} - 20°\text{C}) = (335,200 \text{ J/kg})m$.

Then $Q_{\text{reaction}} = (180 \text{ J/kg} + 335,200 \text{ J/kg})m = (335,380 \text{ J/kg})m$ and

$Q_{\text{reaction}} = m L_{\text{reaction}}$ implies $m L_{\text{reaction}} = (335,380 \text{ J/kg})m$.

The mass m divides out and $L_{\text{reaction}} = 3.4 \times 10^5 \text{ J/kg}$

(Note that the amount of energy converted to work is negligible for the two significant figures to which the answer should be expressed.)

7-43 a) $\gamma = C_p/C_V = (C_V + R)/C_V = 1 + R/C_V = 1.40$

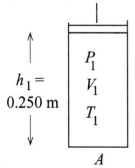
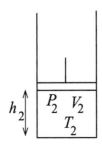

$p_1 = 1.01 \times 10^5 \text{ Pa}$
$p_2 = 4.20 \times 10^5 \text{ Pa} + p_{\text{air}} = 5.21 \times 10^5 \text{ Pa}$
$V_1 = h_1 A$
$V_2 = h_2 A$

adiabatic process: $p_1 V_1^\gamma = p_2 V_2^\gamma$

$p_1 h_1^\gamma A^\gamma = p_2 h_2^\gamma A^\gamma$

$$h_2 = h_1 \left(\frac{p_1}{p_2}\right)^{1/\gamma} = (0.250 \text{ m}) \left(\frac{1.01 \times 10^5 \text{ Pa}}{5.21 \times 10^5 \text{ Pa}}\right)^{1/1.40} = 0.0774 \text{ m}$$

The piston has moved a distance $h_1 - h_2 = 0.250 \text{ m} - 0.0774 \text{ m} = 0.173 \text{ m}$.

b) $T_1 V_1^{\gamma-1} = T_2 V_2^{\gamma-1}$

$T_1 h_1^{\gamma-1} A^{\gamma-1} = T_2 h_2^{\gamma-1} A^{\gamma-1}$

$$T_2 = T_1 \left(\frac{h_1}{h_2}\right)^{\gamma-1} = 300.1 \text{ K} \left(\frac{0.250 \text{ m}}{0.0774 \text{ m}}\right)^{0.40} = 479.7 \text{ K} = 207°\text{C}$$

c) $W = nC_V(T_1 - T_2)$ (Eq.17-25)

$W = (20.0 \text{ mol})(20.8 \text{ J/mol} \cdot \text{K})(300.1 \text{ K} - 479.7 \text{ K}) = -7.47 \times 10^4 \text{ J}$

17-45 a) <u>isothermal</u> ($\Delta T = 0$)

$\Delta U = Q - W$; $W = +300 \text{ J}$

For any process of an ideal gas, $\Delta U = nC_V \Delta T$. So for an ideal gas, if $\Delta T =$ then $\Delta U = 0$ and $Q = W = +300 \text{ J}$.

b) <u>adiabatic</u> ($Q = 0$)

$\Delta U = Q - W$; $W = +300 \text{ J}$

$Q = 0$ says $\Delta U = -W = -300 \text{ J}$

c) <u>isobaric</u> $\Delta p = 0$

$W = p\Delta V = nR \Delta T$; $\Delta T = W/nR$

$Q = nC_p \Delta T$ and for a monatomic ideal gas $C_p = \frac{5}{2}R$

Thus $Q = n\frac{5}{2}R\Delta T = (5Rn/2)(W/nR) = 5W/2 = +750 \text{ J}$.

$\Delta U = nC_V \Delta T$ for any ideal gas process and $C_V = C_p - R = \frac{3}{2}R$.

Thus $\Delta U = 3W/2 = +450 \text{ J}$

17-47 a) initial expansion (state 1 → state 2)

$p_1 = 2.40 \times 10^5 \text{ Pa}$, $T_1 = 355 \text{ K}$, $p_2 = 2.40 \times 10^5 \text{ Pa}$, $V_2 = 2V_1$

$pV = nRT$; $T/V = p/nR = $ constant, so $T_1/V_1 = T_2/V_2$ and $T_2 = T_1(V_2/V_1)$

$355 \text{ K}(2V_1/V_1) = 710 \text{ K}$

$\Delta p = 0$ so $W = p\Delta V = nR \Delta T = (0.250 \text{ mol})(8.3145 \text{ J/mol} \cdot \text{K})(710 \text{ K} - 355 \text{ K})$ $+738 \text{ J}$

$Q = nC_p \Delta T = (0.250 \text{ mol})(29.17 \text{ J/mol} \cdot \text{K})(710 \text{ K} - 355 \text{ K}) = +2590 \text{ J}$

$\Delta U = Q - W = 2590 \text{ J} - 738 \text{ J} = 1850 \text{ J}$

b) At the beginning of the final cooling process (cooling at constant volume), $T = 710$ K. The gas returns to its original volume and pressure, so also to its original temperature of 355 K.

$\Delta V = 0$ so $W = 0$

$Q = nC_V \Delta T = (0.250 \text{ mol})(20.85 \text{ J/mol} \cdot \text{K})(355 \text{ K} - 710 \text{ K}) = -1850$ J

$\Delta U = Q - W = -1850$ J.

c) For any ideal gas process $\Delta U = nC_V \Delta T$. For an isothermal process $\Delta T = 0$, so $\Delta U = 0$.

17-51 **a)** In the adiabatic expansion the pressure decreases.

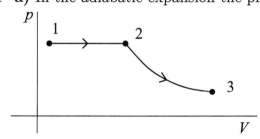

b) process $1 \to 2$ ($\Delta p = 0$)

$pV = nRT$ and p constant implies $V_1/T_1 = V_2/T_2$

$T_2 = T_1(V_2/V_1) = (300 \text{ K})(2V_1/V_1) = 600$ K

process $2 \to 3$ ($Q = 0$)

$T_2 = 600$ K

Temperature returns to its initial value says that $T_3 = T_1 = 300$ K

$Q_{12} = nC_p \Delta T$, since $\Delta p = 0$ for this process.

$Q_{12} = (0.350 \text{ mol})(34.60 \text{ J/mol} \cdot \text{K})(600 \text{ K} - 300 \text{ K}) = 3600$ J

$Q_{23} = 0$ since process $2 \to 3$ is adiabatic.

Then $Q = Q_{12} + Q_{23} = 3600$ J.

c) $\Delta U_{12} = nC_V \Delta T = nC_V(T_2 - T_1)$

$\Delta U_{23} = nC_V \Delta T = nC_V(T_3 - T_2)$

$\Delta U_{\text{tot}} = \Delta U_{12} + \Delta U_{23} = nC_V(T_2 - T_1 + T_3 - T_2) = nC_V(T_3 - T_1) = 0$, since $T_1 = T_3$; $\Delta U_{\text{tot}} = 0$ since the ideal gas ends up with the same temperature as at the start.

d) $\Delta U = Q - W$

$W = Q - \Delta U$ and $\Delta U = 0$ give $W = Q = 3600$ J

e) In process $1 \to 2$ the volume doubles, so $V_2 = 0.0140$ m^3.

Process $2 \to 3$ is adiabatic and we know T_2 and T_3. To find what happens to th
volume use $T_2 V_2^{\gamma-1} = T_3 V_3^{\gamma-1}$.

$V_3^{\gamma-1} = V_2^{\gamma-1}(T_2/T_3)$ so $V_3 = V_2(T_2/T_3)^{1/(\gamma-1)}$

For H$_2$S $\gamma = 1.33$ (from Table 17-1), so $V_3 = (0.0140$ m$^3)(600$ K$/300$ K$)^{1/0.33}$
0.114 m^3.

(Note that $V_3 > V_2$ as it should, since process $2 \to 3$ is an expansion.)

CHAPTER 18

THE SECOND LAW OF THERMODYNAMICS

Exercises

8-3 **a)** $e = \dfrac{\text{work output}}{\text{heat energy input}} = \dfrac{W}{Q_H} = \dfrac{3700 \text{ J}}{16,100 \text{ J}} = 0.23 = 23\%.$

b) $W = Q = |Q_H| - |Q_C|$

Heat discarded is $|Q_C| = |Q_H| - W = 16,100 \text{ J} - 3700 \text{ J} = 12,400 \text{ J}.$

c) Q_H is supplied by burning fuel; $Q_H = mL_c$ where L_c is the heat of combustion.

$$m = \dfrac{Q_H}{L_c} = \dfrac{16,100 \text{ J}}{4.60 \times 10^4 \text{ J/kg}} = 0.350 \text{ g}.$$

d) $W = 3700$ J per cycle

In $t = 1.00$ s the engine goes through 60.0 cycles.

$P = W/t = 60.0(3700 \text{ J})/1.00 \text{ s} = 222 \text{ kW}$

$P = (2.22 \times 10^5 \text{ W})(1 \text{ hp}/746 \text{ W}) = 298 \text{ hp}$

8-7 **a)** Process $a \to b$ is adiabatic so $T_a V_a^{\gamma-1} = T_b V_b^{\gamma-1}$

$V_b = V, \quad V_a = rV$ (as given in Fig.18-3)

$T_a = 22.0°\text{C} = 295 \text{ K}$

$T_b = T_a(V_a/V_b)^{\gamma-1} = (295 \text{ K})(rV/V)^{1.40-1} = (295 \text{ K})(9.50)^{0.4} = 726 \text{ K} = 453°\text{C}$

b) $pV = nRT$, $pV/T = $ constant, so $p_a V_a/T_a = p_b V_b/T_b$

$p_b = p_a(V_a/V_b)(T_b/T_a) = 8.50 \times 10^4 \text{ Pa}(rV/V)(726 \text{ K}/295 \text{ K}) = 1.99 \times 10^6 \text{ Pa}$

8-9 **a)** Performance coefficient $K = Q_C/|W|$ (Eq.18-9)

$|W| = Q_C/K = 3.40 \times 10^4 \text{ J}/2.10 = 1.62 \times 10^4 \text{ J}$

b)

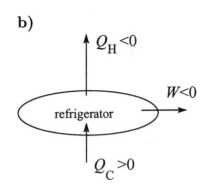

$W = Q_C + Q_H$
$Q_H = W - Q_C$
$Q_H = -1.62 \times 10^4 \text{ J} - 3.40 \times 10^4 \text{ J} = -5.02 \times 10^4$
(negative because heat goes out of the system)

18-11 a)

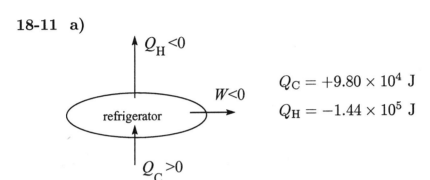

$Q_C = +9.80 \times 10^4 \text{ J}$
$Q_H = -1.44 \times 10^5 \text{ J}$

$W = Q_C + Q_H = +9.80 \times 10^4 \text{ J} - 1.44 \times 10^5 \text{ J} = -4.60 \times 10^4 \text{ J}$
(W negative means power is consumed, not produced, by the device)
$P = W/t = -4.60 \times 10^4 \text{ J}/60.0 \text{ s} = -767 \text{ W}$

b) EER $= (3.413)K$
$K = |Q_C|/|W| = 9.80 \times 10^4 \text{ J}/4.60 \times 10^4 \text{ J} = 2.13$
EER $= (3.413)(2.13) = 7.27$

18-13 a)

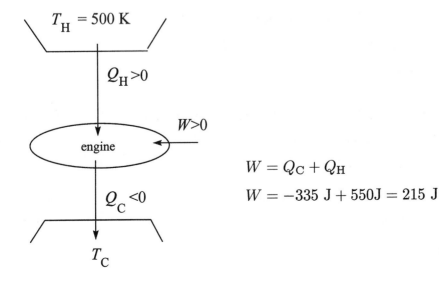

$W = Q_C + Q_H$
$W = -335 \text{ J} + 550 \text{J} = 215 \text{ J}$

b) For a Carnot cycle, $\dfrac{|Q_C|}{|Q_H|} = \dfrac{T_C}{T_H}$ (Eq.18-13)

$$T_C = T_H \frac{|Q_C|}{|Q_H|} = 620 \text{ K} \left(\frac{335 \text{ J}}{550 \text{ J}} \right) = 378 \text{ K}$$

c) $e(\text{Carnot}) = 1 - T_C/T_H = 1 - 378 \text{ K}/620 \text{ K} = 0.390 = 39.0\%$

18-15 a)

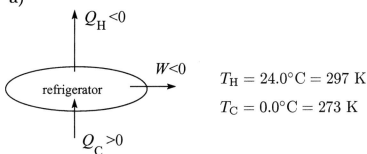

$T_H = 24.0°\text{C} = 297 \text{ K}$

$T_C = 0.0°\text{C} = 273 \text{ K}$

The amount of heat taken out of the water to make the liquid → solid phase change is $Q = -mL_f = -(85.0 \text{ kg})(334 \times 10^3 \text{ J/kg}) = -2.84 \times 10^7 \text{ J}$. This amount of heat must go into the working substance of the refrigerator, so $Q_C = +2.84 \times 10^7 \text{ J}$.

For Carnot cycle $|Q_C|/|Q_H| = T_C/T_H$

$|Q_H| = |Q_C|(T_H/T_C) = 2.84 \times 10^7 \text{ J}(297 \text{ K}/273 \text{ K}) = 3.09 \times 10^7 \text{ J}$

b) $W = Q_C + Q_H = +2.84 \times 10^7 \text{ J} - 3.09 \times 10^7 \text{ J} = -2.5 \times 10^6 \text{ J}$

(W is negative because this much energy must be supplied to the refrigerator rather than obtained from it.)

18-21 First use $Q_{\text{system}} = 0$ to find the final temperature of the mixed water.

The 1.00 kg of water warms from 20.0°C to T:

$Q = mc\,\Delta T = (1.00 \text{ kg})(4190 \text{ J/kg} \cdot \text{K})(T - 20.0°\text{C})$

The 2.00 kg of water cools from 80.0°C to T:

$Q = mc\,\Delta T = (2.00 \text{ kg})(4190 \text{ J/kg} \cdot \text{K})(T - 80.0°\text{C})$

$Q_{\text{system}} = 0$ says

$(1.00 \text{ kg})(4190 \text{ J/kg} \cdot \text{K})(T - 20.0°\text{C}) + (2.00 \text{ kg})(4190 \text{ J/kg} \cdot \text{K})(T - 80.0°\text{C}) = 0$

$(T - 20.0°\text{C}) + 2(T - 80.0°\text{C}) = 0$

$3T = 180°\text{C}$ and $T = 60.0°\text{C}$

Now we can calculate the entropy change.

<u>The 1.00 kg of water:</u>

$\Delta S = \int_1^2 dQ/T$

$dQ = mc\,dT$ so $\Delta S = mc \int_1^2 dT/T = mc\ln(T_2/T_1)$ (Example 18-6).

$$\Delta S = (1.00 \text{ kg})(4190 \text{ J/kg} \cdot \text{K}) \ln\left(\frac{(60.0 + 273.15) \text{ K}}{(20.0 + 273.15) \text{ K}}\right) = 535.9 \text{ J/K}$$

The 2.00 kg of water:

$$\Delta S = mc\ln(T_2/T_1) = (2.00 \text{ kg})(4190 \text{ J/kg} \cdot \text{K}) \ln\left(\frac{(60.0 + 273.15) \text{ K}}{(80.0 + 273.15) \text{ K}}\right) =$$

-488.6 J/K

The entropy change of the system is $\Delta S_{tot} = 535.9 \text{ J/K} - 488.6 \text{ J/K} = 47 \text{ J/K}$.

18-23 The initial and final states are at the same temperature, at the normal boiling point of 4.216 K. Calculate the entropy change for the irreversible process by considering a reversible isothermal process that connects the same two states.

The heat flow for the helium is $Q = -mL_v$, negative since in condensation heat flows out of the helium. The heat of vaporization L_v is given in Table 15-4 and $L_v = 20.9 \times 10^3$ J/kg.

$Q = -mL_v = -(0.130 \text{ kg})(20.9 \times 10^3 \text{ J/kg}) = -2717 \text{ J}$

Since we are considering a reversible isothermal process Eq.(18-18) applies and $\Delta S = Q/T = -2717 \text{ J}/4.216 \text{ K} = -644 \text{ J/K}$.

18-27 Reversible implies $\Delta S = \int_1^2 dQ/T$

Isothermal means T is constant, and then $\Delta S = Q/T$.

Calculate Q:

$\Delta U = Q - W$

$\Delta U = nC_V\,\Delta T$ (any ideal gas process); $\Delta U = 0$ since $\Delta T = 0$

$W = nRT \int_{V_1}^{V_2} dV/V = nRT\ln(V_2/V_1)$

$W = (2.00 \text{ mol})(8.3145 \text{ J/mol} \cdot \text{K})(298 \text{ K})\ln(0.0420 \text{ m}^3/0.0280 \text{ m}^3) = 2.01 \times 10^3$

Then $Q = \Delta U + W = 0 + 2.01 \times 10^3 \text{ J} = 2.01 \times 10^3 \text{ J}$.

$\Delta S = Q/T = 2.01 \times 10^3 \text{ J}/298 \text{ K} = +6.74 \text{ J/K}$

18-29 a) The velocity distribution of Eq.(16-32) depends only on T, so in an isothermal process it does not change.

b) Following the reasoning of Example 18-11, the number of possible positions available to each molecule is altered by a factor of 3 (becomes larger). Hence the number of microscopic states the gas occupies at volume $3V$ is $w_2 = (3)^N w_1$, where N is the number of molecules and w_1 is the number of possible microscopic states at the start of the process, where the volume is V. Then, by Eq.(18-23),

$\Delta S = k \ln(w_2/w_1) = k \ln(3)^N = Nk \ln(3) = nN_A k \ln(3) = nR \ln(3)$
$\Delta S = (2.00 \text{ mol})(8.3145 \text{ J/mol} \cdot \text{K}) \ln(3) = +18.3 \text{ J/K}$

c) $\Delta S = Q/T$

Need to calculate Q:
$\Delta T = 0$ implies $\Delta U = 0$, since system is an ideal gas.
Then by $\Delta U = Q - W$, $Q = W$.
For an isothermal process, $W = \int_{V_1}^{V_2} p \, dV = \int_{V_1}^{V_2} (nRT/V) \, dV = nRT \ln(V_2/V_1)$
Thus $Q = nRT \ln(V_2/V_1)$ and $\Delta S = Q/T = nR \ln(V_2/V_1)$
$\Delta S = (2.00 \text{ mol})(8.3145 \text{ J/mol} \cdot \text{K}) \ln(3V_1/V_1) = +18.3 \text{ J/K}$
This is the same result as obtained in part (b).

18-33 a) Total energy collected in one hour:
$(150 \text{ W/m}^2)(3600 \text{ s})(8.0 \text{ m}^2)(0.60) = 2.592 \times 10^6 \text{ J}$
This amount of energy is transferred to the water, to raise its temperature.

$$Q = mc \, \Delta T \text{ so } m = \frac{Q}{c \, \Delta T} = \frac{2.592 \times 10^6 \text{ J}}{(4190 \text{ J/kg} \cdot \text{K})(55.0°\text{C} - 15.0°\text{C})} = 15.47 \text{ kg}$$

$$V = \frac{m}{\rho} = \frac{15.47 \text{ kg}}{1000 \text{ kg/m}^3} = 0.0155 \text{ m}^3 = 15 \text{ L}.$$

b) The answer in part (a) is for one hour. The volume of 55.0°C water produced in one day (24 hours) is $24(15 \text{ L}) = 360 \text{ L}$.
The number of inhabitants is $360 \text{ L}/(75 \text{ L/inhabitant}) = 5$.

18-35 a) Calculate the mass of water required and then from this
   the volume of water:

$$Q = mc \, \Delta T \text{ so } m = \frac{Q}{c \, \Delta T} = \frac{4.00 \times 10^9 \text{ J}}{(4190 \text{ J/kg} \cdot \text{K})(49.0°\text{C} - 21.0°\text{C})} = 3.409 \times 10^4 \text{ kg}$$

$\rho = m/V$ so $V = m/\rho = (3.409 \times 10^4 \text{ kg})/(1.00 \times 10^3 \text{ kg/m}^3) = 34.1 \text{ m}^3$

b) Repeat the above calculation, but now with Glauber salt in place of water. The essential difference is that Glauber salt undergoes a phase transition in this temperature range (at 32.0°C).
$Q = mc_{\text{solid}}(32.0°\text{C} - 21.0°\text{C}) + mL_{\text{f}} + mc_{\text{liquid}}(49.0°\text{C} - 32.0°\text{C})$
$Q = m[(1930 \text{ J/kg} \cdot \text{K})(11.0 \text{ K}) + 2.42 \times 10^5 \text{ J/kg} + (2850 \text{ J/kg} \cdot \text{K})(17.0 \text{ K})]$
$Q = m[3.117 \times 10^5 \text{ J/kg}]$

Then $m = \dfrac{Q}{3.117 \times 10^5 \text{ J/kg}} = \dfrac{4.00 \times 10^9 \text{ J}}{3.117 \times 10^5 \text{ J/kg}} = 1.283 \times 10^4$ kg.

$V = \dfrac{m}{\rho} = \dfrac{1.283 \times 10^4 \text{ kg}}{1600 \text{ kg/m}^3} = 8.02 \text{ m}^3$

The space requirement is smaller by about a factor of 4 when Glauber salt is used instead of water.

Problems

18-39 $\gamma = 1.40$

$C_V = R/(\gamma - 1) = 20.70$ J/mol · K
$C_p = C_V + R = 29.10$ J/mol · K

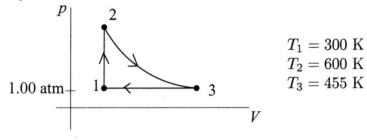

$T_1 = 300$ K
$T_2 = 600$ K
$T_3 = 455$ K

a) <u>point 1</u>

$p_1 = 1.00$ atm $= 1.013 \times 10^5$ Pa (given); $pV = nRT$;

$V_1 = \dfrac{nRT_1}{p_1} = \dfrac{(0.350 \text{ mol})(8.3145 \text{ J/mol} \cdot \text{K})(300 \text{ K})}{1.013 \times 10^5 \text{ Pa}} = 8.62 \times 10^{-3} \text{ m}^3$

<u>point 2</u>

process $1 \rightarrow 2$ at constant volume so $V_2 = V_1 = 8.62 \times 10^{-3}$ m³

$pV = nRT$ and n, R, V constant implies $p_1/T_1 = p_2/T_2$

$p_2 = p_1(T_2/T_1) = (1.00 \text{ atm})(600 \text{ K}/300 \text{ K}) = 2.00$ atm $= 2.03 \times 10^5$ Pa

<u>point 3</u>

Consider the process $3 \rightarrow 1$, since it is simpler than $2 \rightarrow 3$.

Process $3 \rightarrow 1$ is at constant pressure so $p_3 = p_1 = 1.00$ atm $= 1.013 \times 10^5$ Pa

$pV = nRT$ and n, R, p constant implies $V_1/T_1 = V_3/T_3$

$V_3 = V_1(T_3/T_1) = (8.62 \times 10^{-3} \text{ m}^3)(455 \text{ K}/300 \text{ K}) = 13.1 \times 10^{-3} \text{ m}^3$

b) <u>process $1 \rightarrow 2$</u>

constant volume ($\Delta V = 0$)

$Q = nC_V \Delta T = (0.350 \text{ mol})(20.79 \text{ J/mol} \cdot \text{K})(600 \text{ K} - 300 \text{ K}) = 2180$ J

$\Delta V = 0$ and $W = 0$

Then $\Delta U = Q - W = 2180$ J

process $2 \to 3$

adiabatic means $Q = 0$

$\Delta U = nC_V \Delta T$ (any process), so

$\Delta U = (0.350 \text{ mol})(20.70 \text{ J/mol} \cdot \text{K})(455 \text{ K} - 600 \text{ K}) = -1050$ J

Then $\Delta U = Q - W$ gives $W = Q - \Delta U = +1050$ J. (It is correct for W to be positive since ΔV is positive.)

process $3 \to 1$

For constant pressure

$W = p\Delta V = (1.013 \times 10^5 \text{ Pa})(8.62 \times 10^{-3} \text{ m}^3 - 13.1 \times 10^{-3} \text{ m}^3) = -450$ J

or $W = nR\Delta T = (0.350 \text{ mol})(8.3145 \text{ J/mol} \cdot \text{K})(300 \text{ K} - 455 \text{ K}) = -450$ J, which checks. (It is correct for W to be negative, since ΔV is negative for this process.)

$Q = nC_p \Delta T = (0.350 \text{ mol})(29.10 \text{ J/mol} \cdot \text{K})(300 \text{ K} - 455 \text{ K}) = -1580$ J

$\Delta U = Q - W = -1580 \text{ J} - (-450 \text{ K}) = -1130$ J

or $\Delta U = nC_V \Delta T = (0.350 \text{ mol})(20.79 \text{ J/mol} \cdot \text{K})(300 \text{ K} - 455 \text{ K}) = -1130$ J, which checks

c) $W_{\text{net}} = W_{1 \to 2} + W_{2 \to 3} + W_{3 \to 1} = 0 + 1050 \text{ J} - 450 \text{ J} = +600$ J

d) $Q_{\text{net}} = Q_{1 \to 2} + Q_{2 \to 3} + Q_{3 \to 1} = 2180 \text{ J} + 0 - 1580 \text{ J} = +600$ J

Note: For a cycle $\Delta U = 0$, so by $\Delta U = Q - W$ it must be that $Q_{\text{net}} = W_{\text{net}}$ for a cycle. We can also check that $\Delta U_{\text{net}} = 0$:

$\Delta U_{\text{net}} = \Delta U_{1 \to 2} + \Delta U_{2 \to 3} + \Delta U_{3 \to 1} = 2180 \text{ J} - 1050 \text{ J} - 1130 \text{ J} = 0$

e) $e = \dfrac{\text{work output}}{\text{heat energy input}} = \dfrac{W}{Q_{\text{H}}} = \dfrac{600 \text{ J}}{2180 \text{ J}} = 0.275 = 27.5\%.$

$e(\text{Carnot}) = 1 - T_{\text{C}}/T_{\text{H}} = 1 - 300 \text{ K}/600 \text{ K} = 0.500.$

18-41

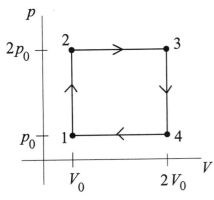

$C_V = 5R/2$

for an ideal gas $C_p = C_V + R = 7R/2$

Calculate Q and W for each process:

process $1 \to 2$

$\Delta V = 0$ implies $W = 0$

$\Delta V = 0$ implies $Q = nC_V \, \Delta T = nC_V(T_2 - T_1)$

But $pV = nRT$ and V constant says $p_1 V = nRT_1$ and $p_2 V = nRT_2$.

Thus $(p_2 - p_1)V = nR(T_2 - T_1)$; $V \Delta p = nR \, \Delta T$ (true when V is constant).

Then $Q = nC_V \, \Delta T = nC_V(V \, \Delta p / nR) = (C_V/R)V \, \Delta p = (C_V/R)V_0(2p_0 - p_0) = (C_V/R)p_0V_0$. $Q > 0$; heat is absorbed by the gas.)

process $2 \to 3$

$\Delta p = 0$ so $W = p \Delta V = p(V_3 - V_2) = 2p_0(2V_0 - V_0) = 2p_0V_0$ (W is positive since V increases.)

$\Delta p = 0$ implies $Q = nC_p \, \Delta T = nC_p(T_2 - T_1)$

But $pV = nRT$ and p constant says $pV_1 = nRT_1$ and $pV_2 = nRT_2$.

Thus $p(V_2 - V_1) = nR(T_2 - T_1)$; $p \Delta V = nR \, \Delta T$ (true when p is constant).

Then $Q = nC_p \, \Delta T = nC_p(p \, \Delta V / nR) = (C_p/R)p \, \Delta V = (C_p/R)2p_0(2V_0 - V_0) = (C_p/R)2p_0V_0$. ($Q > 0$; heat is absorbed by the gas.)

process $3 \to 4$

$\Delta V = 0$ implies $W = 0$

$\Delta V = 0$ so

$Q = nC_V \, \Delta T = nC_V(V \, \Delta p / nR) = (C_V/R)(2V_0)(p_0 - 2p_0) = -2(C_V/R)p_0V_0$ ($Q < 0$ so heat is rejected by the gas.)

process $4 \to 1$

$\Delta p = 0$ so $W = p \Delta V = p(V_1 - V_4) = p_0(V_0 - 2V_0) = -p_0V_0$ (W is negative since V decreases)

$\Delta p = 0$ so $Q = nC_p \, \Delta T = nC_p(p \, \Delta V / nR) = (C_p/R)p \, \Delta V = (C_p/R)p_0(V_0 - 2V_0) = -(C_p/R)p_0V_0$ ($Q < 0$ so heat is rejected by the gas.)

total work performed by the gas during the cycle:

$W_{\text{tot}} = W_{1 \to 2} + W_{2 \to 3} + W_{3 \to 4} + W_{4 \to 1} = 0 + 2p_0V_0 + 0 - p_0V_0 = p_0V_0$

(Note that W_{tot} equals the area enclosed by the cycle in the pV-diagram.)

total heat absorbed by the gas during the cycle (Q_{H}):

Heat is absorbed in processes $1 \to 2$ and $2 \to 3$.

$$Q_{\text{H}} = Q_{1 \to 2} + Q_{2 \to 3} = \frac{C_V}{R}p_0V_0 + 2\frac{C_p}{R}p_0V_0 = \left(\frac{C_V + 2C_p}{R} \right)p_0V_0$$

But $C_p = C_V + R$ so $Q_{\text{H}} = \dfrac{C_V + 2(C_V + R)}{R}p_0V_0 = \left(\dfrac{3C_V + 2R}{R} \right)p_0V_0$.

<u>total heat rejected</u> by the gas during the cycle (Q_C):

Heat is rejected in processes $3 \to 4$ and $4 \to 1$.

$$Q_C = Q_{3\to4} + Q_{4\to1} = -2\frac{C_V}{R}p_0V_0 - \frac{C_p}{R}p_0V_0 = -\left(\frac{2C_V + C_p}{R}\right)p_0V_0$$

But $C_p = C_V + R$ so $Q_C = -\frac{2C_V + (C_V + R)}{R}p_0V_0 = -\left(\frac{3C_V + R}{R}\right)p_0V_0.$

As a check on the calculations note that

$$Q_C + Q_H = -\left(\frac{3C_V + R}{R}\right)p_0V_0 + \left(\frac{3C_V + 2R}{R}\right)p_0V_0 = p_0V_0 = W, \text{ as it should.}$$

<u>efficiency</u>

$$e = \frac{W}{Q_H} = \frac{p_0V_0}{([3C_V + 2R]/R)(p_0V_0)} = \frac{R}{3C_V + 2R} = \frac{R}{3(5R/2) + 2R} = \frac{2}{19}.$$

$e = 0.105 = 10.5\%$

18-43

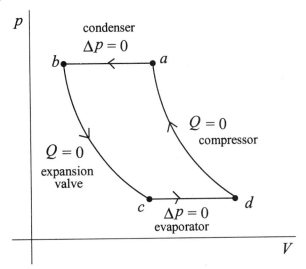

a) <u>process $c \to d$</u>

$\Delta U = U_d - U_c = 1657 \times 10^3 \text{ J} - 1005 \times 10^3 \text{ J} = 6.52 \times 10^5 \text{ J}$

$W = \int_{V_c}^{V_d} p\,dV = p\Delta V$ (since is a constant pressure process)

$W = (363 \times 10^3 \text{ Pa})(0.4513 \text{ m}^3 - 0.2202 \text{ m}^3) = +8.39 \times 10^4 \text{ J}$ (positive since process is an expansion)

$\Delta U = Q - W$ so $Q = \Delta U + W = 6.52 \times 10^5 \text{ J} + 8.39 \times 10^4 \text{ J} = 7.36 \times 10^5 \text{ J}.$

(Q positive so heat goes into the coolant)

b) process $a \rightarrow b$

$\Delta U = U_b - U_a = 1171 \times 10^3 \text{ J} - 1969 \times 10^3 \text{ J} = -7.98 \times 10^5 \text{ J}$

$W = p \Delta V = (2305 \times 10^3 \text{ Pa})(0.00946 \text{ m}^3 - 0.0682 \text{ m}^3) = -1.35 \times 10^5 \text{ J}$

(negative since $\Delta V < 0$ for the process)

$Q = \Delta U + W = -7.98 \times 10^5 \text{ J} - 1.35 \times 10^5 \text{ J} = -9.33 \times 10^5 \text{ J}$

(negative so heat comes out of coolant).

c) The coolant cannot be treated as an ideal gas, so we can't calculate W for the adiabatic processes. But $\Delta U = 0$ (for cycle) so $W_{net} = Q_{net}$.

$Q = 0$ for the two adiabatic processes, so

$Q_{net} = Q_{cd} + Q_{ab} = 7.36 \times 10^5 \text{ J} - 9.33 \times 10^5 \text{ J} = -1.97 \times 10^5 \text{ J}$

Thus $W_{net} = -1.97 \times 10^5$ J (negative since work is done on the coolant, the working substance).

d) $K = Q_C / |W| = (+7.36 \times 10^5 \text{ J}) / (+1.97 \times 10^5 \text{ J}) = 3.74.$

18-49 a) Eq.(18-6): $e = 1 - 1/(r^{\gamma - 1}) = 1 - 1/(10.6^{0.4}) = 0.6111$

$e = (Q_H + Q_C)/Q_H$ and we are given $Q_H = 200$ J; calculate Q_C.

$Q_C = (e - 1)Q_H = (0.6111 - 1)(200 \text{ J}) = -78$ J (negative since corresponds to heat leaving)

Then $W = Q_C + Q_H = -78 \text{ J} + 200 \text{ J} = 122$ J. (Positive, in agreement with Fig.18-3.)

b) For each cylinder of area $A = \pi(d/2)^2$ the piston moves 0.0864 m and the volume changes from rV to V.

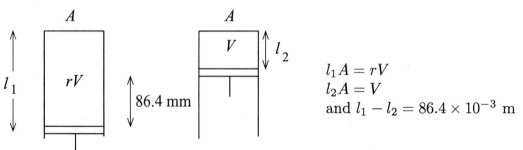

$l_1 A = rV$
$l_2 A = V$
and $l_1 - l_2 = 86.4 \times 10^{-3}$ m

$l_1 A - l_2 A = rV - V$ and $(l_1 - l_2)A = (r - 1)V$

$V = \dfrac{(l_1 - l_2)A}{r - 1} = \dfrac{(86.4 \times 10^{-3} \text{ m})\pi(41.25 \times 10^{-3} \text{ m})^2}{10.6 - 1} = 4.811 \times 10^{-5} \text{ m}^3$

At point a the volume is $rV = 10.6(4.811 \times 10^{-5} \text{ m}^3) = 5.10 \times 10^{-4} \text{ m}^3$.

c) point a: $T_a = 300$ K, $p_a = 8.50 \times 10^4$ Pa, and $V_a = 5.10 \times 10^{-4} \text{ m}^3$

<u>point b</u>:

$V_b = V_a/r = 4.81 \times 10^{-5} \text{ m}^3$

process $a \rightarrow b$ is adiabatic, so $T_a V_a^{\gamma-1} = T_b V_b^{\gamma-1}$

$T_a(rV)^{\gamma-1} = T_b V^{\gamma-1}$

$T_b = T_a r^{\gamma-1} = 300 \text{ K}(10.6)^{0.4} = 771 \text{ K}$

$pV = nRT$ so $pV/T = nR = $ constant, so $p_a V_a/T_a = p_b V_b/T_b$

$p_b = p_a(V_a/V_b)(T_b/T_a) = (8.50 \times 10^4 \text{ Pa})(rV/V)(771 \text{ K}/300 \text{ K}) = 2.32 \times 10^6 \text{ Pa}$

<u>point c</u>

process $b \rightarrow c$ is at constant volume, so $V_c = V_b = 4.81 \times 10^{-5} \text{ m}^3$

$Q_H = nC_V \, \Delta T = nC_V(T_c - T_b)$. The problem specifies $Q_H = 200$ J; use to calculate T_c. First use the p, V, T values at point a to calculate the number of moles n.

$$n = \frac{pV}{RT} = \frac{(8.50 \times 10^4 \text{ Pa})(5.10 \times 10^{-4} \text{ m}^3)}{(8.3145 \text{ J/mol} \cdot \text{K})(300 \text{ K})} = 0.01738 \text{ mol}$$

Then $T_c - T_b = \dfrac{Q_H}{nC_V} = \dfrac{200 \text{ J}}{(0.01738 \text{ mol})(20.5 \text{ J/mol} \cdot \text{K})} = 561.3 \text{ K}$, and

$T_c = T_b + 561.3 \text{ K} = 771 \text{ K} + 561 \text{ K} = 1332 \text{ K}$

$p/T = nR/V = $ constant so $p_b/T_b = p_c/T_c$

$p_c = p_b(T_c/T_b) = (2.32 \times 10^6 \text{ Pa})(1332 \text{ K}/771 \text{ K}) = 4.01 \times 10^6 \text{ Pa}$

<u>point d</u>:

$V_d = V_a = 5.10 \times 10^{-4} \text{ m}^3$

process $c \rightarrow d$ is adiabatic, so $T_d V_d^{\gamma-1} = T_c V_c^{\gamma-1}$

$T_d(rV)^{\gamma-1} = T_c V^{\gamma-1}$

$T_d = T_c/r^{\gamma-1} = 1332 \text{ K}/10.6^{0.4} = 518 \text{ K}$

$p_c V_c/T_c = p_d V_d/T_d$

$p_d = p_c(V_c/V_d)(T_d/T_c) = (4.01 \times 10^6 \text{ Pa})(V/rV)(518 \text{ K}/1332 \text{ K}) = 1.47 \times 10^5 \text{ Pa}$

Can look at process $d \rightarrow a$ as a check:

$Q_C = nC_V(T_a - T_d) = (0.01738 \text{ mol})(20.5 \text{ J/mol} \cdot \text{K})(300 \text{ K} - 518 \text{ K}) = -78 \text{ J}$, which agrees with part (a)

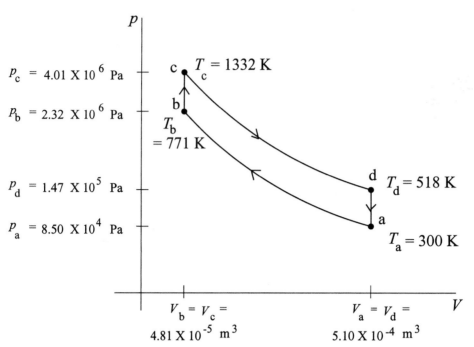

$P_c = 4.01 \times 10^6$ Pa

$P_b = 2.32 \times 10^6$ Pa

$P_d = 1.47 \times 10^5$ Pa

$P_a = 8.50 \times 10^4$ Pa

$T_c = 1332$ K

$T_b = 771$ K

$T_d = 518$ K

$T_a = 300$ K

$V_b = V_c = 4.81 \times 10^{-5}$ m^3

$V_a = V_d = 5.10 \times 10^{-4}$ m^3

d) From part (a) the efficiency of this Otto cycle is $e = 0.611 = 61.1\%$.

The efficiency of a Carnot cycle operating between 1332 K and 300 K is
$e(\text{Carnot}) = 1 - T_C/T_H = 1 - 300\text{ K}/1332\text{ K} = 0.775 = 77.5\%$, which is larger.

18-51 First use the methods of Chapter 15 to calculate the final temperature T of
the system:

0.600 kg of water (cools from 45.0°C to T)
$Q = mc\,\Delta T =$
$(0.600\text{ kg})(4190\text{ J/kg}\cdot\text{K})(T - 45.0°\text{C}) = (2514\text{ J/K})T - 1.1313 \times 10^5$ J

0.0500 kg of ice (warms to 0°C, melts, and water warms from 0°C to T)
$Q = mc_{\text{ice}}(0°\text{C} - (-15.0°\text{C})) + mL_f + mc_{\text{water}}(T - 0°\text{C})$
$Q = 0.0500\text{ kg}[(2100\text{ J/kg}\cdot\text{K})(15.0°\text{C}) + 334 \times 10^3\text{ J/kg} + (4190\text{ J/kg}\cdot\text{K})(T - 0°\text{C})]$
$Q = 1575\text{ J} + 1.67 \times 10^4\text{ J} + (209.5\text{ J/K})T = 1.828 \times 10^4\text{ J} + (209.5\text{ J/K})T$

$Q_{\text{system}} = 0$ gives $(2514\text{ J/K})T - 1.1313 \times 10^5\text{ J} + 1.828 \times 10^4\text{ J} + (209.5\text{ J/K})T = 0$

$(2.724 \times 10^3\text{ J/K})T = 9.485 \times 10^4$ J
$T = (9.485 \times 10^4\text{ J})/(2.724 \times 10^3\text{ J/K}) = 34.82°\text{C} = 308$ K

Now we can calculate the entropy changes:

ice:
The process takes ice at $-15°$C and produces water at $34.8°$C. Calculate ΔS for

a reversible process between these two states, in which heat is added very slowly. ΔS is path independent, so ΔS for a reversible process is the same as ΔS for the actual (irreversible) process as long as the initial and final states are the same.

$\Delta S = \int_1^2 dQ/T$, where T must be in kelvins

For a temperature change $dQ = mc\,dT$ so $\Delta S = \int_{T_1}^{T_2}(mc/T)\,dT = mc\ln(T_2/T_1)$.

For a phase change, since it occurs at constant T,

$\Delta S = \int_1^2 dQ/T = Q/T = \pm mL/T$.

Therefore $\Delta S_{\text{ice}} = mc_{\text{ice}}\ln(273\text{ K}/258\text{ K}) + mL_{\text{f}}/273K + mc_{\text{water}}\ln(308\text{ K}/273\text{ K})$

$\Delta S_{\text{ice}} = (0.0500\text{ kg})[(2100\text{ J/kg} \cdot \text{K})\ln(273\text{ K}/258\text{ K}) + (334 \times 10^3\text{ J/kg})/273\text{ K} + (4190\text{ J/kg} \cdot \text{K})\ln(308\text{ K}/273\text{ K})]$

$\Delta S_{\text{ice}} = 5.93\text{ J/K} + 61.17\text{ J/K} + 25.27\text{ J/K} = 92.4\text{ J/K}$

water

$\Delta S_{\text{water}} = mc\ln(T_2/T_1) = (0.600\text{ kg})(4190\text{ J/kg} \cdot \text{K})\ln(308\text{ K}/318\text{ K}) = -80.3\text{ J/K}$

For the system, $\Delta S = \Delta S_{\text{ice}} + \Delta S_{\text{water}} = 92.4\text{ J/K} - 80.3\text{ J/K} = +12\text{ J/K}$

Note: Our calculation gives $\Delta S > 0$, as it must for an irreversible process of an isolated system.

CHAPTER 19

MECHANICAL WAVES

Exercises 1, 5, 7, 11, 15, 17, 19, 23, 25, 29
Problems 31, 35, 41, 43

Exercises

19-1   **a)** The distance between crests is the wavelength; $\lambda = 6.0$ m.

Time between largest upward displacement to largest downward displacement is one-half the period; $T/2 = 2.5$ s so $T = 5.0$ s.

$f = 1/T = 1/5.0$ s $= 0.20$ Hz

Then $v = f\lambda = (0.20$ Hz$)(6.0$ m$) = 1.2$ m/s.

b) The distance between the highest and lowest points is twice the amplitude, so $2A = 0.62$ m and $A = 0.31$ m.

c) v unchanged but A becomes 0.15 m.

d) No. Expect longitudinal as well as transverse surface waves.

19-5   **a)** $v = 344$ m/s; $v = f\lambda$ so $\lambda = v/f$

$f = 20.0$ Hz: $\lambda = (344$ m/s$)/(20.0$ Hz$) = 17.2$ m

$f = 20,000$ Hz: $\lambda = (344$ m/s$)/(20,000$ Hz$) = 0.0172$ m

b) $v = 1480$ m/s

$f = 20.0$ Hz: $\lambda = (1480$ m/s$)/(20.0$ Hz$) = 74.0$ m

$f = 20,000$ Hz: $\lambda = (1480$ m/s$)/(20,000$ Hz$) = 0.0740$ m

19-7 $v = 8.00$ m/s, $A = 0.0700$ m, $\lambda = 0.320$ m

a) $v = f\lambda$ so $f = v/\lambda = (8.00$ m/s$)/(0.320$ m$) = 25.0$ Hz

$T = 1/f = 1/25.0$ Hz $= 0.0400$ s

$k = 2\pi/k = 2\pi$ rad$/0.320$ m $= 19.6$ rad/m

b) For a wave traveling in the $-x$-direction,

$y(x, t) = A\sin 2\pi(t/T + x/\lambda)$ (Eq.(19-8)).

At $x = 0$, $y(0, t) = A\sin 2\pi(t/T)$, so $y = 0$ at $t = 0$ and y is positive for t slightly greater than zero. This equation describes the wave specified in the problem.

Substitute in numerical values:

$y(x, t) = (0.0700$ m$)\sin(2\pi(t/0.0400$ s $+ x/0.320$ m$))$.

Or, $y(x,t) = (0.0700 \text{ m}) \sin((157 \text{ rad/s})t + (19.6 \text{ m}^{-1})x)$.

c) From part (b), $y = (0.0700 \text{ m}) \sin(2\pi(t/0.0400 \text{ s} + x/0.320 \text{ m}))$.

Plug in $x = 0.360$ m and $t = 0.150$ s:

$y = (0.0700 \text{ m}) \sin(2\pi(0.150 \text{ s}/0.0400 \text{ s} + 0.360 \text{ m}/0.320 \text{ m}))$

$y = (0.0700 \text{ m}) \sin[2\pi(4.875 \text{ rad})] = -0.0495 \text{ m} = -4.95 \text{ cm}$

d) In part (c) $t = 0.150$ s.

$y = 0$ means $\sin 2\pi(t/T + x/\lambda) = 0$

$\sin\theta = 0$ for $\theta = 0, \pi, 2\pi, \ldots = n\pi$ for $n = 0, 1, 2, \ldots$

So $y = 0$ when $2\pi(t/T + x/\lambda) = n\pi$ or $t/T + x/\lambda = n/2$

$t = T(n/2 - x/\lambda) = (0.0400 \text{ s})(n/2 - 0.360 \text{ m}/0.320 \text{ m}) = (0.0400 \text{ s})(n/2 - 1.125)$

For $n = 9$, $t = 0.1350$ s (before the instant in part (c))

For $n = 10$, $t = 0.1550$ s (the first occurrence of $y = 0$ after the instant in part (c))

Thus the elapsed time is $0.1550 \text{ s} - 0.1500 \text{ s} = 0.0050 \text{ s}$.

e) Time between successive occurrences of $y = 0$ is $T/2 = 0.0200$ s; this much additional time elapses until $y = 0$ again.

19-11 a) The maximum y is 4 mm (read from graph).

b) For either x the time for one full cycle is 0.040 s; this is the period.

c) Since $y = 0$ for $x = 0$ and $t = 0$ and since the wave is traveling in the $+x$-direction then $y(x,t) = A\sin[2\pi(t/T - x/\lambda)]$.

From the graph, if the wave is traveling in the $+x$-direction and if $x = 0$ and $x = 0.090$ m are within one wavelength the peak at $t = 0.01$ s for $x = 0$ moves so that it occurs at $t = 0.035$ s (read from graph so is approximate) for $x = 0.090$ m. The peak for $x = 0$ is the first peak past $t = 0$ so corresponds to the first maximum in $\sin[2\pi(t/T - x/\lambda)]$ and hence occurs at $2\pi(t/T - x/\lambda) = \pi/2$. If this same peak moves to $t_1 = 0.035$ s at $x_1 = 0.090$ m, then

$2\pi(t_1/T - x_1/\lambda) = \pi/2$

Solve for λ: $t_1/T - x_1/\lambda = 1/4$

$x_1/\lambda = t_1/T - 1/4 = 0.035 \text{ s}/0.040 \text{ s} - 0.25 = 0.625$

$\lambda = x_1/0.625 = 0.090 \text{ m}/0.625 = 0.14 \text{ m}$.

Then $v = f\lambda = \lambda/T = 0.14 \text{ m}/0.040 \text{ s} = 3.5 \text{ m/s}$.

d) If the wave is traveling in the $-x$-direction, then $y(x,t) = A\sin(2\pi(t/T + x/\lambda))$ and the peak at $t = 0.050$ s for $x = 0$ corresponds to the peak at $t_1 = 0.035$ s for $x_1 = 0.090$ m. This peak at $x = 0$ is the second peak past the origin so corresponds to $2\pi(t/T + x/\lambda) = 5\pi/2$. If this same peak moves to $t_1 = 0.035$ s for

$x_1 = 0.090$ m, then $2\pi(t_1/T + x_1/\lambda) = 5\pi/2$.

$t_1/T + x_1/\lambda = 5/4$

$x_1/\lambda = 5/4 - t_1/T = 5/4 - 0.035$ s$/0.040$ s $= 0.375$

$\lambda = x_1/0.375 = 0.090$ m$/0.375 = 0.24$ m.

Then $v = f\lambda = \lambda/T = 0.24$ m$/0.040$ s $= 6.0$ m/s.

e) No. Wouldn't know which point in the wave at $x = 0$ moved to which point a
$x = 0.090$ m.

19-15 a) The tension F in the rope is the weight of the hanging mass:

$F = mg = (1.50$ kg$)(9.80$ m/s$^2) = 14.7$ N

$v = \sqrt{F/\mu} = \sqrt{14.7 \text{ N}/(0.0550 \text{ kg/m})} = 16.3$ m/s

b) $v = f\lambda$ so $\lambda = v/f = (16.3$ m/s$)/120$ Hz $= 0.136$ m.

c) $v = \sqrt{F/\mu}$, where $F = mg$. Doubling m increases v by a factor of $\sqrt{2}$.

$\lambda = v/f$. f remains 120 Hz and v increases by a factor of $\sqrt{2}$, so λ increases by a
factor of $\sqrt{2}$.

19-17 Calculate the wave speed for transverse waves on the rubber tube.

The tension in the tube is equal to the weight of the suspended mass:

$F = mg = (7.50$ kg$)(9.80$ m/s$^2) = 73.5$ N.

The mass per unit length for the tube is

$\mu = m/L = 0.800$ kg$/14.0$ m $= 0.0571$ kg/m.

$$v = \sqrt{\frac{F}{\mu}} = \sqrt{\frac{73.5 \text{ N}}{(0.0571 \text{ kg/m})}} = 35.9 \text{ m/s}$$

The time is the distance traveled divided by the speed, so $t = 14.0$ m$/(35.9$ m/s$) =$
0.390 s

19-19 a) Calculate the wave speed:

$v = f\lambda = (400$ Hz$)(8.00$ m$) = 3200$ m/s.

$v = \sqrt{B/\rho}$ (Eq.19-21) so $B = \rho v^2 = (1300$ kg/m$^3)(3200$ m/s$)^2 = 1.33 \times 10^{10}$ Pa

b) The wave speed is $v = 1.50$ m$/3.90 \times 10^{-4}$ s $= 3846$ m/s

$v = \sqrt{Y/\rho}$ so $Y = \rho v^2 = (6400$ kg/m$^3)(3846$ m/s$)^2 = 9.47 \times 10^{10}$ Pa

19-23 a) The speed of sound v at the altitude of the plane satisfies the equation

850 km/h $= 0.85v$.

$v = (850 \text{ km/h})/0.85 = (1000 \text{ km/h})(1000 \text{ m}/1 \text{ km})(1 \text{ h}/3600 \text{ s}) = 277.8 \text{ m/s}$

If the air is assumed to be an ideal gas, then v is related to T by Eq.(19-27):

$v = \sqrt{\gamma RT/M}$ and $T = Mv^2/\gamma R$

For air, $\gamma = 1.40$ and $M = 28.8 \times 10^{-3}$ kg/mol (Example 19-7).

Then $T = \dfrac{(28.8 \times 10^{-3} \text{ kg/mol})(277.8 \text{ m/s})^2}{1.40(8.3145 \text{ J/mol} \cdot \text{K})} = 190.0 \text{ K} = -83°\text{C}$

b) No. $p = nRT/V = mRT/MV = \rho RT/M$. We know T but we don't know the density ρ of the air at this altitude.

19-25 Calculate the time it takes each sound wave to travel the $L = 80.0$ m length of the pipe:

wave in air: $t = 80.0 \text{ m}/(344 \text{ m/s}) = 0.2326 \text{ s}$

wave in the metal: $v = \sqrt{\dfrac{Y}{\rho}} = \sqrt{\dfrac{9.0 \times 10^{10} \text{ Pa}}{8600 \text{ kg/m}^3}} = 3235 \text{ m/s}$

$t = \dfrac{80.0 \text{ m}}{3235 \text{ m/s}} = 0.0247 \text{ s}$

The time interval between the two sounds is $\Delta t = 0.2326 \text{ s} - 0.0247 \text{ s} = 0.208 \text{ s}$

19-29 a) $v = \sqrt{\dfrac{Y}{\rho}} = \sqrt{\dfrac{11 \times 10^{10} \text{ Pa}}{8.9 \times 10^3 \text{ kg/m}^3}} = 3516 \text{ m/s}$

$v = f\lambda$ so $\lambda = \dfrac{v}{f} = \dfrac{3516 \text{ m/s}}{220 \text{ Hz}} = 16 \text{ m}$

b) $I = P_{\text{av}}/\pi r^2 = \frac{1}{2}\sqrt{\rho Y}\omega^2 A^2$ (Eq.19-35)

$\omega = 2\pi f = 1382 \text{ rad/s}$

$A = \left(\dfrac{2P_{\text{av}}}{\pi r^2 \omega^2 \sqrt{\rho Y}}\right)^{1/2} = 3.3 \times 10^{-8} \text{ m}$

c) From Eq.(19-9), $v_{\text{max}} = \omega A = (1382 \text{ rad/s})(3.3 \times 10^{-8} \text{ m}) = 4.6 \times 10^{-5} \text{ m/s}$

Problems

19-31 $A = 2.50 \times 10^{-3} \text{ m}$, $\lambda = 1.80 \text{ m}$, $v = 36.0 \text{ m/s}$

a) $v = f\lambda$ so $f = v/\lambda = (36.0 \text{ m/s})/1.80 \text{ m} = 20.0 \text{ Hz}$

$\omega = 2\pi f = 2\pi(20.0 \text{ Hz}) = 126 \text{ rad/s}$

$k = 2\pi/\lambda = 2\pi$ rad$/1.80$ m $= 3.49$ rad/m

b) For a wave traveling to the right, $y(x,t) = A\sin(\omega t - kx)$.

This equation gives that the $x = 0$ end of the string moves upward just after $t = 0$.

Put in the numbers: $y(x,t) = (2.50 \times 10^{-3}$ m$)\sin((126$ rad/s$)t - (3.49$ rad/m$)x)$.

c) The left hand end is located at $x = 0$. Put this value into the equation of part (b): $y(0,t) = +(2.50 \times 10^{-3}$ m$)\sin((126$ rad/s$)t)$.

d) Put $x = 1.35$ m into the equation of part (b):

$y(1.35$ m$, t) = (2.50 \times 10^{-3}$ m$)\sin((126$ rad/s$)t - (3.49$ rad/m$)(1.35$ m$))$.

$y(1.35$ m$, t) = (2.50 \times 10^{-3}$ m$)\sin((125$ rad/s$)t - 4.71$ rad$)$

4.71 rad $= 3\pi/2$ so $y(1.35$ m$, t) = (2.50 \times 10^{-3}$ m$)\sin((125$ rad/s$)t - 3\pi/2$ rad$)$

e) $y = A\sin(\omega t - kx)$ (part (b))

The transverse velocity is given by

$$v_y = \frac{\partial y}{\partial t} = A\frac{\partial}{\partial t}\sin(\omega t - kx) = +A\omega\cos(\omega t - kx).$$

The maximum v_y is $A\omega = (2.50 \times 10^{-3}$ m$)(126$ rad/s$) = 0.315$ m/s.

f) $y(x,t) = (2.50 \times 10^{-3}$ m$)\sin((126$ rad/s$)t - (3.49$ rad/m$)x)$

$t = 0.0625$ s and $x = 1.35$ m gives

$y = (2.50 \times 10^{-3}$ m$)\sin((126$ rad/s$)(0.0625$ s$) - (3.49$ rad/m$)(1.35$ m$)) = 0$.

$v_y = +A\omega\cos(\omega t - kx) = +(0.315$ m/s$)\cos((126$ rad/s$)t - (3.49$ rad/m$)x)$

$t = 0.0625$ s and $x = 1.35$ m gives

$v_y = (0.315$ m/s$)\cos((125$ rad/s$)(0.0625$ s$) - (3.49$ rad/m$)(1.35$ m$)) = -0.314$ m/s

19-35 a) $\omega = 2\pi f$, $\quad f = v/\lambda$, and $v = \sqrt{F/\mu}$

These equations combine to give $\omega = 2\pi f = 2\pi(v/\lambda) = (2\pi/\lambda)\sqrt{F/\mu}$.

But also $\omega = \sqrt{k'/m}$. Equating these expressions for ω gives

$k' = m(2\pi/\lambda)^2(F/\mu)$

But $m = \mu\,\Delta x$ so $k' = \Delta x(2\pi/\lambda)^2 F$

b) The "force constant" k' is independent of the amplitude A and mass per unit length μ, just as is the case for a simple harmonic oscillator. The force constant is proportional to the tension in the string F and inversely proportional to the wavelength λ. The tension supplies the restoring force and the $1/\lambda^2$ factor represents the dependence of the restoring force on the curvature of the string.

19-41 a) Eq.(19-33): $P_{av} = \frac{1}{2}\sqrt{\mu F}\omega^2 A^2$

$v = \sqrt{F/\mu}$ says $\sqrt{\mu} = \sqrt{F}/v$ so $P_{av} = \frac{1}{2}(\sqrt{F}/v)\sqrt{F}\omega^2 A^2 = \frac{1}{2}F\omega^2 A^2/v$

$\omega = 2\pi f$ so $\omega/v = 2\pi f/v = 2\pi/\lambda = k$ and
$P_{av} = \frac{1}{2}Fk\omega A^2$, as was to be shown.

b) Use Eq.(19-33) since it involves just ω, not k: $P_{av} = \frac{1}{2}\sqrt{\mu F}\omega^2 A^2$.

P_{av}, μ, A all constant so $\sqrt{F}\omega^2$ is constant, and $\sqrt{F_1}\omega_1^2 = \sqrt{F_2}\omega_2^2$.

$\omega_2 = \omega_1(F_1/F_2)^{1/4} = \omega_1(F_1/4F_1)^{1/4} = \omega_1(4)^{-1/4} = \omega_1/\sqrt{2}$

ω must be changed by a factor of $1/\sqrt{2}$ (decreased)

In the equation derived in part (a), $P_{av} = \frac{1}{2}Fk\omega A^2$, if P_{av} and A are constant then $Fk\omega$ must be constant, and $F_1 k_1 \omega_1 = F_2 k_2 \omega_2$.

$$k_2 = k_1 \left(\frac{F_1}{F_2}\right)\left(\frac{\omega_1}{\omega_2}\right) = k_1 \left(\frac{F_1}{4F_1}\right)\left(\frac{\omega_1}{\omega_1/\sqrt{2}}\right) = k_1 \frac{\sqrt{2}}{4} = k_1 \sqrt{\frac{2}{16}} = k_1/\sqrt{8}$$

k must be changed by a factor of $1/\sqrt{8}$ (decreased).

19-43 $v_1 = \sqrt{F/\mu}$, $v_1^2 = F/\mu$ and $F = \mu v_1^2$

The length and hence μ stay the same but the tension decreases by
$\Delta F = -Y\alpha A\,\Delta T$.
$v_2 = \sqrt{(F + \Delta F)/\mu} = \sqrt{(F - Y\alpha A\,\Delta T)/\mu}$
$v_2^2 = F/\mu - Y\alpha A\,\Delta T/\mu = v_1^2 - Y\alpha A\,\Delta T/\mu$
And $\mu = m/L$ so $A/\mu = AL/m = V/m = 1/\rho$. ($A$ is the cross-sectional area of the wire, V is the volume of a length L.)
Thus $v_1^2 - v_2^2 = \alpha(Y\,\Delta T/\rho)$ and

$$\alpha = \frac{v_1^2 - v_2^2}{(Y/\rho)\,\Delta T}$$

CHAPTER 20
WAVE INTERFERENCE AND NORMAL MODES

Exercises 3, 5, 9, 11, 15, 21, 23
Problems 27, 29, 31, 37, 39, 43

Exercises

20-3 Eq.(20-1): $y = (A_{\text{SW}} \sin kx) \cos \omega t$

a) At a node $y = 0$ for all t. This requires that $\sin kx = 0$ and this occurs fo
$kx = n\pi$, $n = 0, 1, 2, \ldots$

$$x = n\pi/k = \frac{n\pi}{0.750\pi \text{ rad/m}} = (1.33 \text{ m})n, \ n = 0, 1, 2, \ldots$$

b) At an antinode $\sin kx = \pm 1$ so y will have maximum amplitude.
This occurs when $kx = (n + \frac{1}{2})\pi$, $n = 0, 1, 2, \ldots$

$$x = (n + \tfrac{1}{2})\pi/k = (n + \tfrac{1}{2})\frac{\pi}{0.750\pi \text{ rad/m}} = (1.33 \text{ m})(n + \frac{1}{2}), \ n = 0, 1, 2, \ldots$$

20-5 **a)** $y = (A_{\text{SW}} \sin kx) \cos \omega t$ (Eq.20-1)

$A_{\text{SW}} = 0.850$ cm

$T = 0.0750$ s, $f = 1/T = 1/0.0750$ s $= 13.33$ Hz, $\omega = 2\pi f = 26.6\pi$ rad/s

antinode to antinode distance is $\lambda/2 = 15.0$ cm, so $\lambda = 30.0$ cm

$k = 2\pi/\lambda = 2\pi/30.0$ cm $= \pi/15.0$ cm

$y(x,t) = (0.850 \text{ cm}) \sin(\pi x/15.0 \text{ cm}) \cos((26.6\pi \text{ s}^{-1})t)$

b) $v = f\lambda = (13.33 \text{ Hz})(0.300 \text{ m}) = 4.00$ m/s

c) The amplitude of the simple harmonic motion at a particular x is
$A_{\text{SW}} \sin kx = (0.850 \text{ cm}) \sin(\pi x/15.0 \text{ cm})$, where x is measured from the node a
the fixed end of the string.

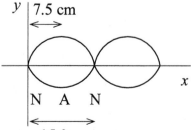

3.0 cm to the right of the first antinode so
$x = 7.50$ cm $+ 3.0$ cm $= 10.5$ cm

$x = 10.5$ cm implies $A_{\text{SW}} \sin kx = (0.850 \text{ cm}) \sin(\pi(10.5 \text{ cm})/15.0 \text{ cm}) = 0.688$ cm

0-9 a) fundamental

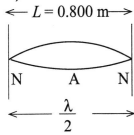

$\leftarrow L = 0.800 \text{ m} \longrightarrow$

$f = 60.0$ Hz

From the sketch,

$\lambda/2 = L$ so

$\lambda = 2L = 1.60$ m

fundamental

$v = f\lambda = (60.0 \text{ Hz})(1.60 \text{ m}) = 96.0$ Hz

b) The tension is related to the wave speed by Eq.(19-13):

$v = \sqrt{F/\mu}$ so $F = \mu v^2$.

$\mu = m/L = 0.0400 \text{ kg}/0.800 \text{ m} = 0.0500$ kg/m

$F = \mu v^2 = (0.0500 \text{ kg/m})(96.0 \text{ m/s})^2 = 461$ N.

0-11 a) $f = 440$ Hz when a length $L = 0.600$ m vibrates;

use this information to calculate the speed v of waves on the string.

For the fundamental $\lambda/2 = L$ so $\lambda = 2L = 2(0.600 \text{ m}) = 1.20$ m.

Then $v = f\lambda = (440 \text{ Hz})(1.20 \text{ m}) = 528$ m/s.

Now find the length $L = x$ of the string that makes $f = 587$ Hz.

$$\lambda = \frac{v}{f} = \frac{528 \text{ m/s}}{587 \text{ Hz}} = 0.900 \text{ m}$$

$L = \lambda/2 = 0.450$ m, so $x = 0.450$ m $= 45.0$ cm.

b) No retuning means same wave speed as in part (a). Find the length of vibrating string needed to produce $f = 392$ Hz.

$$\lambda = \frac{v}{f} = \frac{528 \text{ m/s}}{392 \text{ Hz}} = 1.35 \text{ m}$$

$L = \lambda/2 = 0.675$ m; string is shorter than this. No, not possible.

0-15 a) The placement of the displacement nodes and antinodes along

the pipe is as sketched below. The open ends are displacement antinodes.

| fundamental | 1st overtone | 2nd overtone |

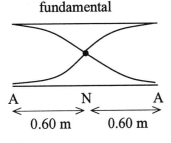

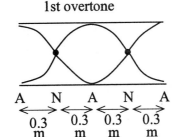

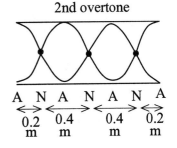

A   N   A

0.60 m   0.60 m

A   N   A   N   A

0.3   0.3   0.3   0.3

m   m   m   m

A N A N A N A

0.2   0.4   0.4   0.2

m   m   m   m

Location of the displacement nodes (N) measured from the left end:
fundamental 0.60 m
1st overtone 0.30 m, 0.90 m
2nd overtone 0.20 m, 0.60 m, 1.00 m

Location of the pressure nodes (displacement antinodes (A)) measured from t
left end: fundamental 0, 1.20 m
1st overtone 0, 0.60 m, 1.20 m
2nd overtone 0, 0.40 m, 0.80 m, 1.20 m

b) The open end is a displacement antinode and the closed end is a displaceme
node.

fundamental	1st overtone	2nd overtone
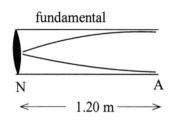	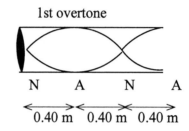	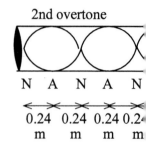
N A	N A N A	N A N A N
←——— 1.20 m ———→	0.40 m 0.40 m 0.40 m	0.24 0.24 0.24 0.2
		m m m m

Location of the displacement nodes (N) measured from the closed end:
fundamental 0
1st overtone 0, 0.80 m
2nd overtone 0, 0.48 m, 0.96 m

Location of the pressure nodes (displacement antinodes (A)) measured from t
closed end: fundamental 1.20 m
1st overtone 0.40 m, 1.20 m
2nd overtone 0.24 m, 0.72 m, 1.20 m

20-21

A   B   Q

◁←————————→◁←————→●
      2.00 m   1.00 m

a) Path difference from points A and B to point Q is 3.00 m − 1.00 m = 2.00 m
Constructive interference implies path difference = $n\lambda$, $n = 1, 2, 3, \ldots$
2.00 m = $n\lambda$ so $\lambda = 2.00$ m/n

$$f = \frac{v}{\lambda} = \frac{nv}{2.00 \text{ m}} = \frac{n(344 \text{ m/s})}{2.00 \text{ m}} = n(172 \text{ Hz}), \quad n = 1, 2, 3, \ldots$$

The lowest frequency for which constructive interference occurs is 172 Hz.

b) Destructive interference implies path difference = $(n/2)\lambda$, $n = 1, 3, 5, \ldots$
2.00 m = $(n/2)\lambda$ so $\lambda = 4.00$ m/n

$$f = \frac{v}{\lambda} = \frac{nv}{4.00 \text{ m}} = \frac{n(344 \text{ m/s})}{4.00 \text{ m})} = n(86 \text{ Hz}), \; n = 1, 3, 5, \ldots$$

The lowest frequency for which destructive interference occurs is 86 Hz.

0-23 a)

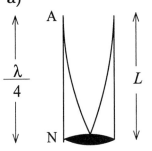

$$\lambda/4 = L$$
$$\lambda = 4L = 4(0.140 \text{ m}) = 0.560 \text{ m}$$

$$f = \frac{v}{\lambda} = \frac{344 \text{ m/s}}{0.560 \text{ m}} = 614 \text{ Hz}$$

b) Now the length L of the air column becomes $\frac{1}{2}(0.140 \text{ m}) = 0.070 \text{ m}$ and $\lambda = 4L = 0.280 \text{ m}$.

$$f = \frac{v}{\lambda} = \frac{344 \text{ m/s}}{0.280 \text{ m}} = 1230 \text{ Hz}$$

Problems

0-27 a) For reflection from a free end of a string the reflected wave is <u>not</u>
inverted, so $y(x, t) = y_1(x, t) + y_2(x, t)$, where
$y_1(x, t) = A \sin(\omega t + kx)$ (traveling to the left)
$y_2(x, t) = A \sin(\omega t - kx)$ (traveling to the right)
Thus $y(x, t) = A[\sin(\omega t + kx) + \sin(\omega t - kx)]$.
Apply the trig identity $\sin(a \pm b) = \sin a \cos b \pm \cos a \sin b$ with $a = \omega t$ and $b = kx$:
$\sin(\omega t + kx) = \sin \omega t \cos kx + \cos \omega t \sin kx$ and
$\sin(\omega t - kx) = \sin \omega t \cos kx - \cos \omega t \sin kx$.
Then $y(x, t) = (2A \cos kx) \sin \omega t$ (the other two terms cancel)

b) For $x = 0$, $\cos kx = 1$ and $y(x, t) = 2A \sin \omega t$. The amplitude of the simple harmonic motion at $x = 0$ is $2A$, which is the maximum for this standing wave, so $x = 0$ is an antinode.

c) $y_{max} = 2A$ from part (b).

$$v_y = \frac{\partial y}{\partial t} = \frac{\partial}{\partial t}[(2A \cos kx) \sin \omega t] = 2A \cos kx \frac{\partial \sin \omega t}{\partial t} = 2A\omega \cos kx \cos \omega t.$$

At $x = 0$, $v_y = 2A\omega \cos \omega t$ and $(v_y)_{max} = 2A\omega$

$$a_y = \frac{\partial^2 y}{\partial t^2} = \frac{\partial v_y}{\partial t} = 2A\omega \cos kx \frac{\partial \cos \omega t}{\partial t} = -2A\omega^2 \cos kx \sin \omega t$$

At $x = 0$, $a_y = -2A\omega^2 \sin\omega t$ and $(a_y)_{\max} = 2A\omega^2$.

20-29 a) Plank oscillates with maximum amplitude at its center says that it is oscillatin
in its fundamental mode.

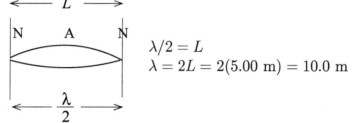

$\lambda/2 = L$
$\lambda = 2L = 2(5.00 \text{ m}) = 10.0 \text{ m}$

Student jumps upward two times per second so $f = 2.00$ Hz.
$v = f\lambda = (2.00 \text{ Hz})(10.0 \text{ m}) = 20.0 \text{ m/s}$.

b) There now must be an antinode 1.25 m (a distance of $L/4$) from one end. T
nodal structure for the standing wave must be:

Now $\lambda = L = 5.00$ m.

v depends on the properties of the plank, so it is the same as in part (a).
$$f = \frac{v}{\lambda} = \frac{20.0 \text{ m/s}}{5.00 \text{ m}} = 4.00 \text{ Hz}$$
The student now has to jump four times each second.

20-31 a) For an open pipe the successive harmonics are $f_n = nf_1$, $n = 1, 2, 3, \ldots$.
For a stopped pipe the successive harmonics are $f_n = nf_1$, $n = 1, 3, 5, \ldots$.
If the pipe is open and these harmonics are successive, then $f_n = nf_1 = 1372$ H
and $f_{n+1} = (n+1)f_1 = 1764$ Hz.
Subtract the first equation from the second: $(n+1)f_1 - nf_1 = 1764 \text{ Hz} - 1372 \text{ H}$
This gives $f_1 = 392$ Hz.

Then $n = \dfrac{1372 \text{ Hz}}{392 \text{ Hz}} = 3.5$. But n must be an integer, so the pipe can't be open.

If the pipe is stopped and these harmonics are successive, then $f_n = nf_1 = 13$
Hz and $f_{n+2} = (n+2)f_1 = 1764$ Hz (in this case succesive harmonics differ in
by 2).
Subtracting one equation from the other gives $2f_1 = 392$ Hz and $f_1 = 196$ Hz.
Then $n = 1372 \text{ Hz}/f_1 = 7$ so 1372 Hz $= 7f_1$ and 1764 Hz $= 9f_1$.

The solution gives integer n as it should; the pipe is stopped.

b) From part (a) these are the 7th and 9th harmonics.

c) From part (a) $f_1 = 196$ Hz.

For a stopped pipe $f_1 = \dfrac{v}{4L}$ and $L = \dfrac{v}{4f_1} = \dfrac{344 \text{ m/s}}{4(196 \text{ Hz})} = 0.439$ m.

0-37 a) There is a node at the piston, so the distance the piston moves is the node to node distance, $\lambda/2$.
$\lambda/2 = 37.5$ cm, so $\lambda = 2(37.5 \text{ cm}) = 75.0$ cm $= 0.750$ m.
$v = f\lambda = (500 \text{ Hz})(0.750 \text{ m}) = 375$ m/s

b) $v = \sqrt{\gamma RT/M}$ (Eq.19-27)

$$\gamma = \frac{Mv^2}{RT} = \frac{(28.8 \times 10^{-3} \text{ kg/mol})(375 \text{ m/s})^2}{(8.3145 \text{ J/mol} \cdot \text{K})(350 \text{ K})} = 1.39.$$

c) There is a node at the piston so when the piston is 18.0 cm from the open end the node is inside the pipe, 18.0 cm from the open end. The node to antinode distance is $\lambda/4 = 18.8$ cm, so the antinode is 0.8 cm beyond the open end of the pipe.

0-39 a) The tension F is related to the wave speed by $v = \sqrt{F/\mu}$ (eq.(19-13), so use the information given to calculate v:

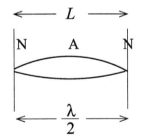

$\lambda/2 = L$
$\lambda = 2L = 2(0.600 \text{ m}) = 1.20$ m

fundamental
$v = f\lambda = (65.4 \text{ Hz})(1.20 \text{ m}) = 78.5$ m/s
$\mu = m/L = 14.4 \times 10^{-3} \text{ kg}/0.600 \text{ m} = 0.024$ kg/m
Then $F = \mu v^2 = (0.024 \text{ kg/m})(78.5 \text{ m/s})^2 = 148$ N.

b) $F = \mu v^2$ and $v = f\lambda$ give $F = \mu f^2 \lambda^2$.
μ is a property of the string so is constant.
λ is determined by the length of the string so stays constant.
μ, λ constant implies $F/f^2 = \mu\lambda^2 =$ constant, so $F_1/f_1^2 = F_2/f_2^2$

$$F_2 = F_1 \left(\frac{f_2}{f_1}\right)^2 = (148 \text{ N}) \left(\frac{73.4 \text{ Hz}}{65.4 \text{ Hz}}\right)^2 = 186 \text{ N}.$$

The percent change in F is $\dfrac{F_2 - F_1}{F_1} = \dfrac{186 \text{ N} - 148 \text{ N}}{148 \text{ N}} = 0.26 = 26\%.$

20-43 $\lambda = v/f = 344 \text{ m/s}/784 \text{ Hz} = 0.4388 \text{ m}$

a)

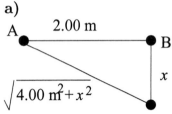

path difference $= \sqrt{4.00 \text{ m}^2 + x^2} - x$

The condition for destructive interference is that path difference $= (n + \frac{1}{2})\lambda$, n
$0, 1, 2, \ldots$ so $\sqrt{4.00 \text{ m}^2 + x^2} - x = (n + \frac{1}{2})\lambda$
Square both side of the equation: $4.00 \text{ m}^2 + x^2 = x^2 + (2n + 1)\lambda x + (n + \frac{1}{2})^2\lambda^2$
$$x = \frac{4.00 \text{ m}^2 - (n + \frac{1}{2})^2\lambda^2}{(2n + 1)\lambda}$$
Values of n from $n = 0$ through $n = 4$ give the following values of x: 9.00 m, 2.
m, 1.28 m, 0.534 m, 0.026 m. For $n \geq 5$ the equation would give $x < 0$ and this
not possible.

b) The condition for constructive interference is that path difference $= n\lambda$, n
$0, 1, 2, \ldots$ so $\sqrt{4.00 \text{ m}^2 + x^2} - x = n\lambda$
Square both side of the equation: $4.00 \text{ m}^2 + x^2 = x^2 + 2n\lambda x + n^2\lambda^2$
$$x = \frac{4.00 \text{ m}^2 - n^2\lambda^2}{2n\lambda}$$
Values of n from $n = 1$ through $n = 4$ give the following values of x: 4.34 m, 1.
m, 0.861 m, 0.262 m. For $n \geq 5$ the equation would give $x < 0$ and this is r
possible. $n = 0$ gives $x \to \infty$ so there also is constructive interference in the lir
that the microphone is very far from B.

c) The condition for destructive interference is that path difference $= (n + \frac{1}{2}$
$n = 0, 1, 2, \ldots$ so the smallest path difference that give destructive interferer
is $\lambda/2$. The maximum path length difference is 2.00 m, when the microphone
very close to B. Thus the longest wavelength for which destructive interference
possible is given by 2.00 m $= \lambda/2$ and $\lambda = 4.00$ m.

This corresponds to a frequency of $f = v/\lambda = (344 \text{ m/s})/4.00 \text{ m} = 86 \text{ Hz}$. This
the smallest frequency for which destructive intererence is still possible.

CHAPTER 21
SOUND AND HEARING

Exercises

1-1 **a)** $\lambda = v/f = (344 \text{ m/s})/1000 \text{ Hz} = 0.344 \text{ m}$

b) $p_{max} = BkA$ and Bk is constant gives $p_{max1}/A_1 = p_{max2}/A_2$

$$A_2 = A_1 \left(\frac{p_{max2}}{p_{max1}} \right) = 1.2 \times 10^{-8} \text{ m} \left(\frac{30 \text{ Pa}}{3.0 \times 10^{-2} \text{ Pa}} \right) = 1.2 \times 10^{-5} \text{ m}$$

c) $p_{max} = BkA = 2\pi BA/\lambda$

$p_{max}\lambda = 2\pi BA = $ constant so $p_{max1}\lambda_1 = p_{max2}\lambda_2$ and

$$\lambda_2 = \lambda_1 \left(\frac{p_{max1}}{p_{max2}} \right) = (0.344 \text{ m}) \left(\frac{3.0 \times 10^{-2} \text{ Pa}}{1.5 \times 10^{-3} \text{ Pa}} \right) = 6.9 \text{ m}$$

$f = v/\lambda = (344 \text{ m/s})/6.9 \text{ m} = 50 \text{ Hz}$

1-3 $p_{max} = BkA$ Eq.(21-5). As computed in Example 21-1 the adiabatic bulk modulus for air is $B = 1.42 \times 10^5$ Pa.

a) $f = 150$ Hz

Need to calculate k: $\lambda = v/f$ and $k = 2\pi/\lambda$ so

$k = 2\pi f/v = (2\pi \text{ rad})(150 \text{ Hz})/344 \text{ m/s} = 2.74 \text{ rad/m}$.

Then $p_{max} = BkA = (1.42 \times 10^5 \text{ Pa})(2.74 \text{ rad/m})(0.0200 \times 10^{-3} \text{ m}) = 7.78$ Pa.

This is below the pain threshold of 30 Pa.

b) f is larger by a factor of 10 so $k = 2\pi f/v$ is larger by a factor of 10, and $p_{max} = BkA$ is larger by a factor of 10. $p_{max} = 77.8$ Pa, above the pain threshold.

c) There is again an increase in f, k, and p_{max} of a factor of 10, so $p_{max} = 778$ Pa, far above the pain threshold.

1-5 **a)** $\omega = 2\pi f = (2\pi \text{ rad})(150 \text{ Hz}) = 942.5 \text{ rad/s}$

$$k = \frac{2\pi}{\lambda} = \frac{2\pi f}{v} = \frac{\omega}{v} = \frac{942.5 \text{ rad/s}}{344 \text{ m/s}} = 2.74 \text{ rad/m}$$

$B = 1.42 \times 10^5$ Pa (Example 21-1)

Then $p_{max} = BkA = (1.42 \times 10^5 \text{ Pa})(2.74 \text{ rad/m})(5.00 \times 10^{-6} \text{ m}) = 1.95$ Pa.

b) Eq.(21-6): $I = \frac{1}{2}\omega BkA^2$

$I = \frac{1}{2}(942.5 \text{ rad/s})(1.42 \times 10^5 \text{ Pa})(2.74 \text{ rad/m})(5.00 \times 10^{-6} \text{ m})^2 = 4.58 \times 10^{-3}$ W/m^2.

c) Eq.(21-11): $\beta = (10 \text{ dB})\log(I/I_0)$, with $I_0 = 1 \times 10^{-12}$ W/m^2.

$\beta = (10 \text{ dB})\log((4.58 \times 10^{-3} \text{ W/m}^2)/(1 \times 10^{-12} \text{ W/m}^2)) = 96.6$ dB.

21-9   Let 1 refer to the mother and 2 to the father.

From Example 21-7, $\beta_2 - \beta_1 = (10 \text{ dB})\log(I_2/I_1)$

Eq.(21-10): $I_1/I_2 = r_2^2/r_1^2$ or $I_2/I_1 = r_1^2/r_2^2$

$\Delta\beta = \beta_2 - \beta_1 = (10 \text{ dB})\log(I_2/I_1) = (10 \text{ dB})\log(r_1/r_2)^2 = (20 \text{ dB})\log(r_1/r_2)$

$\Delta\beta = (20 \text{ dB})\log(1.50 \text{ m}/0.30 \text{ m}) = 14.0$ dB.

21-11   **a)** From Example 21-7 $\Delta\beta = (10 \text{ dB})\log(I_2/I_1)$

13.0 dB $= 10$ dB $\log(I_2/I_1)$ so $1.3 = \log(I_2/I_1)$ and $I_2/I_1 = 20.0$.

b) According to the equation in part (a) difference in two sound intensity levels determined by the ratio of the sound intensities. So you don't need to know ⌐ just the ratio I_2/I_1.

21-17   Note: When the source is at rest $\lambda = \dfrac{v}{f_S} = \dfrac{344 \text{ m/s}}{400 \text{ Hz}} = 0.860$ m.

a) Eq.(21-15): $\lambda = \dfrac{v - v_S}{f_S} = \dfrac{344 \text{ m/s} - 25.0 \text{ m/s}}{400 \text{ Hz}} = 0.798$ m

b) Eq.(21-16): $\lambda = \dfrac{v + v_S}{f_S} = \dfrac{344 \text{ m/s} + 25.0 \text{ m/s}}{400 \text{ Hz}} = 0.922$ m

c) $f_L = v/\lambda$ (since $v_L = 0$). so $f_L = (344 \text{ m/s})/0.798 \text{ m} = 431$ Hz

d) $f_L = v/\lambda = (344 \text{ m/s})/0.922 \text{ m} = 373$ Hz

21-19   **a)** The positive direction is from the listener toward the source.

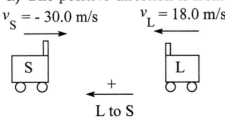

$$f_L = \left(\frac{v + v_L}{v + v_S}\right) f_S = \left(\frac{344 \text{ m/s} + 18.0 \text{ m/s}}{344 \text{ m/s} - 30.0 \text{ m/s}}\right)(262 \text{ Hz}) = 302 \text{ Hz}$$

Listener and source are approaching and $f_L > f_S$.

b)

$v_L = -18.0$ m/s $\qquad v_S = +30.0$ m/s

L to S

$$f_L = \left(\frac{v + v_L}{v + v_S}\right) f_S = \left(\frac{344 \text{ m/s} - 18.0 \text{ m/s}}{344 \text{ m/s} + 30.0 \text{ m/s}}\right)(262 \text{ Hz}) = 228 \text{ Hz}$$

Listener and source are moving away from each other and $f_L < f_S$.

Problems

1-23 **a)** Use the intensity level β to calculate I at this distance.

$\beta = (10 \text{ dB}) \log(I/I_0)$

$52.0 \text{ dB} = (10 \text{ dB}) \log(I/(10^{-12} \text{ W/m}^2))$

$\log(I/(10^{-12} \text{ W/m}^2)) = 5.20$ implies $I = 1.585 \times 10^{-7} \text{ W/m}^2$

Then use Eq.(21-9) to calculate p_{max}:

$$I = \frac{p_{max}^2}{2\rho v} \text{ so } p_{max} = \sqrt{2\rho v I}$$

From Example 21-3, $\rho = 1.20 \text{ kg/m}^3$ for air at 20°C.

$$p_{max} = \sqrt{2\rho v I} = \sqrt{2(1.20 \text{ kg/m}^3)(344 \text{ m/s})(1.585 \times 10^{-7} \text{ W/m}^2)} = 0.0114 \text{ Pa}$$

b) Eq.(21-5): $p_{max} = BkA$ so $A = \dfrac{p_{max}}{Bk}$

For air $B = 1.42 \times 10^5$ Pa (Example 21-1).

$$k = \frac{2\pi}{\lambda} = \frac{2\pi f}{v} = \frac{(2\pi \text{ rad})(587 \text{ Hz})}{344 \text{ m/s}} = 10.72 \text{ rad/m}$$

$$A = \frac{p_{max}}{Bk} = \frac{0.0114 \text{ Pa}}{(1.42 \times 10^5 \text{ Pa})(10.72 \text{ rad/m})} = 7.49 \times 10^{-9} \text{ m}$$

c) $\beta_2 - \beta_1 = (10 \text{ dB}) \log(I_2/I_1)$ (Example 21-7).

Eq.(21-10): $I_1/I_2 = r_2^2/r_1^2$ so $I_2/I_1 = r_1^2/r_2^2$

$\beta_2 - \beta_1 = (10 \text{ dB}) \log(r_1/r_2)^2 = (20 \text{ dB}) \log(r_1/r_2)$.

$\beta_2 = 52.0$ dB and $r_2 = 5.00$ m. Then $\beta_1 = 30.0$ dB and we need to calculate r_1.

52.0 dB $- 30.0$ dB $= (20$ dB$)\log(r_1/r_2)$

22.0 dB $= (20$ dB$)\log(r_1/r_2)$

$\log(r_1/r_2) = 1.10$ so $r_1 = 12.6r_2 = 63.0$ m.

21-25 a) The amplitude of the oscillations is ΔR.

$I = \frac{1}{2}\sqrt{\rho B}(2\pi f)^2 A^2 = 2\sqrt{\rho B}\pi^2 f^2(\Delta R)^2$

b) $P = I(4\pi R^2) = 8\pi^3\sqrt{\rho B}f^2 R^2(\Delta R)^2$

c) $I_R/I_d = d^2/R^2$

$I_d = (R/d)^2 I_R = 2\pi^2\sqrt{\rho B}(Rf/d)^2(\Delta R)^2$

$I = p_{max}^2/2\sqrt{\rho B}$ so

$p_{max} = \sqrt{(2\sqrt{\rho B}I)} = 2\pi\sqrt{\rho B}(Rf/d)\,\Delta R$

$A = \dfrac{p_{max}}{Bk} = \dfrac{p_{max}\lambda}{B2\pi} = \dfrac{p_{max}v}{B2\pi f} = v\sqrt{\rho/B}(R/d)\,\Delta R$

But $v = \sqrt{B/\rho}$ so $v\sqrt{\rho/B} = 1$ so $A = (R/d)\,\Delta R$.

21-27 a) $\lambda = \dfrac{v}{f} = \dfrac{1482 \text{ m/s}}{22.0 \times 10^3 \text{ Hz}} = 0.0674$ m

b) The Problem-Solving Strategy in the text describes how to do this problem. T
frequency of the directly radiated waves is $f_S = 22,000$ Hz. The moving whale fi
plays the role of a moving listener, receiving waves with frequency f_L'. The wh
then acts as a moving source, emitting waves with the same frequency, $f_S' =$
with which they are received. Let the speed of the whale be v_W.

whale receives waves

$v_L = +v_W$

$f_L' = f_S\left(\dfrac{v + v_L}{v + v_S}\right) = f_S\left(\dfrac{v + v_W}{v}\right)$

whale re-emits the waves

$$f_L = f_S\left(\frac{v+v_L}{v+v_S}\right) = f_S'\left(\frac{v}{v-v_W}\right)$$

But $f_S' = f_L'$ so $f_L = f_S\left(\dfrac{v+v_W}{v}\right)\left(\dfrac{v}{v-v_W}\right) = f_S\left(\dfrac{v+v_W}{v-v_W}\right)$.

Then $\Delta f = f_S - f_L = f_S\left(1 - \dfrac{v+v_W}{v-v_W}\right) = f_S\left(\dfrac{v-v_W-v-v_W}{v-v_W}\right) = \dfrac{-2f_S v_W}{v-v_W}$.

$$\Delta f = \frac{-2(2.20\times 10^4 \text{ Hz})(4.95 \text{ m/s})}{1482 \text{ m/s} - 4.95 \text{ m/s}} = 147 \text{ Hz}.$$

Listener and source toward away from each other so frequency is raised.

1-33 a) From Problem 21-31, $f_R = f_S(1 - v/c)$

v is negative since the source is approaching:

$v = -(42.0 \text{ km/h})(1000 \text{ m/1 km})(1 \text{ h}/3600 \text{ s}) = -11.67 \text{ m/s}$

Approaching means that the frequency is increased.

$$\Delta f = f_S\left(-\frac{v}{c}\right) = 2800\times 10^6 \text{ Hz}\left(-\frac{-11.67 \text{ m/s}}{3.00\times 10^8 \text{ m/s}}\right) = 109 \text{ Hz}$$

b) Approaching, so the frequency is increased. The frequency of the waves received and reflected by the water is very close to 2880 MHz, so get an additional shift of 109 Hz and the total shift in frequency is $2(109 \text{ Hz}) = 218 \text{ Hz}$.

1-35 Eq.(21-19): $\sin\alpha = v/v_S$

From Fig.21-16 $\tan\alpha = h/v_S T$. And $\tan\alpha = \dfrac{\sin\alpha}{\cos\alpha} = \dfrac{\sin\alpha}{\sqrt{1-\sin^2\alpha}}$.

Combining these equations we get $\dfrac{h}{v_S T} = \dfrac{v/v_S}{\sqrt{1-(v/v_S)^2}}$ and $\dfrac{h}{T} = \dfrac{v}{\sqrt{1-(v/v_S)^2}}$.

$$1 - (v/v_s)^2 = \frac{v^2 T^2}{h^2} \text{ and } v_S^2 = \frac{v^2}{1 - v^2 T^2/h^2}$$

$$v_S = \frac{hv}{\sqrt{h^2 - v^2 T^2}} \text{ as was to be shown.}$$